Martin C. Moore-Ede is Associate Professor
of Physiology, Harvard Medical School;
Frank M. Sulzman, Assistant Professor of
Biology, State University of New York,
Binghamton; and Charles A. Fuller, Assis-
tant Professor of Biomedical Sciences,
University of California, Riverside.

The Clocks That Time Us

This volume is published as part of a long-standing cooperative program between Harvard University Press and the Commonwealth Fund, a philanthropic foundation, to encourage the publication of significant and scholarly books in medicine and health.

The Clocks That Time Us

Physiology of the Circadian Timing System

Martin C. Moore-Ede,
Frank M. Sulzman, and
Charles A. Fuller

 A Commonwealth Fund Book

Harvard University Press
Cambridge, Massachusetts, and London, England
1982

Martin C. Moore-Ede
Department of Physiology
Harvard Medical School

Frank M. Sulzman
Department of Biological Sciences
State University of New York at Binghamton

Charles A. Fuller
Division of Biomedical Sciences
University of California at Riverside

Library of Congress Cataloging in Publication Data

Moore-Ede, Martin C., 1945-
    The clocks that time us.

    "A Commonwealth Fund book."
    Includes bibliographical references and index.
    1. Circadian rhythms. I. Sulzman, Frank M., 1944 - II.
Fuller, Charles A., 1949-   III. Title.
    QP84.6.M66   599.01′882   81-6780
    ISBN 0-674-13580-6   AACR2

To Donna, Pat, and Kelly

In appreciation of their understanding
of our time-consuming interest in clocks

Foreword

In the nineteenth century and the first half of the twentieth century, several works appeared on the topic of daily physiological cycles in plants and animals. The claims of certain authors that "endogenous daily," or "endogenous diurnal," rhythms might be involved in these cycles were often rejected as sheer mysticism by their critics. At that time the view was widely held that all organisms, from unicellulars to humans, were much like a clean slate at birth and therefore could not contain in their genetic heritage information on the spatial and temporal structure of their environment. Even after 1955, when biological and medical congresses began to deliberate on physiological rhythms, many biologists had misgivings. Against this background, it was believed appropriate to substitute for "endogenous rhythm" the more cautious term "circadian rhythm."

Well over a thousand papers now appear annually on circadian rhythms, many dealing specifically with their importance for the development and behavior of unicellulars, plants, and animals. Gradually, however, the medical implications of circadian rhythms are gaining attention, as the proceedings of various symposia bear witness. In such proceedings, by and large, the work of the participating research teams is discussed and summarized. Novices in the field or those who want to delve deeper into the subject of circadian timing often find such reports one-sided, representing isolated viewpoints. Such accounts focus even less on the interrelatedness of basic biological knowledge in the field and various medical aspects. Yet an awareness of such close interrelatedness is vital for fruitful research in medicine.

I know of no other book on the circadian timing system that proceeds on the basis of known fundamental biological principles and does such justice to the many-sided medical aspects in a comparative vein. For all of us working in this area, it is fascinating to watch how this marvelous timing system has arisen in the course of evolution, starting from the clock already "invented" and used by the single-celled flagellates.

Erwin Bünning

Tübingen,
West Germany

Preface

This book is the outgrowth of seven years of collaboration which started in the physiological laboratories at Harvard Medical School. We came from different backgrounds: one of us a physician and mammalian physiologist, another a biologist, and the third a neuroscientist. Although each of us was interested in timekeeping processes, we found that our preconceptions and diverse scientific backgrounds led to many a long debate on the mechanisms of biological clocks and the ways that mammalian behavior and physiology are scheduled within the 24-hour day.

In the course of our discussions it became clear that just as mammalian physiology can be conceptually divided into cardiovascular, respiratory, and other systems, so can a *circadian timing system* be defined. This system, concerned with the timing and coordination of events within an approximately 24-hour time scale, has its own set of specialized structures, mechanisms, and functions. The circadian timing system can thus be analyzed using the same intellectual rubric as the other, better-established organ systems.

Knowledge about the structure and function of the circadian timing system has been rapidly expanding over the last twenty-five years; its enormous import for physiological regulation and its implications for clinical medicine and occupational health are now readily apparent to those who work in this scientific field. Yet the majority of biologists, physicians, and occupational health specialists are only vaguely aware of the impact of circadian rhythmicity upon their disciplines. Thus in 1981 the editors of *Nature* (vol. 293, p. 531) wrote, in a major review of the neurosciences, "Although the phenomenon of diurnal metabolic activity is not well attested in human beings, it does appear that insects such as the locust can keep time with the Sun"—apparently unaware of the thousands of papers on human circadian rhythms published over the last hundred years.

How could this research on the clocks that time our sleep and wakefulness, our metabolic, endocrine, and neural functions be so easily ignored? The answer lies partly in the unfamiliar vocabulary used by scientists in this field, partly in the lack of exposure of most biologists and medical scientists to the concepts of oscillator theory and dynamic systems, and partly in the fact that the specific structures that act as circadian clocks in the mammalian brain were not identified until 1972. Thus the major biological and medical courses and textbooks do not yet discuss the physiological mechanisms responsible for circadian rhythmicity, or they provide only a cursory and often inaccurate account. Fortunately, the situation is changing, and many universities have initiated courses on biological clocks at the graduate and undergraduate levels.

We took on the challenge of writing this book because we could find no book to give to our colleagues or recommend to our students that clearly explained the structure and function of circadian clocks in mammals or provided a basis for understanding the medical and public health implications of circadian rhythmicity. There is much elegant and important work in this field that cannot readily be appreciated without an introductory book to unlock the door to circadian language and oscillator concepts.

We have learned much from writing it and have been stimulated by the process of having to clarify for the reader issues we had come to take for granted. We hope the book will reflect the excitement of those working in the field and yet explain clearly what is known and why it is important. It is impossible for us to trace where each idea presented in these pages came from; we owe a debt to so many of our colleagues that it is difficult to thank them all. In particular, Charles A. Czeisler, Richard E. Kronauer, David A. Kass, Ralph Lydic, H. Elliott Albers, Philippa H. Gander, and Margaret L. Moline have contributed much in criticism, current data and, most important, ideas. We have much to thank them for.

We have also taken the opportunity to tell the history of this field while the major players, some now in their eighties, are still active. We owe much to Jürgen Aschoff, Erwin Bünning, Patricia J. DeCoursey, Leland N. Edmunds, Franz Halberg, J. Woodland Hastings, Michael Menaker, Robert Y. Moore, Colin S. Pittendrigh, Curt P. Richter, Benjamin Rusak, Elliot D. Weitzman, Rütger A. Wever, and Arthur T. Winfree. We have taken efforts to sort out from their first-hand accounts the evolution of this field. Each of them has read chapters, or even the whole book, and provided invaluable criticism, setting us straight on many points. Any errors that persist are ours, not theirs. Others, including Wolfgang Engelmann, J. Allan Hobson, Robert W. McCarley, Michael Terman, and Torsten N. Wiesel have helped us at various points along the way. Still others have allowed us to reproduce their published data, for which we are much appreciative. We would especially like to acknowledge the kind permission of several investigators to reproduce their unpublished data in figures, including Robert Y. Moore, Ralph Lydic, H. Elliott Albers, Janet M. Zimmerman, and Philippa H. Gander.

The task of writing, referencing, indexing, and preparing figures would have been totally unmanageable without the editorial assistance of Louise C. Kilham, Joan Finkelstein, and Barbara E. Bunker; the artwork of Suzanne Lawson, Pamela P. Blum, Dennis Young, and Karen Watson; the photography of Steven Borack, Robert H. Rubin, and Howard Cook; and the technical support of Wendy S. Schmelzer, Meredith M. MacKenzie-Lamb, Christine J. Harling, Sharon M. Eagan, Beverly J. Tepper, Samuel F. Pilato, David Goldman, Christine M. Collins, Corey Goldman, and Mark S. Menser. The staff of the Harvard University Press—William Patrick, Susan Wallace, and Peg Anderson, in particular—have helped us write what we intended to say.

Finally we wish to acknowledge the generous support of the Commonwealth Fund in aiding the production of this volume, and of the agencies, through the following grants, which have supported our own research: NIH Grants NS13921, GN22085, MH28460, BRD-RR09070, GM29327, GM28265; NSF Grants PCM76-19943, BNS-7924412; AFOSR

Grant 78-3560; NASA Grants NAS9-14249, NSG-9054, NCA2-0R284, NAS2-10547, NAS9-15975, NAS2-10621, NAS2-10536. Martin C. Moore-Ede was the recipient of NIH Career Development Award NS-00247.

Contents

The Clocks That Time Us

Symbols and Abbreviations

τ period of biological rhythm

T   period of zeitgeber

ϕ phase of biological rhythm

Φ phase of zeitgeber

ψ phase relationship between zeitgeber and biological rhythm

α length of daily activity

ρ length of daily rest ($\alpha + \rho = \tau$)

a   amplitude of biological rhythm (mean to maximum or mean to minimum)

A   amplitude of zeitgeber cycle

CT   circadian time, a time scale covering one full circadian period. The zero point is usually defined as lights-on when the circadian rhythm is entrained by a light-dark cycle, but is defined arbitrarily when the rhythm is free-running.

DD   continuous darkness

LD   light–dark cycle. The respective durations of light and dark (in hours) may also be indicated, for example LD 14:10, indicating that the cycle is 14 hours of light and 10 hours of darkness.

LL   continuous illumination

r   range of biological rhythm (maximum to minimum)

R   range of zeitgeber cycle

ZT   zeitgeber time, a time scale covering one full zeitgeber period. The zero point is defined arbitrarily.

A Physiological System Measuring Time

The expression "as sure as night follows day" reflects the stability of certain cycles in our environment. The earth, spinning on its axis approximately once every 24 hours, submits plants and animals to highly predictable daily rhythms of light and temperature (Fig. 1.1). The availability of food and the activity of predators are in turn affected by these periodic variations. It is not surprising, therefore, to discover that the behavior and metabolism of most organisms follow a 24-hour schedule.

The most obvious explanation for such 24-hour rhythms is that plants and animals passively respond to the cycles in their environment. However, when an organism is isolated from all environmental time cues—when light, food, temperature, and sound are kept constant around the clock—the majority of its rhythms persist with an independent period. This "free-running" period is usually close to but not exactly 24 hours.

It is little wonder that living systems should generate periodic oscillations in a wide variety of physiological variables. Engineers know that complex control systems tend to oscillate in their performance; indeed, much of their job is to minimize such oscillations when they interfere with the function of the machine. The remarkable development, however, is that evolution has selected the frequency of certain oscillations in biological systems so that they can serve as an organism's internal clocks.

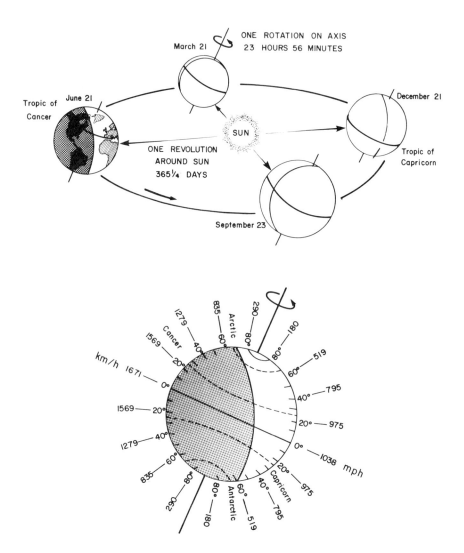

Fig. 1.1 The earth's yearly revolution about the sun and its daily rotation on its axis determine the light–dark patterns to which we are exposed. Seasonal changes occur because the earth's axis of rotation is tilted with respect to its plane of revolution. The north pole is tilted toward the sun from March to September, so the northern hemisphere receives more sunlight per day than the southern hemisphere. Then from September to

The physiological system responsible for measuring time and synchronizing an organism's internal processes with the daily events in its environment is known as the circadian timing system. The word *circadian* (Latin: *circa*=about; *dies*=day) was coined by Franz Halberg in 1959 to describe the approximately 24-hour cycles that are endogenously generated by an organism. Although virtually all plants and animals have circadian timing systems, this book will focus on mammals, because there is consistency in the physiological strategies of this class and the most direct parallels to clinical medicine can be drawn. However, from time to time we will refer to other organisms that can help to explain mammalian mechanisms.

In addition to circadian processes, a very wide range of biological rhythms, with periods of less than a second to more than a year, has been demonstrated in mammals, as Figure 1.2 illustrates. Some rhythms are normal products of mammalian physiology, while others become apparent only in disease states. Many of the observed periodicities correspond to periodicities in the environment, such as the solar year. Circadian events are the focus of this book, but we will consider the interrelationships between the circadian timing system and other biological rhythms, such as the estrous cycle and circannual rhythms.

March the southern hemisphere receives more sunlight per day than the northern hemisphere because the north pole is tilted away from the sun. In midsummer at the north pole there is continuous light, and in midwinter there is continuous darkness. In contrast, at the equator there is no seasonal change; there are always 12 hours of light and 12 hours of darkness. In the regions between the equator and the poles the duration of the light period continuously changes during the year.

As the earth rotates in an easterly direction (*lower panel*), the earth's shadow and the time of dawn and dusk move in a westerly direction across the earth's surface. The earth's shadow moves most quickly at the equator and most slowly at the poles, simply because the earth's circumference is greater at the equator so that the shadow has a greater distance to travel.

The true period of rotation of the earth is not 24.0 hours, but 23 hours and 56 minutes. But because the earth is revolving about the sun in the same direction that it is rotating, it must turn for an extra 4 minutes each day to complete a rotation. (Copyright 1982 by Moore-Ede, Sulzman, and Fuller.)

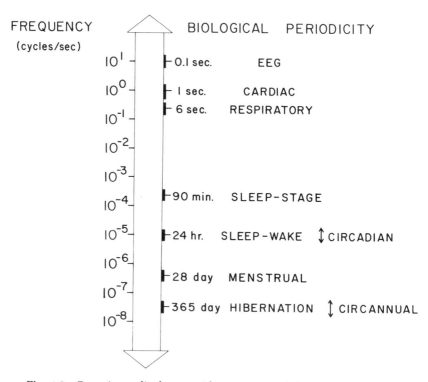

FREQUENCY (cycles/sec)

BIOLOGICAL PERIODICITY

Frequency	Period	
10^1	0.1 sec.	EEG
10^0	1 sec.	CARDIAC
10^{-1}	6 sec.	RESPIRATORY
10^{-2}		
10^{-3}		
10^{-4}	90 min.	SLEEP-STAGE
10^{-5}	24 hr.	SLEEP-WAKE ↕ CIRCADIAN
10^{-6}	28 day	MENSTRUAL
10^{-7}	365 day	HIBERNATION ↕ CIRCANNUAL
10^{-8}		

Fig. 1.2 Organisms display a wide spectrum of frequencies in their rhythmic processes. The left side shows the various frequencies, and the right side shows the periods (that is, 1/frequency) of selected examples. High-frequency events include the electrical activity of the brain (EEG: electroencephalogram), heart beat (cardiac), rate of breathing (respiratory), and sleep-stage (rate of progression through different levels of sleep). These cycles are often referred to as *ultradian* rhythms (processes having periods much less than 24 hours). *Circadian* rhythms are often arbitrarily defined as having periodicities between 20 and 28 hours. Two examples of low-frequency cycles are the menstrual cycle of women and the seasonal (circannual) cycle of hibernation. Such long-period rhythms are referred to as *infradian* rhythms. (Copyright 1982 by Moore-Ede, Sulzman, and Fuller.)

We will also discuss the many analogous features of the various biological timekeeping systems, because these provide a useful theoretical base for looking at biological clocks in all their apparent diversity.

Historical Milestones

Twenty-four-hour rhythms in the activities of plants and animals must have been recognized from the earliest times. The tendency for some organisms to sleep at night and some during the day would have been obvious to man in his earliest contemplations of the temporal order of nature. Written records going back to the marches of Alexander the Great in the fourth century B.C. document the daily movements of leaves and flower petals. Androsthenes reported that the tamarind tree (*Tamarindus indicus*), known at that time primarily for its laxative fruit, opened its leaves during the day and closed them at night (Bretzl, 1903). However, in these early writings there is no sign that these diurnal rhythms were interpreted as anything other than passive responses to a cyclic environment.

ENDOGENOUS ORIGIN OF CIRCADIAN RHYTHMS

The apparently quite reasonable assumption that biological rhythms were a direct consequence of a periodic environment remained untested until Jean Jacques d'Ortous de Mairan, an astronomer by training, conducted a critical experiment in 1729. In a brief communication to L'Academie Royale des Sciences of Paris (Fig. 1.3), transmitted by M. Marchant, it was reported how de Mairan had studied the leaf movements of a "sensitive" heliotrope plant (probably *Mimosa pudica*). This plant opens its leaves and pedicels during the day and folds them at night. When de Mairan moved this plant to a place where sunlight could not reach it, he found that the plant still opened its leaves during the day and folded them for the entire night (Fig. 1.4). Thus, the persistence of circadian rhythms in the absence of environmental time cues was first demonstrated.

De Mairan recognized that these rhythms were related to the sleeping patterns of bedridden patients, which persist on a circadian schedule even when the patients are unaware of day and night. He also suggested that other plants be examined for the same phenomenon and that the role of environmental temperature as a synchronizer be tested. De Mairan proposed that another interesting experiment would be to reverse light and dark and examine the response of the plant but he concluded that his everyday chores did not permit him to pursue

OBSERVATION BOTANIQUE.

ON fçait que la Senfitive eft *heliotrope*, c'eft-à-dire que fes rameaux & fes feüilles fe dirigent toûjours vers le côté d'où vient la plus grande lumiére, & l'on fçait de plus qu'à cette propriété qui lui eft commune avec d'autres Plantes, elle en joint une qui lui eft plus particuliére, elle eft Senfitive à l'égard du Soleil ou du jour, fes feüilles & leurs pédicules fe replient & fe contraétent vers le coucher du Soleil, de la même maniére dont cela fe fait quand on touche la Plante, ou qu'on l'agite. Mais M. de Mairan a obfervé qu'il n'eft point néceffaire pour ce phénoméne qu'elle foit au Soleil ou au grand air, il eft feulement un peu moins marqué lorfqu'on la tient toûjours enfermée dans un lieu obfcur, elle s'épanoüit encore très-fenfiblement pendant le jour, & fe replie ou fe refferre réguliérement le foir pour toute la nuit. L'expérience a été faite fur la fin de l'Eté, & bien répétée. La Senfitive fent donc le Soleil fans le voir en aucune maniére ; & cela paroît avoir rapport à cette malheureufe délicateffe d'un grand nombre de Malades, qui s'apperçoivent dans leurs Lits de la différence du jour & de la nuit.

Il feroit curieux d'éprouver fi d'autres Plantes, dont les feüilles ou les fleurs s'ouvrent le jour, & fe ferment la nuit, conferveroient comme la Senfitive cette propriété dans des lieux obfcurs ; fi on pourroit faire par art, par des fourneaux plus ou moins chauds, un jour & une nuit qu'elles fentiffent ; fi on pourroit renverfer par là l'ordre des phénomenes du vrai jour & de la vraye nuit, &c. Mais les occupations ordinaires de M. Mairan ne lui ont pas permis de pouffer les expériences jufque-là, & il fe contente d'une fimple invitation aux Botaniftes & aux Phificiens, qui pourront eux-mêmes avoir d'autres chofes à fuivre. La marche de la véritable Phifique, qui eft l'Expérimentale, ne peut être que fort lente.

E ij

M. Marchant a lû la Defcription De l'*Althoea* Diofc. & Plin. C. B. Pin. 315. *Guimauve*, avec la Critique des Auteurs Botaniftes fur cette Plante.

De la *Mitella Americana, florum foliis fimbriatis*. Inft. Raii Herb. 242.

Et de la *Sanicula*, feu *Cortufa Americana, altera, flore minuto, fimbriato*. Hort. Reg. Par.

Fig. 1.3 The original communication of Jean Jacques d'Ortous de Mairan on the persistence of the daily rhythm of a plant's leaf movements in constant conditions, reported to L'Academie Royale des Sciences in Paris by M. Marchant in 1729.

his experiment that far. Finally, he extended an invitation to botanists and physicists to pursue this research, noting that the progress of true experimental science can only be very slow.

With this last statement, he was also to prove his considerable foresight. Thirty years passed before his experiments were even repeated, by Duhamel Du Monceau (1759), who showed that the rhythm of leaf movements was not caused by variations in environmental temperature, and by Zinn (1759). Furthermore, the fact that circadian timekeeping is an endogenous property of animals and plants was not generally accepted until almost 250 years later.

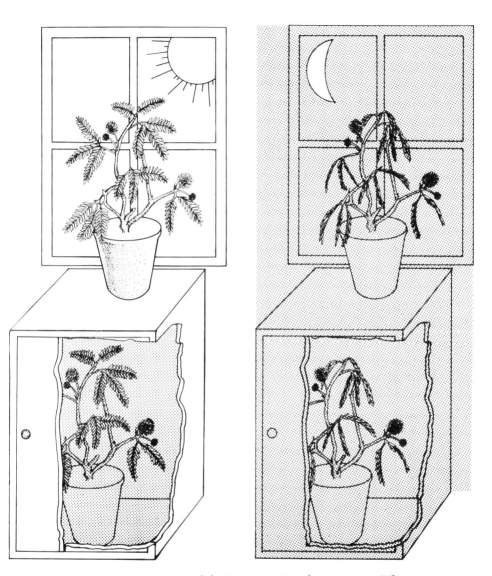

Fig. 1.4 A representation of de Mairan's original experiment. When exposed to sunlight during the day (*upper left*), the leaves of the plant were open, and during the night (*upper right*) the leaves were folded. De Mairan showed that sunlight was not necessary for these leaf movements by placing the plant in total darkness; even under these constant conditions, the leaves opened during the day (*lower left*) and folded during the night (*lower right*). (Copyright 1982 by Moore-Ede, Sulzman, and Fuller.)

In 1832 Augustin de Candolle discovered that the daily leaf movements of *Mimosa pudica* not only persisted in constant darkness but that the leaves opened an hour or two earlier each day, so that they displayed a periodicity of 22 to 23 hours. This was the first demonstration that circadian clocks would "free-run" with their own endogenous period when they were no longer synchronized to a 24-hour light–dark cycle. Although not fully appreciated at first, this finding made it most improbable that some other day–night variation, such as room temperature, was continuing to drive the rhythm even though the light–dark cycle was eliminated. In de Candolle's words, his findings showed "an inherent tendency of plants to show periodic movements."

Wilhelm Pfeffer, a plant physiologist well respected for his introduction of exact physicochemical methods, at first disputed the conclusion that the circadian leaf movements of plants were endogenously produced. In his book *Physiologische Untersuchungen*, published in 1873, he suggested that the persistence of rhythms was a result of light leaks in the dark rooms used in earlier studies. In subsequent studies Pfeffer clearly proved otherwise and reversed his position, but this reversal was not recognized by the general scientific community. Pfeffer's extensive later experiments were unfortunately published in an obscure journal named, with Germanic literality, *Abhandlungen der Mathematisch-Physischen Klasse der Koeniglich Saechsischen Akadamie der Wissenschaften* (Pfeffer, 1915). As a consequence, Pfeffer's earlier doubts were remembered instead of his later findings (Bünning and Chandrashekeran, 1975). However, the case for the inherent nature of diurnal rhythms was not completely ignored. Darwin, in his book *On the Power of Movement in Plants* (1880), indicated that he believed the diurnal periodicity in leaf movements was an inherent property of plants.

CIRCADIAN OSCILLATIONS AS BIOLOGICAL CLOCKS

The conceptual advance that stimulated modern scientific activity in this field was the recognition that circadian rhythms are the outputs of a system whose main function is to measure time. This shift in scientific emphasis from circadian rhythms as a biological curiosity to circadian systems as critical time-measuring devices can be traced back to the chance observations of the Swiss physician August Forel in 1910. When he and his family were taking breakfast one morning on the terrace of his summer home in the Swiss Alps, Forel noticed a few worker bees from a hive located about 125 meters from the house arrived to sample some marmalade on the table. After a few days, he found that the bees often appeared on the terrace just before breakfast was served, as if they knew it was time for the food to arrive. Finally, finding it impossible to eat outside, the family moved inside, only to notice that for several days the bees continued to arrive outside exactly at breakfast time and walk around the terrace table as though they expected to find food. Because the bees appeared only at that time of day and did not come back at any other time, Forel suggested that the bees might possess a *zeitgedächtnis*—a memory for time.

That organisms could truly measure the passage of time was demonstrated by von Frisch and his graduate student, Beling, in 1929. Beling's technique was to mark bees individually and then offer them sugar water at an artificial feeding place for several days—always at the same time of day (Beling, 1929). Then, on the day of study, no food was placed in the dish and the time of arrival of each bee was observed (Fig. 1.5). The bees arrived within the training time, or very close to it, in almost all situations. Even conducting such experiments in the constant conditions of a salt mine, 180 meters below the surface of the earth, did not alter the bees' capacity for timekeeping. Evidence that an internal circadian system was responsible was provided by the finding that the bees could be trained to arrive only when food was placed at intervals very close to 24 hours; however, if food was presented at 19-hour or 48-hour intervals, the bees showed no capability for recognizing the pattern and entraining to it.

Renner (1955) conducted a definitive experiment to test whether bees were sensing something in their environment which the experimenters had missed. In his laboratory in France, he trained 40 bees to collect sugar water between 8:15 P.M. and 10:15 P.M. each evening in a closed room. He then transported the bees overnight to New York City, where they were placed in a similarly organized laboratory. The next day the bees again arrived at the feeding table at 8:15 to 10:15 P.M. *French time.* These experiments showed that neither cosmic rays nor

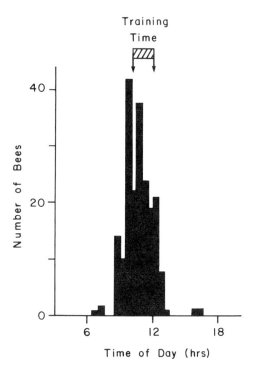

Fig. 1.5 A graph of the number of bees visiting a site where they had been fed sugar water on previous days although none was available on the day of study. The time at which food had been presented (the training time) is shown at the top. The distribution of the arrival times is slightly skewed to the left, showing the bees' tendency to arrive early. Beling (1929) showed that bees could not be trained to arrive at the time of feeding if the food was provided at 19- or 48-hour intervals. (After Renner, 1960.)

any other unscreened environmental time cue was transmitting temporal information to the bees.

The internal circadian timing system thus enables bees to avoid wasting energy in futile visits to flowers which, following a circadian schedule themselves, offer their nectar or pollen only at restricted times of day. As first extensively documented by Linnaeus (1751), each flower species has a characteristic time of opening and closing its petals. Once a bee has identified a flower with nectar available at a given time of day, the bee and others from the same hive can return each day at the appropriate time whether or not the sun is shining or other temporal cues are present. Indeed, because of the predictability of the opening and closing of certain flowers, Linnaeus devised a clock, depicted in Figure 1.6, by cultivating in the appropriate areas of a garden flowers which open and close their petals at different specific times of day (von Marilaun, 1895). By inspecting his garden he could tell the time of day from his flower clock, a prospect apparently greeted with some concern by the clockmakers of his day!

Beginning in the first part of this century, Erwin Bünning laid the foundations for much of our current understanding of the properties of circadian systems. For example, he demonstrated that plants (1932) and insects (1935b) still displayed circadian rhythms after they or their parents were raised in constant conditions. He went on to show that the free-running period was genetically inherited when strains of plants with different endogenous periods were crossed (1935a). Furthermore, he was the first to recognize that circadian clocks measure the length of the day as well as the total circadian cycle (1936), and he pointed out the adaptive advantages to organisms of being able to detect seasonal changes in day length (see Chapter 6). In the years since, Bünning and a large number of other investigators have extensively documented the properties of circadian clocks and demonstrated their generality in organisms ranging from single-celled algae to man (Bünning, 1960, 1973, 1977).

Since the 1950s Colin Pittendrigh has done much to convince biologists of the importance of circadian clocks. In 1954 he published the first of an elegant series of papers demonstrating that the time of day that the fruit fly (*Drosophila*) emerged from its pupa (eclosion) was controlled by a circadian clock which was little influenced by variations in environmental temperature. Although most metabolic processes speed up considerably with increases in body temperature, the period of circadian clocks does not. Pittendrigh showed that circadian clocks are temperature-compensated, an essential feature that makes the clock a viable timekeeping device. Pittendrigh's work also demonstrated that circadian rhythms were not learned phenomena, as Beling and associates had implied by their use of the word *zeitgedächtnis*—memory for time. *Drosophila* pupae raised in constant conditions did

Fig. 1.6 A representation of the flower clock proposed by Linnaeus in 1751, showing the characteristic times of petal opening and closing for various species of flowers. The 12 hours of the clock run from 6 A.M. to 6 P.M. Drawing by Ursula Schleicher-Benz. (From *Lindauer Bilderbogen* no. 5, edited by Friedrich Boer, Jan Thorbecke Verlag, Sigmaringen, West Germany.)

not need to be exposed to a 24-hour light–dark cycle to show circadian rhythmicity; a brief flash of light was all that was necessary to trigger the rhythm. His forceful arguments about the endogenous timekeeping properties of circadian systems are well summarized in Pittendrigh (1960).

The importance of circadian timekeeping to animals was also demonstrated by Kramer (1952), who showed that migratory birds can use the sun as a compass even though it moves across the daytime sky. The changing position of the sun throughout the day is compensated for by the birds' circadian clocks, which adjust the direction of flight with respect to the sun. Hence if a bird in the northern hemisphere was migrating south, it would continue to do so throughout the day, although at first the sun would be on the bird's left, then later directly ahead, and finally in the evening on the bird's right.

Despite the growing evidence of the endogenous nature of circadian timekeeping processes, some researchers still viewed circadian rhythms as a product of environmental oscillations. Brown (1957, 1972) has been one of the last proponents of this concept. There was a certain plausibility to the argument that circadian timekeeping processes could be the products of some geophysical oscillation, the nature of which was unknown to the investigator. Much of Brown's work has been to demonstrate that organisms are responsive to weak electromagnetic fields in their environment. Although he has shown that such fields could potentially transmit circadian information, he has not proved that they actually do so.

A number of studies have attempted to rule out the effects of unrecognized cues which might be produced by the earth's rotation. Hamner and co-workers (1962), for example, showed the persistence of circadian rhythmicity in hamsters (*Mesocricetus auratus*), fruit flies (*Drosophila*), and a fungus (*Neurospora*) when placed on a table rotating counter to the earth's rotation at the South Pole. Through complex arguments, however, Brown still claimed that some unrecognized 24-hour cue was not excluded; for example, the electrical generator for the South Pole Station might be on a 24-hour timer and hence create periodic electromagnetic fields. Space flight studies in which organisms can be isolated from 24-hour cues would seem an obvious environment to test the exogenous hypothesis. However, up to now animals on space missions (the monkey on NASA's Biosatellite III, for example) have been maintained on a 24-hour light–dark cycle in the capsule. We plan to test the exogenous hypothesis during NASA's first Spacelab mission by flying *Neurospora* in constant conditions well away from the earth's 24-hour periodicity (Sulzman, Fuller, and Moore-Ede, 1978).

More problematic for the argument that circadian rhythms are exogenously generated are the extensive data showing a genetic basis for circadian rhythmicity and the comprehensive studies demonstrating the unique free-running periods for each species and for each individual within the species (Chapter 2). Together with the study of organisms in space, these investigations are placing the final nails in the coffin of the exogenous theory.

The endogenous–exogenous debate proved to have a braking effect on progress in circadian research, since it was intellectually less interesting to study a phenomenon that might be a simple response to the 24-hour rhythms in the environment. Out of this debate, however, came the awareness that particular subtle environmental events such as barometric pressure (Hayden and Lindberg, 1969) and electrostatic fields (Dowse and Palmer, 1969) can in certain limited circumstances provide discrete synchronizing stimuli to circadian systems, and these stimuli must be controlled in experiments. Such effects are usually of concern only in experiments where a circadian rhythm shows a precise 24-hour period despite the exclusion of all known time cues from the organism's environment. This period length may sometimes occur by chance, but one must make sure that the environment is not contributing an unrecognized 24-hour cue.

CIRCADIAN RHYTHMICITY IN MAMMALS

The endogenous nature of circadian rhythms in mammals appears to have been ignored until William Ogle in 1866 undertook some careful observations of the daily body temperature rhythm in man. In his own words, "There is a rise in the early mornings while we are still asleep, and a fall in the evening while we are still awake, which cannot be explained by reference to any of the hitherto mentioned influences. They are not due to variations in light; they are probably produced by periodic variations in the activity of the organic functions."

Ogle demonstrated that this rhythm was not directly dependent on either obvious environmental influences or the sleep–wake cycle of his subjects, but it was not until Simpson and Galbraith's studies of monkeys that the rhythm of mammalian body temperature was conclusively demonstrated to be endogenous. In incredibly laborious studies, without benefit of automation, Simpson and Galbraith (1906) measured the rectal and axillary temperature of five monkeys every two hours day and night for 60 days. They demonstrated not only that there was a body temperature rhythm but that it persisted in constant darkness or constant light with an endogenous periodicity. Furthermore, this rhythm appeared to be synchronized by the 24-hour light–dark cycle, since the rhythm gradually became inverted when the schedule of light and dark was reversed.

The properties of circadian clocks in mammals were first extensively characterized by Curt Richter. In his Ph.D. thesis, published in 1922, he demonstrated the endogenous nature of circadian activity rhythms in the rat and showed that they were synchronized by light–dark cycles and the time of feeding. Over the intervening 60 years he has continued to be a pioneer in the determination of the properties of circadian clocks in a wide variety of mammalian species.

The endogenous nature of the human circadian system was not confirmed until much more recently. Aschoff and Wever (1962) isolated individuals in a sealed cellar below a Munich hospital for 8–19 days in the absence of any environmental time cues, and Siffre (1964) lived alone in an underground cavern for two months in 1962. Both experiments revealed that the spontaneous or free-running period of the human activity–rest cycle was greater than 24 hours, averaging 25 hours, indicating that humans, like other animals, have an endogenous circadian system. Since that time several hundred human subjects have been studied in isolation from environmental time cues in a number of different laboratories.

Another issue that concerned investigators in the 1960s was how widespread were circadian rhythms. A host of papers was published documenting circadian rhythms in hundreds, if not thousands, of biological variables. Particular attention was paid to circadian rhythms in humans because of their potential importance in medicine. Compendia such as *Human Circadian Rhythms* by Conroy and Mills (1970) documented circadian rhythms in hundreds of variables. Halberg and his collaborators, in particular, have devoted considerable effort to plotting phase maps of a wide variety of rhythms (for example, Scheving, Halberg, and Kanabrocki, 1977). As the documentation of rhythms in various human physiological systems proceeded apace, it became apparent that it was often more significant to find no circadian rhythm in a physiological variable than to find one. In a highly rhythmic animal in which multiple control systems show circadian rhythms, a particularly elaborate physiological strategy is required to maintain a variable absolutely constant throughout day and night. Rhythmic systems must counterbalance each other to achieve the nonrhythmic function. One example is blood pressure in the squirrel monkey (Fig. 1.7), which shows no detectable circadian rhythm but is regulated by endocrine systems and renal salt and water excretion mechanisms that are highly rhythmic in their function (Kass, 1980).

THE ANATOMY AND PHYSIOLOGY OF CIRCADIAN TIMEKEEPING

Particular interest has been focused recently on the identification of circadian pacemakers and the study of their functions. As long ago as 1918 neuropathological studies in humans had indicated that an area in the anterior hypothalamus was important in the regulation of sleep (von Economo, 1929). Nauta (1946) further localized this "sleep center" to the suprachiasmatic-preoptic area on the basis of lesioning studies in rodents. The first direct search for a circadian pacemaker in mammals, however, was not undertaken until the 1960s. Curt Richter, in an extensive series of experiments, placed lesions in various locations throughout the central nervous system of rodents and removed a number of endocrine glands in a search for "the biological clock." His findings were never fully published, but his brief reports suggest that the only lesions that would disrupt circadian rhythmicity in wheel-running behavior in rodents were those placed in the anterior-ventral hypothalamus (Richter, 1965, 1967). Then in 1972 two independent groups identified a small, bilateral pair of nuclei in the anterior hypothalamus, the *suprachiasmatic nuclei* (SCN), as a potential circadian pacemaker (Moore and Eichler, 1972; Stephan and Zucker, 1972). Lesions that destroyed the SCN were found to eliminate circadian rhythmicity in many physiological and behavioral variables, including sleep and wakefulness. This finding has stimulated considerable research activity in the last ten years.

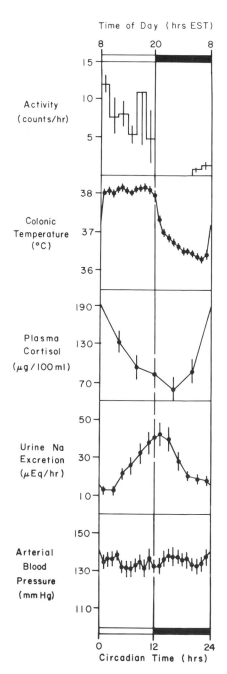

Fig. 1.7 Variation in physiological functions with the time of day in squirrel monkeys maintained in a light–dark cycle. Each curve represents the mean value of several days of data plus the standard error of the mean. There were prominent diurnal rhythms in each of these variables except arterial blood pressure. Although the monkeys were in a periodic environment and showed rhythms in many functions that influence blood pressure, no significant diurnal rhythmicity was seen in this variable. (Copyright 1982 by Moore-Ede, Sulzman, and Fuller.)

We now know that circadian systems in mammals are composed of more than one potentially independent oscillator. Considerable efforts are under way to map the locations of these oscillators and to identify the neural and endocrine modes of temporal communication among them. Concurrent with these anatomical and physiological studies have been studies of the biochemical mechanisms that may be involved in generating circadian rhythms. Early studies indicated that relatively few chemical agents could interfere with circadian systems (Hastings, 1960), although it now appears that certain classes of chemicals, such as the methylxanthines (caffeine is the most common substance in this family of chemicals), can reset the timing of circadian clocks (Mayer and Scherer, 1975; Ehret et al., 1975). The search for the biochemical basis of circadian timekeeping has not been as productive as the anatomical studies, although a number of new genetic approaches offer some hope for the future.

MEDICAL IMPLICATIONS

As the diverse and significant role that circadian timekeeping plays in biological systems became apparent, the implications for human health and disease have become more widely recognized. For example, many drugs show circadian rhythms in both their toxic and therapeutic effects (Halberg, 1960; Reinberg, 1967; Moore-Ede, 1973), and these circadian rhythms have been manipulated to provide the most effective and least toxic timing regimes for cancer chemotherapy and many other treatments. In Chapter 7 we will discuss these issues as well as the problems of disparity between environmental and body time that occur in jet lag and shift work. We will also look at new areas of research which indicate that the circadian timing system (like other body systems) can malfunction and lead to discrete pathological syndromes. For example, certain insomnias (Czeisler, Richardson, Coleman, et al., 1981) and affective disorders (Wehr et al., 1979) appear to be the result of disorders in the circadian timekeeping system.

These studies of the significance of circadian rhythms for human health and disease should be aided by the recent identification of a putative circadian pacemaker in man (Lydic, Schoene, et al., 1980; Moore-Ede et al., 1980). A cluster of neurons in the human hypothalamus was identified which appears to be homologous to the suprachiasmatic nuclei in other mammals, including nonhuman primates. It is fascinating that this nuclear group is located in a region of the hypothalamus where pathological damage has been correlated with disruptions of the circadian sleep–wake cycle (Fulton and Bailey, 1929; von Economo, 1929).

Measurement and Interpretation

The uninitiated person starting to read the literature on circadian rhythmicity, and indeed the experienced circadian clockwatcher, may feel bombarded by a profusion of unfamiliar terms. Only a few of the terms have entered common usage, and we will try to keep the specialized descriptive terms to a minimum as we discuss the characteristics of circadian clocks. Most terms are used across many scientific disciplines and need little explanation. Figure 1.8 provides an introduction to the parameters which describe rhythms, time cues, and reference time scales. These and other terms are defined again in the text and discussed in more detail in the Glossary.

CLOCK MECHANISM VERSUS HANDS

As we start to delve into the workings of biological clocks, it is useful to distinguish between the "mechanism" and the "hands" of the clock. Most of the circadian rhythms that are observed in a wide variety of behavioral, physiological, and biochemical variables are analogous to only the hands of the clock. If we interfere with the rhythmic expression of some biological variable—to take an extreme example, if we tie the limbs of an animal so that it cannot show its circadian pattern of activity—the underlying circadian clock keeps on ticking although its hands (and feet) are tied. Once the animal is released, it will resume activity at the circadian time determined by the mechanism of the clock, which has been running without interruption throughout. Only if the clock mechanism is manipulated does the animal's activity rhythm show any lasting effect.

Although the definition of a circadian rhythm is straightforward, we must ask how one would recognize a circadian clock. Two key properties must be demonstrated for any biological structure to be considered a clock. First, the structure must measure the passage of time indepen-

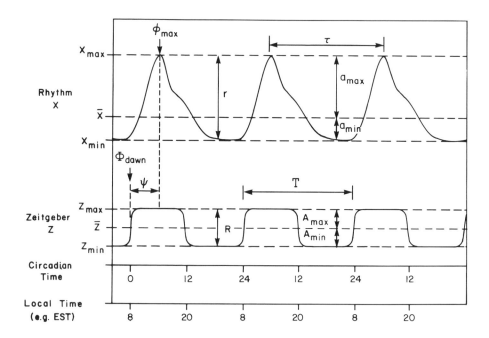

Fig. 1.8 The important parameters of biological rhythms, *zeitgebers* (time cues; literally, "time givers"), and reference time scales. *Top panel,* rhythm of a biological variable, x; *bottom panel,* cycle of an environmental zeitgeber, Z, which entrains the rhythm (that is, resets the rhythm every day to a 24-hour period). These variables are plotted against two different time scales. The *local time scale* indicates the time of day at which the biological measurement was made (for example, Eastern Standard Time in Boston, or Greenwich Mean Time in London). The *circadian time scale* standardizes the relationship of the biological variable to the zeitgeber cycle by defining circadian time (CT) 0 as dawn (CT 24 = CT 0). The parameters for the zeitgeber are given in capital letters, and those for the rhythm in lower case. Identified for rhythm and zeitgeber cycles are the mean values of the variable ($\bar{x}$ and $\bar{Z}$), the maximum and minimum values (x-max and x-min, Z-max and Z-min), the ranges of the oscillations (r and R), the periods of the oscillations (τ and T), the amplitudes (a-max and a-min, A-max and A-min), and the reference phases (ϕ and Φ). The figure also shows the phase relationship (ψ) between dawn on the zeitgeber cycle and the maximum of the rhythm, but ψ can be defined between any two reference points on the zeitgeber and rhythm waveforms. (Copyright 1982 by Moore-Ede, Sulzman, and Fuller.)

dently of any periodic input from its environment. It must, therefore, convert a nonperiodic source of energy into a self-sustaining periodic output. Tissues that display circadian rhythms but cannot sustain circadian oscillations on their own do not contain a clock; they are therefore driven by a circadian clock elsewhere in the organism. If the circadian timekeeping function is the result of interaction among several discrete structures, none of which can independently sustain rhythmicity, then the clock is the minimum combination of essential structures. As we search for circadian clocks, we are therefore looking for the smallest entity, whether it be an organ, a cell, or a subcellular fraction, that can measure circadian time in the absence of time cues in the environment.

Second, for a tissue displaying periodicity to be considered a clock, it must be used to time biological events. Thus, an oscillation in a biological system caused perhaps by some instability in a control system cannot be considered a clock if it does not time other events within the organism.

Let us consider for a moment what we mean by timekeeping. To keep time, a clock must have some degree of both resolution and uniformity (Cannon and Jensen, 1975). The *resolution* of a clock is a measure of its ability to detect the temporal order of two events closely spaced in time. If the events are closer together than a clock can resolve, the clock will not be able to distinguish reliably which event precedes the other. The *uniformity* of a clock is a measure of the regularity of its period and, consequently, the clock's ability to predict the occurrence of other regularly timed phenomena in the universe. Of course, in theory one can never be certain whether the tested clock or the reference clock has the lower uniformity or resolution. However, this is rarely a problem in biology because modern atomic clocks have far higher resolution and uniformity than any biological clock.

Compared to other biological clocks, circadian clocks have relatively high resolution; that is, they can measure time intervals considerably shorter than their own period. Their uniformity is also relatively high, since circadian clocks show little variation in period from cycle to cycle. These circadian periods also do not deviate much (often less than 5%) from the environmental cycles they are used to predict. And because circadian clocks are normally "reset" each day by cues in the environment, this error is not cumulative. (The daily resetting of the clock is referred to as *entrainment*.)

To study the characteristics of a circadian clock, one must be able to accurately record its output. Most behavioral, physiological, and biochemical variables in animals show circadian rhythms, but there are considerable problems in determining which rhythm is the output of which clock and how accurately each rhythm reflects the clock's behavior. A fundamental problem is that in no case has the exact biochemical or biophysical structure of a circadian clock been identified. We can identify specific cells or groups of cells which appear to contain circadian oscillators, but we are far from a complete understanding of their mechanism.

Ideally we would monitor the clock directly, although we would be content to monitor any rhythm that accurately represents the behavior of that clock. Such a rhythm must be governed by that clock alone and must not be influenced by exogenous inputs or by other clocks within the animal (except when these other influences reset the timing of the clock under study). The timing of the rhythm (that is, when it peaks and troughs) with respect to the timing of the clock's own oscillations (the *phase relationship*) is unimportant as long as it is stable.

In practice, such a rhythm is extremely difficult to find and even more difficult to measure. The ideal rhythmic variable should be measurable within an individual animal at many times per circadian cycle for many successive cycles. Ideally the behavior or physiology of the animal should not be influenced by the process of measurement. Furthermore, the process of measurement and recording must be automatable unless a large group of collaborators is available to work shifts around the clock.

An example of a rhythm that meets most of these criteria is the circadian rhythm of wheel-running activity in rodents, which was first studied by Richter (1922) using the running wheels originally developed by Stewart (1898) and Slonaker (1908). Hamsters, for example, will run with a highly reproducible pattern at the same time each day (Fig. 1.9). Many successive circadian cycles can be monitored for the animal's whole lifetime if necessary. Other rhythms that can readily be monitored are feeding and drinking behaviors and body temperature via telemetry from an implanted capsule. Much of the information

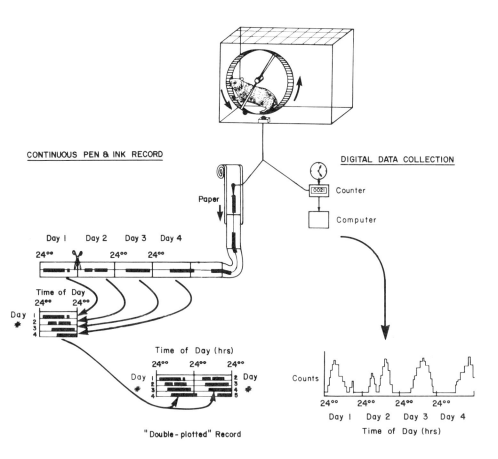

Fig. 1.9 Methods of recording the wheel-running activity of a hamster. In the traditional method, each wheel rotation activates a sideways movement of a pen on a moving sheet of paper. When the 24-hour strips are arranged one under the other, the human eye can easily pick out rhythmic behavior in the daily patterns of activity. The 24-hour strips are often double-plotted so that the continuity of the rhythmic pattern can be visualized (*lower left*); on the first line, day 1 plus day 2 are plotted, and on the next line day 2 plus day 3, and so on. For more objective analyses, a digital record can be made. The amount of activity can then be plotted against time (*lower right*) and analyzed statistically. Rodents may run for miles during a night, and a digital record may be necessary to distinguish between different amounts of activity. (Copyright 1982 by Moore-Ede, Sulzman, and Fuller.)

on the overall properties of the circadian timing system has been gained using such rhythms as markers for the whole system.

Unfortunately, the most easily measured rhythms may be only loosely coupled to the clock under study. To obtain a more direct measure of the circadian clock it may be necessary to measure a less accessible variable. Especially when we attempt to uncover the anatomical and physiological basis of circadian timekeeping, we often find it necessary to use invasive techniques to measure rhythms. Sometimes it is possible to use implanted arterial or venous catheters if the rhythm in some blood constituent is to be measured (Moore-Ede et al., 1977). However, often the animal must be sacrificed to obtain the measurement needed.

If a variable cannot be measured more than once in a given animal, such as when the animal is sacrificed, many pitfalls await the investigator. Since the characteristics of the circadian timing system vary not only among species but also among individuals within a species, pooling the results from several animals can lead to special problems. If the animals in a group have different free-running periods and are placed in a situation with no external time cues, then the timing (*phase*) of each of their rhythms will gradually disperse "around the clock." Figure 1.10 shows that the mean results from such a group may have no visible rhythm, even though each individual animal (contributing toward only one data point at one time of day) may have a prominent rhythm in the variable under study.

For example, the documentation of rhythms in pineal enzyme activity requires sacrificing several rats at each phase of the day. Pohl and Gibbs (1978) have demonstrated that the rhythm of pineal N-acetyltransferase (NAT) activity has a prominent peak in the middle of the night, but it appears to be arrhythmic in blinded rats if one looks at results pooled by solar time. However, when results are pooled by each individual animal's *subjective* time as indicated by the phase of the circadian rest–activity cycle when it was sacrificed, the prominent rhythm again becomes apparent. One can imagine the interpretative problems for an unaware investigator searching for the mechanisms driving the NAT rhythm.

There are several approaches to this problem. First, sacrificing the animals shortly after their release into constant conditions lessens the time in which the free-running rhythms of individual animals can disperse. Thus the mean rhythm of the group will more closely represent the rhythms of the individual animals. Second, one can use another rhythm as a marker for each animal and express the results with respect to the phase of the marker rhythm. In the example cited above, Pohl and Gibbs (1978) found a prominent rhythm in NAT when the data were correlated with the rat's rest–activity cycle. The problem with this approach is that it assumes internal synchrony between the rhythm under study and the marker rhythm, a condition that may not always occur.

In any biological experiment, and particularly in the study of mammals, the investigator must constantly wrestle with the problem of making reliable measurements without disturbing the myriad of variables that influence the function under study. The problems are compounded for the circadian clockwatcher because of the long time course of his experiments. Events which occurred many weeks, months, or even years before may have altered the current behavior or timing of a biological clock, as we will discuss in Chapter 2. But it is this feature of biological clocks, so complicating to the experimenter's life, which enables these clocks to serve as an important source of temporal memory for the animal.

ACTIVE AND PASSIVE COMPONENTS

The adage that "one person's noise is another person's signal" is only too true for biologists attempting to sort out the complexity of concurrent events. Circadian oscillations in biological variables have often been treated as a troublesome variation in "baseline" values by investigators who have been interested only in direct responses to stimuli; circadian rhythms in their data were consequently treated as one more source of error that had to be controlled by undertaking experiments at the same time of day. Unfortunately, such an approach does not always work, because in certain conditions circadian rhythms can free-run with a period different from 24 hours. Thus the rhythmic variation cannot be predicted from the clock hour, and the effect of circadian variation may not be successfully minimized.

In contrast, for those who are studying timekeeping mechanisms,

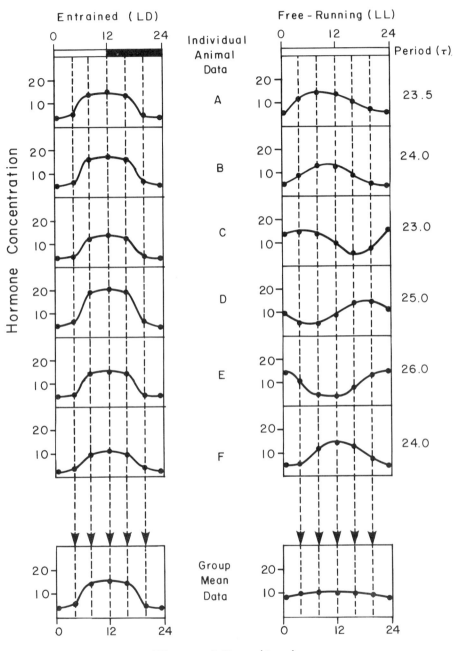

Time of Day (hrs)

Fig. 1.10 Hypothetical averaged hormone rhythms calculated from groups of entrained and free-running animals. The animals on the left are all entrained to a 24-hour period by a cycle of 12 hours of light and 12 hours of darkness (LD 12:12). These rhythms are in phase with each other: low values occur at CT 0 (circadian time) and high values at CT 12. The group mean data (*lower left*), accurately represent the average rhythmic pattern of the group. The animals on the right are all free-running in constant light (LL); their endogenous periods are listed to the right of each record. Because the individual periods are different, after several days of constant conditions the hormone rhythms have become out of phase. Some animals show peaks of activity in the first half of the day, and others in the second half. Because of these phase differences, the group mean data (*lower right*), does not accurately reflect the patterns of the individual animals. In this example the mean pattern shows no rhythm, even though each of the six animals was rhythmic. (Copyright 1982 by Moore-Ede, Sulzman, and Fuller.)

the rhythmic characteristics of biological variables are, of course, the observation of critical interest. The direct stimulus-response effects that have been the concern of most biologists then become the "noise" (or *masking* or *exogenous* effects) for the circadian physiologist. Such effects can alter the observed rhythm without influencing the underlying circadian clock that times the rhythm. As a result, they may obscure the behavior of the clock under study.

This problem is diagrammatically illustrated in Figure 1.11. The rhythm that is measured is the product of both the circadian clock and direct random influences from the environment. For example, humans have a well-defined body temperature rhythm that is actively generated by endogenous circadian clocks (Aschoff, 1970). The rhythm has a maximum during late afternoon and a minimum in the early hours of the morning. Nevertheless, a hot bath at any time of day or night will directly cause a rise in body temperature that may alter the observed body temperature rhythm and thereby obscure experimental measurements of the clock's behavior. As we examine the characteristics of circadian clocks, it will become apparent that these passive effects can be separated from the output of the circadian clocks in a number of ways. Part of the discussion in the following chapters will be concerned, therefore, with the experimental strategies that can be used to isolate circadian phenomena from other events.

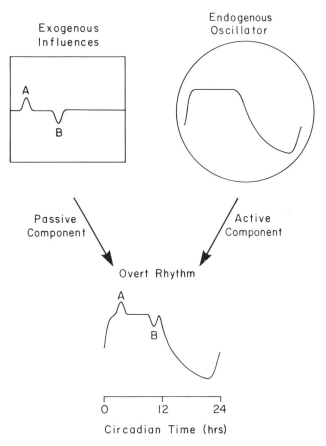

Fig. 1.11 The contributions of exogenous influences and endogenous oscillations to an overt rhythm such as body temperature. *A* represents an increase in body temperature due to a hot shower in the morning and *B* represents a drop in body temperature due to eating a liter of ice cream. These passive components will change body temperature whenever they occur. *Right,* the rhythmic pattern of body temperature as timed by endogenous circadian oscillators. The measured overt rhythm (*below*) reflects both the endogenous rhythm and the exogenous influences. (Copyright 1982 by Moore-Ede, Sulzman, and Fuller.)

THE STUDY of the anatomy and physiology of the circadian timing system in mammals is young. Much is not yet known or is incompletely understood, though the general principles of circadian organization are now apparent. We will try to make clear the boundaries between established fact, hypothesis, and speculation, in the hope of stimulating further research. Some of our postulates will undoubtedly prove wrong, but often most is learned on those occasions when research yields unexpected findings. Having in mind a framework that can guide one to the most critical experimental questions is far better than having no scheme and thus never taking the risk of being wrong.

Characteristics
of Circadian Clocks

When de Mairan, 250 years ago, demonstrated that circadian rhythms persist even when an organism is isolated from time cues, he originated a technique that has been of considerable value in studying the characteristics of circadian clocks. By taking an animal out of its natural environment and placing it in a chamber where the levels of light, temperature, food, and sound are kept constant, the animal's direct responses to events in its environment, which ordinarily may obscure endogenous circadian rhythms, can largely be eliminated. The circadian rhythms that emerge under these constant conditions thus more directly reflect the behavior of the clock (or clocks) under study.

Although some long-term changes occur in the behavior of the circadian timing system when it is freed from the restraints imposed by the environment, the advantages to the experimenter of being able to isolate the system normally outweigh the interpretive problems that may arise. Besides facilitating the analysis of circadian clocks, studies of animals free of environmental time cues can also provide direct insights into situations that occur in nature. Animals living deep in dark caves or in polar regions during the summer or winter solstices, or animals that become trapped during bad weather, may all be deprived of effective circadian time cues. For example, beavers living under the thick ice covering Canadian lakes during the winter have been shown to have free-running circadian rhythms rather than being entrained to a 24-hour period (Bovet and Oertli, 1974), whereas in the summer they become synchronized to a precise 24-hour day. The study of the circadian timing system when it is isolated from effective time cues also has clinical relevance. As we will discuss in Chapter 7, diseases may interrupt entrainment by environmental time cues (for example, in a patient who has become blind) or patients may be isolated from the regular alternation of day and night, as in intensive care units, where there may be little or no contrast between the daytime and nighttime environments.

The Free-Running Clock

Most mammals show persisting circadian rhythms when isolated from environmental time cues. The rest–activity cycle of each of the species shown in Figure 2.1, from rodents to humans, progressively drifts with respect to clock time and thus free-runs with a period either a little shorter or longer than 24.0 hours. Such free-running periods are one indication that circadian rhythms are driven by autonomous clocks within the organism and are not the product of some undetected environmental stimulus.

Each species shows a characteristic free-running period (Fig. 2.2), with the periods of the individual animals normally distributed around the species mean. Most species have free-running periods displaced somewhat from exactly 24.0 hours. This is not because evolution failed to produce a clock mechanism of sufficient precision. Pittendrigh and Daan (1976b) have shown instead that stable entrainment to the natural 24-hour light–dark cycle becomes more difficult the closer an animal's natural period is to 24.0 hours. Hence there may have been selection pressure to evolve circadian clocks with free-running periods deviating somewhat from 24 hours.

Some animals, such as cats and dogs, do not demonstrate clearly defined circadian activity rhythms. As Figure 2.3 shows, when a cat is placed in a room with no time cues, it will show bursts of activity for an hour or two followed by an hour or so of rest, an irregular pattern that is repeated throughout the day and night (Hawking et al., 1971). While there are some indications of a weak daily organization of activity around dawn and dusk in cats kept on a rigid 24-hour schedule (Kavanau, 1971; Lucas, 1978), investigators who have maintained cats in time-cue-free environments have reported little evidence of self-

Time of Day (hrs)

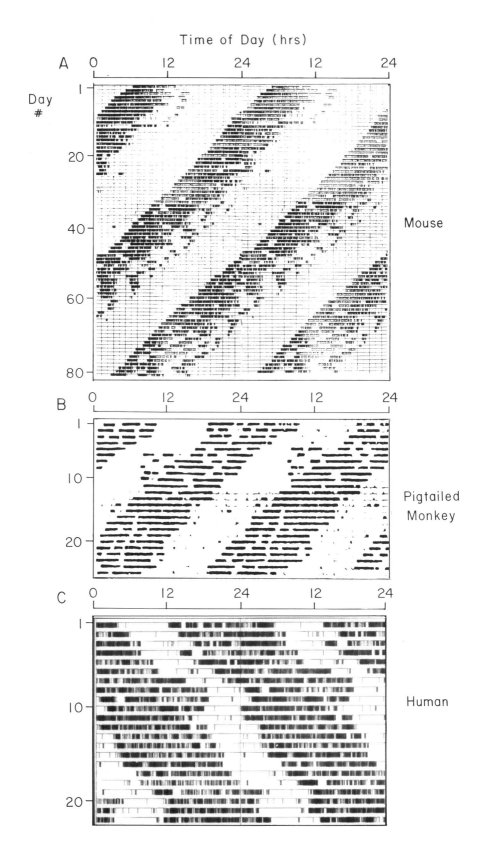

A

0   12   24   12   24

Day #

1

20

40

60

80

Mouse

B

0   12   24   12   24

1

10

20

Pigtailed Monkey

C

0   12   24   12   24

1

10

20

Human

Fig. 2.1 Free-running activity rhythms of individual animals from three mammalian species living without time cues: (A) mouse (*Mus musculus*); (B) pigtailed monkey (*Macaca nemestrina*); (C) human. The time of activity onset (and offset) occurred earlier (*A, B*) or later (*C*) each day. Thus each individual had a free-running period either shorter (*A, B*) or longer (*C*) than 24.0 hours. (*A* reprinted from Ebihara, Tsuji, and Kondo, 1978, reprinted by permission of Pergamon Press; *B* from Aschoff, 1979a, reprinted by permission of Raven Press; *C* kindly provided by J. Zimmerman, Department of Neurology, Montefiore Hospital, New York; copyright 1982 by Zimmerman et al.)

sustained circadian rhythms (Sterman et al., 1965; Hawking et al., 1971), although spectral analysis techniques are sometimes able to detect weak circadian rhythms (Moore-Ede et al., 1981). This virtual absence of circadian rhythmicity appears to be a function of the cat's natural behavior as a predator rather than a consequence of centuries of domestication. Kavanau (1971) has shown that some predators captured from the wild, such as bobcats (*Felis rufus*), similarly show a lack of clear circadian organization, except for bursts of activity during twilight.

The loss of activity rhythms in constant conditions does not necessarily mean that cats and dogs have no circadian clocks. Rather it appears that the coupling of their activity rhythms to the clocks has been lost. Other variables in cats, such as the hormones vasopressin and oxytocin in the cerebrospinal fluid, may continue to show prominent circadian rhythms (Reppert, Artman, et al., 1981). However, those who use cats and dogs in physiological studies should be aware that they do not possess the same circadian organization as humans and most other mammalian species.

PERSISTENCE UNDER CONSTANT CONDITIONS

Self-sustained circadian rhythms will normally persist for considerable lengths of time in an environment without any time cues. Several investigators have reported that circadian rhythms can persist for many years, even for the lifetime of the animal. For example, Richter (1968) blinded the squirrel monkey in Figure 2.4 in July 1964, and the animal showed a persisting circadian rhythm of activity no longer entrained to the environment for as long as it was studied, through September 1967. Richter has studied other individual animals for as long

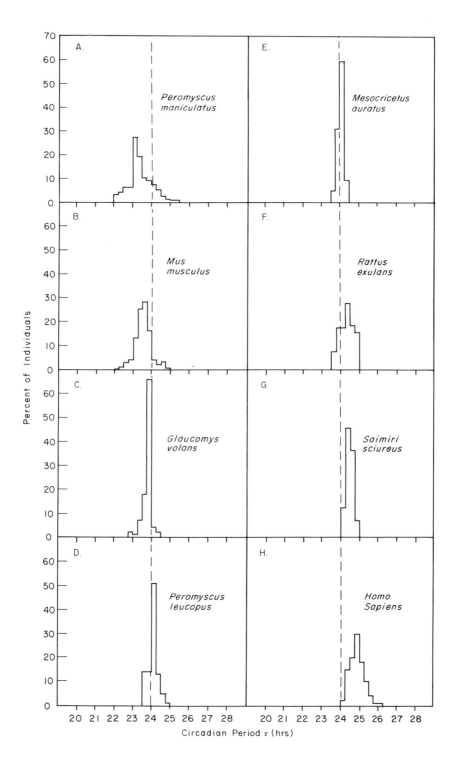

Fig. 2.2 The normal distributions of free-running periods among individuals of eight mammalian species: (*A*, *B*, *D*) three species of mice; (*C*) flying squirrel; (*E*) hamster; (*F*) Polynesian rat; (*G*) squirrel monkey; (*H*) human. All experiments except those with humans were conducted in constant darkness. Each species showed a characteristic distribution, with the mean period displaced one side or the other from 24.0 hours. (*A*, *B*, *D*, *E* from Pittendrigh and Daan, 1976a; *C* from DeCoursey, 1960a, copyright 1960 by the American Association for the Advancement of Science; *F* kindly provided by P. Gander, copyright 1982 by Gander; *G* copyright 1982 by Moore-Ede, Sulzman, and Fuller; *H* from Wever, 1979.)

as ten years. Once, only half in jest, he suggested that he could predict the time the monkey would start its activity the next Christmas Day, the clock was so precise! Maybe an even more impressive demonstration of the persistence of rhythmicity has been provided by Aschoff (1960), who reared several successive generations of mice in isolation from environmental time cues. These animals continued to show circadian rhythms generation after generation, despite the fact that every mouse in subsequent generations had never seen a light–dark cycle.

Not all circadian rhythms persist for that length of time in constant conditions, however. Some rhythms damp out after a few days or weeks, especially in constant light. For example, Honma and Hiroshige (1978) have shown that the rhythm of locomotor activity in Wistar rats becomes less distinct with time and is eventually dominated by ultradian rhythms (rhythms with shorter than circadian periods called "ultra" because their frequency is higher). This is more common in some strains of rodents than in others (Oliverio and Malorni, 1979).

The cause of such damping is often difficult to determine. With lower organisms, in which it is easier to study such phenomena, a circadian clock may stop functioning when the temperature is too low or the light intensity too high, a phenomenon which has been termed *conditionality* (Njus, McMurry, and Hastings, 1977). However, no counterpart has as yet been demonstrated in mammals. Alternatively, it is possible that several clocks are contributing to the rhythm and that in constant conditions they uncouple and free-run independently; hence no coherent circadian rhythm is seen (see Chapter 3). Finally, we must recognize that the clock driving the rhythm may still be running, but the rhythmic events we are monitoring (the "hands" of the

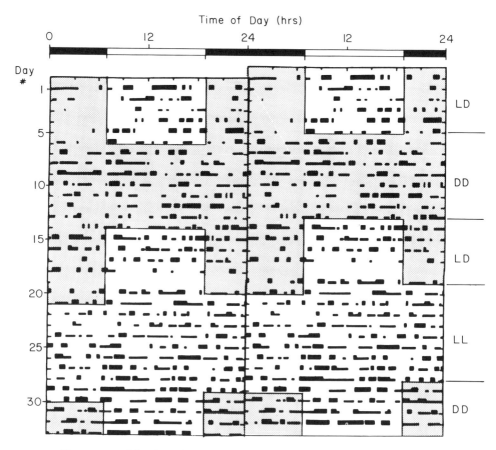

Fig. 2.3 Double-plotted activity record of a cat studied in isolation with various light–dark regimens. Days 1–6, light–dark cycle (LD 12:12); days 7–14, constant darkness; days 15–20, LD 12:12; days 21–28, constant light; and days 29–32, LD 12:12. Shaded areas represent darkness. Although there was some temporal organization of activity in a light–dark cycle, the cat appeared to have no endogenous circadian activity rhythm in constant darkness or constant light. (From Hawking et al., 1971.)

clock) may have become uncoupled from the clock, so that no overt rhythm is seen.

UNIFORMITY

Free-running circadian clocks vary in their uniformity—the stability of their period. Some are timed with impressive precision; others have

Time of Day (hrs)

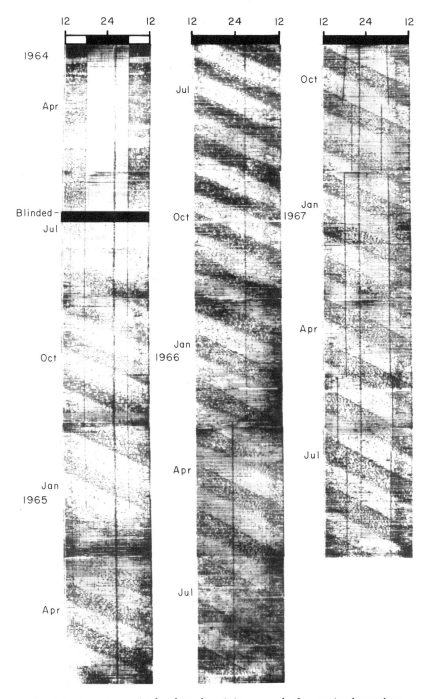

Fig. 2.4 Long-term single-plotted activity record of a squirrel monkey. From February to June 1964 the animal was maintained in an LD 12:12 cycle. Activity was confined to lights-on each day. After the animal was blinded in June 1964, a clear, free-running rhythm was seen (July 1964 to September 1967). Little change was seen in the approximately 25-hour free-running period over the three years of study. (From Richter, 1968.)

less stable periods that drift somewhat with time. This is a key characteristic of a clock, since its usefulness as a timekeeper is determined by its uniformity and therefore its ability to predict the occurrence of other regularly spaced events in its environment.

But measuring the uniformity of a circadian rhythm is not the same as measuring the uniformity of the circadian clock that drives it. Even if a circadian pacemaker emitted perfectly regular signals, the physiological transmission processes that couple the observed rhythm to the clock might have a variable time delay. Consider the circadian activity rhythm. The clock may emit a perfectly regular signal that initiates the waking-up process and starts the animal's daily bout of activity. However, the time to actually wake up on different days might be highly variable and the animal would consequently have a "sloppy" rhythm, with the interval between activity onsets on successive days (that is, the *circadian period*) varying considerably. In other words, the coupling mechanism between internal clock and observed rhythm may contribute to an apparent imprecision in timing.

How can one distinguish between the uniformity of the clock and the uniformity of the transmission process? Pittendrigh and Daan (1976a) found that if the period of the rest–activity cycle in mice on any given day was longer than average, there was a greater than 50% probability that the period of the next day would be less than average. Similarly, if the period was shorter than average, the next day the period would tend to be longer. Their analysis of mice showed that the clock's period (τ_c) was about twice as uniform as the already remarkably uniform period of the activity rhythm (τ_r) which it drove. That is, the clock had a standard deviation of about 9 minutes in estimating 24 hours (0.6% of τ_c), whereas the rhythm had a standard deviation of about 18 minutes (1.2% of τ_r). Furthermore, the difference between individual mice with sloppy rhythms and those with highly uniform rhythms was caused not by a difference in the uniformity of the underlying circadian clock but, instead, by variability in the transmission processes from the clock to the overt rhythm.

TEMPERATURE COMPENSATION

The uniformity of circadian clocks from day to day has been enhanced by the evolution of mechanisms that compensate for changes in environmental temperature. Such compensation is essential, for if the periodicity of circadian clocks were to depend on temperature, their usefulness as timepieces would be limited. In fact, this was a problem that had to be faced by the early builders of man-made clocks, for an increase in temperature would lengthen the pendulum, causing it to swing more slowly, thereby slowing the clock down. It was not until the eighteenth century that clockmakers learned to compensate for this thermal expansion and therefore could begin to construct accurate clocks.

Biological clocks face the same problem. The rate of most biochemical processes changes two- or threefold with each 10° C change in temperature (that is, the $Q_{10} = 2$–3). If the circadian clock depended on a simple metabolic process, the animal would estimate time incorrectly when the environmental temperature changed, an error that could jeopardize the animal's welfare.

Most of the work on the temperature compensation of circadian clocks has been done with plants and insects that do not thermoregulate, so the effects of environmental temperature can be readily studied. In these species the ratio between the period of their circadian rhythms before and after a 10° C increase (the Q_{10}) is typically in the range of 0.85–1.3 (Sweeney and Hastings, 1960). This appears to be the result of active compensation rather than the clock's insensitivity to temperature (Hastings and Sweeney, 1957). The exact mechanism of temperature compensation is as yet poorly understood, but a number of effective methods can easily be envisaged. For example, timekeeping could be the net result of two reactions, one producing a substance that inhibits the other, and both reactions speeding up with increasing temperature. There is evidence in microorganisms that this temperature compensation process is genetically inherited (Feldman, 1971).

Mammals are protected from environmental temperature changes by the fact that they are homeotherms; except when they hibernate, their body temperature is tightly regulated within narrow limits. Have mammals also inherited temperature-compensated circadian clocks, even though they thermoregulate? In hibernating mammals the requirement for temperature compensation is obvious, but even in a nonhibernator the performance of circadian clocks may be enhanced. Body temperature in many mammals has a prominent circadian rhythm with a range (from daily maximum to minimum) of about 2° C. If an animal had an uncompensated pacemaker with a Q_{10} of 2–3, the free-running period could lengthen transiently by several hours during each temperature trough.

While mammals will tolerate only relatively small increases in body temperature, they will survive considerable cooling, in some species to as low as 0° C (Andjus and Smith, 1955). Such cooling techniques have made possible studies of the temperature dependence of mammalian circadian clocks (Rawson, 1960; Richter, 1975; Gibbs, 1981), and have indicated that the clocks are indeed temperature compensated. Nonhibernators such as the rat (Q_{10} = 1.2–1.4) appear to be less well compensated than hibernators such as hamsters and bats (Q_{10} = 1.01–1.13) with the period of circadian clocks more precisely compensated at temperatures near normal body temperature (Gibbs, 1981). However, during hypothermia at temperatures near freezing (0–5° C) Richter (1975) found that circadian clocks in animals can stop altogether, although some animals apparently can continue to measure time even when heartbeat and respiration have ceased. Although temperature compensation is not perfect, the remarkable uniformity normally shown by circadian clocks in mammals attests to the role that temperature compensation and perhaps other homeostatic compensations do play (Pittendrigh and Caldarola, 1973; Albers, 1981).

LIGHT-INTENSITY EFFECTS

One determinant of the length of the free-running period of a circadian clock in a constant environment is the intensity of the light to which the animal is exposed. In 1960 Aschoff showed that more intense light shortens the free-running period in diurnal organisms but lengthens it in nocturnal organisms. The influence of light intensity on the free-running period was actually first observed by Johnson (1939), but it was not until Aschoff systematically studied this effect in both nocturnal species, mainly rodents, and diurnal species, mainly birds, that their distinctly different behaviors were made apparent.

Aschoff (1960) further concluded that under more intense light the time an animal is active, compared with the time it is at rest, increases in diurnal species but decreases in nocturnal species. Also, in diurnal species the total amount of activity during a free-running period increases with light intensity, while the reverse is true in nocturnal species. Aschoff was less confident about a third generalization—that the free-running period is longer than 24.0 hours for diurnal animals in constant darkness and shorter than 24.0 hours for nocturnal species under the same conditions.

In Figure 2.5 (A, B) the responses of the nocturnal deermouse (*Peromyscus maniculatus*) to changes in light intensity can be seen to conform to Aschoff's three rules. The transfer of the mouse from 200 lux to darkness, or vice versa, produces a sharp change in the free-running period, with longer periods in the brighter light (Pittendrigh, 1967). At the same time, the activity/rest ratio is decreased and the total daily activity reduced, with a shorter daily episode of activity in bright light. The deermice in Figure 2.5 also conform to Aschoff's third proposition, since their free-running period is typically less than 24 hours in constant darkness. Most other nocturnal mammals similarly conform to Aschoff's rules (see Fig. 2.6A).

However, diurnal mammals show much less consistent behavior with different light intensities. Aschoff's rules apply to some species, such as the diurnal ground squirrel (*Ammospermophilus leucurus*) shown in Figure 2.5C. With increasing light intensity this animal shows a decrease in the free-running period, an increase in the activity/rest ratio, and an increase in total activity. Furthermore, in constant darkness its free-running period is greater than 24 hours (Kramm, 1971).

Many other diurnal mammals do not conform to Aschoff's original rule. For example, when we subjected squirrel monkeys (*Saimiri sciureus*) to a decrease in light intensity from 1,800 to 1 lux, the free-running period was decreased by 0.5 hours, while the activity/rest ratio followed Aschoff's predictions (Fig. 2.5D; Sulzman, Fuller, and Moore-Ede, 1979a). Martinez (1972) examined the activity rhythm of rhesus monkeys over a wide range of lighting conditions from darkness to 1,250 lux, including five intermediate intensities. The data indicated that the period was constant up to about 50 lux, then increased with brighter light. The activity/rest ratios did not change in the different lighting conditions. Farrer and Ternes (1969), measuring the feeding rhythm of a chimpanzee in isolation, found that at 1 lux the period was 23.8 hours and at 85 lux it was 25.1 hours. Visual examination of their published data indicates that the activity/rest ratio decreased as the light became brighter. Hoffman (1965) reported that the free-run-

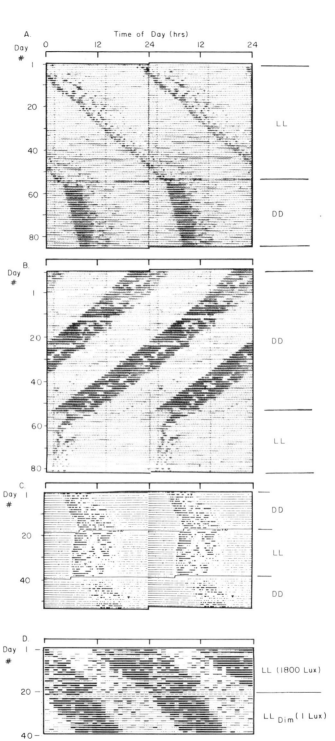

A.

Time of Day (hrs)

Day #

B.

Day #

C.

Day #

D.

Day #

Fig. 2.5 Light-intensity effects on circadian free-running period. (A) The nocturnal deermouse conforms to Aschoff's rules. A change from constant light (200 lux) to darkness resulted in a shortening of the free-running period and a lengthening of the daily activity bout. (B) Conversely, transferring a mouse from constant darkness into constant light (200 lux) lengthened the free-running period and reduced the length and amount of activity per day. (C) The diurnal rodent (*Ammospermophilus leucurus*) obeys Aschoff's rule for diurnal species, showing a decrease in period in constant light as compared with constant darkness and an increased length and amount of activity per day. (D) However, the diurnal squirrel monkey does not conform to Aschoff's original rule. A reduction in light intensity from 1,800 lux to 1 lux shortened the free-running period, although it also decreased the amount and length of activity per day, as Aschoff's rule predicts. (A, B reprinted from Pittendrigh, 1967a, by permission of Pergamon Press; C from Kramm, 1971; D from Sulzman, Fuller, and Moore-Ede, 1979a.)

ning period in a prosimian, *Tupaia belangeri,* also contradicted Aschoff's rules, though no information was given about the activity/rest ratio.

In Wever's (1969) experiments with isolated humans, the free-running period increased with light intensity in seven cases and decreased in three cases, while there was no correlation for one subject. The remaining five subjects showed internal desynchronization (see Chapter 3). Although these results are mixed, they suggest that the free-running period of humans also does not follow Aschoff's rule.

Twenty years after he first formulated these rules, Aschoff (1979) surveyed the extensive studies of light effects that have examined his proposals. Certain nonmammalian classes follow the rules well, as do most night-active mammals (Fig. 2.6A), but diurnal mammals (Fig. 2.6B) show either little response to light intensity or an effect in the reverse direction. However, significant differences do exist between nocturnal and diurnal mammals. For example, the slope of the relationship between free-running period (τ) and light intensity is much steeper in nocturnal animals.

AFTEREFFECTS

The free-running period of circadian rhythms in a constant environment may be influenced for 100 days or more by the conditions to which the animal was previously subjected. These history-dependent

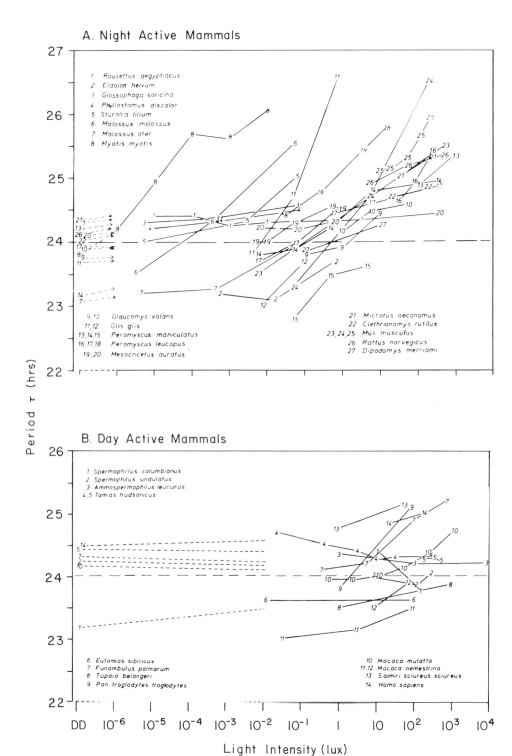

A. Night Active Mammals

1 Rousettus aegyptiacus
2 Eidolon helvum
3 Glossophaga soricina
4 Phyllostomus discolor
5 Sturnira lilium
6 Molossus molossus
7 Molossus ater
8 Myotis myotis

9,10 Glaucomys volans
11,12 Glis glis
13,14,15 Peromyscus maniculatus
16,17,18 Peromyscus leucopus
19,20 Mesocricetus auratus

21 Microtus oeconomus
22 Clethrionomys rutilus
23,24,25 Mus musculus
26 Rattus norvegicus
27 Dipodomys merriami

B. Day Active Mammals

1 Spermophilus columbianus
2 Spermophilus undulatus
3 Ammospermophilus leucurus
4,5 Tamias hudsonicus

6 Eutamias sibiricus
7 Funambulus palmarum
8 Tupaia belangeri
9 Pan troglodytes troglodytes

10 Macaca mulatta
11,12 Macaca nemestrina
13 Saimiri sciureus sciureus
14 Homo sapiens

Period τ (hrs)

Light Intensity (lux)

Fig. 2.6 Summary of the effects of light intensity on the free-running period of night-active (A) and day-active (B) mammals. Most night-active mammals obey Aschoff's original rule, showing a lengthening of free-running period under increased light intensity. While many day-active animals do not conform to Aschoff's rule they still behave differently from the night-active species and are mostly not very responsive to changes in light intensity. (From Aschoff, 1979b.)

phenomena, first recognized by Pittendrigh (1960), are termed *aftereffects*. For example, it may take many cycles for the free-running period to stabilize after a change in light intensity. Hamsters increase their period by about half an hour after light intensity is increased from constant darkness to constant light, but it takes 60 days to achieve the new, stable free-running period because the period during constant darkness has an aftereffect on the period in constant light. Such aftereffects of course complicate experiments testing Aschoff's rules and other properties of circadian clocks.

Similarly, after an animal has been released into constant conditions after being entrained to environmental time cues, the period of the free-running circadian rhythm is influenced by the previous entrained period for many subsequent cycles. In Figure 2.7, mouse A was entrained by a 20-hour, and mouse B by a 28-hour, light–dark cycle before each was released into constant darkness (Pittendrigh and Daan, 1976a). Both mice showed long-term changes in period which gradually approached each other. When two groups of mice were studied, one released from a 20-hour and the other from a 28-hour LD cycle, it was found that over 100 days passed before the mean periods of the two groups merged (Fig. 2.7C).

The number of hours of light in the light–dark cycle entraining an animal also has an aftereffect on the free-running period after release into constant conditions (Fig. 2.8). When the *photoperiod* (length of lights-on) is long, for example 18 hours of light, 6 of darkness (LD 18:6), the subsequent free-running period is much shorter than it is when the animal is released from a light–dark cycle with a short photoperiod (for example, LD 1:23) (Pittendrigh, 1974).

Aftereffects in an animal's free-running period are also observed following a *phase shift* of a circadian clock (a change in the timing of the clock with respect to a reference time scale such as the 24-hour geo-

A.

Time of Day (hrs)

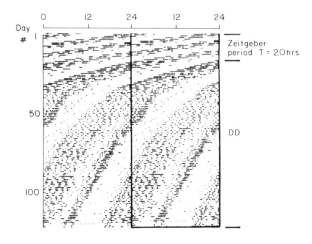

B.

Time of Day (hrs)

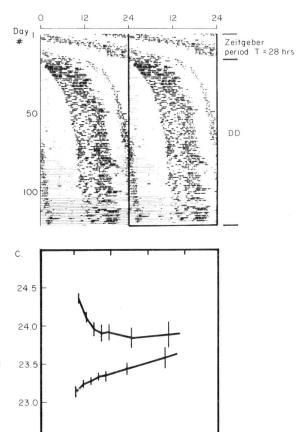

C.

Fig. 2.7 Aftereffects of prior entrainment. Mice (Mus musculus) en-trained to a 20-hour light–dark cycle (A) initially showed a shorter free-running period in wheel-running activity when released into constant darkness (DD) than when they were previously entrained (B) to a 28-hour cycle. (C) The mean periods and standard error of the mean (SEM) from the two groups of animals. It takes more than 100 days in constant dark-ness after the transition from the 20-hour or 28-hour light–dark cycle for the periods of the two groups to merge. (From Pittendrigh and Daan, 1976a.)

physical day). For example, if the clock is forced to transiently in-crease its period to resynchronize with a light–dark cycle whose tim-ing has been delayed, the aftereffect of that increase in period will be to lengthen the free-running period of the clock. If the animal is subse-quently released into constant conditions, then its circadian rhythms will run with a longer period for a number of cycles before regaining their original period.

These aftereffects indicate that events which impose a period on a circadian clock alter the clock's properties so that residual effects of the imposed period remain for many cycles. Such aftereffects suggest that not only can environmental time cues entrain circadian clocks to their own period, they can also to a certain extent mold the properties of circadian clocks so they can more readily entrain. Thus, a clock with a 23-hour endogenous period which becomes entrained to a 24-hour light–dark cycle may have its endogenous period modified to 23.5 hours, thereby facilitating entrainment.

Given all of these aftereffects, there are obviously some problems in knowing an animal's "true" free-running period. The 100–200 days necessary to overcome aftereffects is a considerable fraction of the lifespan of an animal such as a mouse, and one must also take into ac-count the effects of aging on circadian clocks, as we will see. The most useful way to control for aftereffects in experiments is to repeatedly reentrain the animals to the same light–dark cycle and to release them into constant conditions before each study.

The formulation of these behaviors of circadian pacemakers is largely due to the considerable work of Colin Pittendrigh. However, his work in mammals has focused mainly on a few nocturnal rodent species. Do these generalizations hold for diurnal mammals, which are normally exposed to very different lighting? Already we have seen

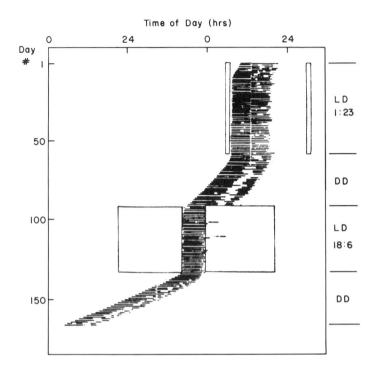

Fig. 2.8 Aftereffects of prior photoperiod on the free-running period of wheel-running activity in a mouse (*Peromyscus leucopus*). After a short photoperiod (1 hour of light per day), the free-running period in constant darkness was longer than after a long photoperiod (18 hours of light per day). (From Pittendrigh, 1974.)

that the behavior of many diurnal mammals is not well described by Aschoff's original rules. Similarly, as Pittendrigh and Daan (1976a) have pointed out, the aftereffects of phase shifts in the diurnal antelope ground squirrel (*Ammospermophilus leucurus*) do not correspond to those of nocturnal rodents. Kramm (1971) showed that after a transient lengthening of period during a phase shift, the free-running rhythm had a shortened period; after a transient shortening of period, a lengthened free-running period was seen. Furthermore, some species show much less marked aftereffects than others. These properties of circadian clocks deserve further attention, especially in diurnal mammals, although their resolutions may require considerable patience on the part of the experimenter because of the long time constants of these responses.

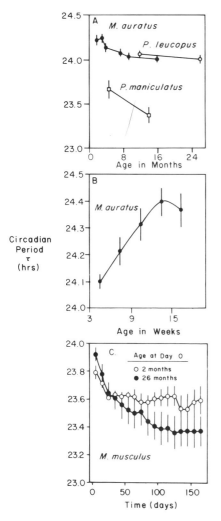

Fig. 2.9 Effect of age on the free-running period. (A) Shortening of the free-running period with age in three rodent species. (B) Lengthening of the free-running period in young hamsters during the first 15 weeks of life. After this age the free-running period in this species started to shorten, as (A) shows. (C) Separation of aftereffects from age effects. Mice aged 2 months (open circles) and 26 months (solid circles) were released into constant darkness. The younger mice consistently showed a longer free-running period than the older mice, even though both showed the aftereffects of entrainment to a 24-hour light–dark cycle. (A and C from Pittendrigh and Daan, 1976a; B from Davis and Menaker, 1980.)

AGE

Circadian clocks, unlike their owners, often run factor with advanc-ing age, so that the free-running period slowly shortens over the ani-mal's lifespan (Pittendrigh and Daan, 1974). Figure 2.9A shows the in-fluence of aging in three species of rodents studied in constant darkness. Over the course of a year each of these species showed a de-crease in the free-running period of 0.2 hours to 0.4 hours. However, as Figure 2.9B shows, newborn hamsters show a steadily increasing period, which reverses around three months of age, the point where Pittendrigh's observations began (Davis and Menaker, 1980); other mammalian species may show no changes in period with advancing age (Kenagy, 1978; Gander, 1980).

Pittendrigh and Daan's (1974) 14 months of observations of hamsters ruled out the possibility that these changes were caused by seasonal variations in the circadian period. Furthermore, age effects do not appear to be aftereffects of prior entrainment conditions, which would depend solely on the length of time after release into constant conditions. When such effects were controlled for by releasing animals of the same strain but two different ages into constant darkness, the free-running period was shorter in the older group throughout the experiment (Fig. 2.9C).

It may be that these effects of aging are not caused by physical changes in the pacemaker (master clock regulating the circadian timing of the organism) per se. Hormones, for example, could modify the pacemaker's function. Both testosterone (Daan et al., 1975) and estrogens (Morin, Fitzgerald, and Zucker, 1977) shorten the period in some species, and it seems tempting to ascribe the age-dependent shortening of period to this effect. The story is not so simple, however, for at puberty, when changes in reproductive endocrinology of the hamster are rampant, the free-running period is longest, and the shortening continues into advanced age, when male and female reproductive hormones are on the wane.

GENETIC BASIS OF CIRCADIAN CLOCKS

Now that we have discussed the influences of environmental temperature, light intensity, and aftereffects on circadian clocks, it is appropriate to raise the nature-versus-nurture question. To what extent are clock properties inherited rather than determined by the animal's, or its parents', exposure to environmental conditions? Ever since Bünning's (1935a) early studies on the genetics of circadian rhythms in plants, the answer has been clear. The contribution of environmental influences is minor compared to that of the genetically programmed determinants of clock properties. The exposure of organisms to abnormal light–dark cycles or constant conditions has no influence on the free-running period of their offspring. For example, Hemmingsen and Krarup (1937), who exposed rats to LD 8:8 cycles, observed no changes in the circadian rhythmicity of the following generation; and Aschoff

(1960), who exposed mice for five successive generations to a constant level of light intensity, saw no changes in the activity rhythms of the offspring.

When plants are selected and then crossed on the basis of the length of their free-running period (Bünning, 1935a) or when fruit flies are mated on the basis of the timing of their rhythms with respect to an entraining light–dark cycle (Pittendrigh, 1967b), these characteristics are preserved in succeeding generations. Furthermore, as we have already mentioned (Fig. 2.2), different mammalian species show quite characteristic free-running periods. Even different strains of a species such as the mouse show consistently different free-running periods and other circadian characteristics (Ebihara, Tsuji, and Kondo, 1978; Oliverio and Malorni, 1979).

Most systematic studies of circadian genetics have been done in lower organisms with short lifespans in which new mutants can be more readily generated. In these studies mutants with different free-running periods have been obtained using X-radiation, ultraviolet light, or chemical mutagens, and these free-running periods are inherited by the offspring. Using fruit flies (*Drosophila melanogaster,* Konopka and Benzer, 1971; Konopka, 1979), fungi (*Neurospora crassa,* Feldman and Hoyle, 1973), and algae (*Chlamydomonas reinhardi,* Bruce, 1972), it has been possible to map the chromosomal location of each mutation. In *Drosophila,* for example, mutants with 19- or 28-hour periods and an arrhythmic mutant have been derived from the wild-type strain, which has a 24.2-hour period. Each mutation has been mapped to the same genetic locus on the X chromosome (Konopka and Benzer, 1971), suggesting that specific "clock genes" exist. Such techniques open new doors to the study of circadian clock mechanisms. It is now becoming possible to identify the proteins and biological structures that are synthesized by the clock genes (Konopka and Wells, 1980), and eventually, through recombinant DNA techniques, it may become feasible to study the molecular biology of mammalian clocks.

Entrainment by Environmental Time Cues

Since circadian clocks, which free-run under constant conditions, are entrained to a 24-hour period in natural environments, it is clear that circadian systems are sensitive to some environmental time cue.

Studies of animals in rigorous isolation have made it apparent that only certain specific environmental variables are capable of acting as circadian time cues. In 1951 Aschoff coined the term *zeitgeber,* from the German, meaning "time-giver," to describe such periodic environmental cycles (Aschoff, 1951, 1954). Although zeitgeber is the most frequently used term, a number of others appear in the literature, including "synchronizer," "entraining agent," and "time cue."

ONTOGENY

A mammal's first exposure to a cyclic environment is in the uterus. The concentration of nutrients and hormones crossing over the placenta into the bloodstream of the fetus reflects the mother's circadian rhythmicity (Reppert et al., 1979). This is apparently a strong enough signal to entrain the circadian pacemakers of the fetus. Deguchi (1975) found that the phase of the rhythm of a pineal enzyme, N-acetyltransferase (NAT), in newborn rat pups was determined by the phase of the mother before birth, whether or not the mother was entrained to a light–dark cycle. This entrainment *in utero* is much stronger than the entrainment by the mother immediately after birth, even though she is the sole source of nutrients; rat pups transferred to a foster mother at two days of age retain the phase of their natural mother rather than completely resynchronizing to the foster parent. However, after ten days of life, the nursing mother does begin to exert a synchronizing influence, apparently related to the timing of nursing, because when the rat pups are later weaned the mother no longer acts as a zeitgeber (Deguchi, 1978).

Although the circadian pacemakers of a newborn animal appear to be keeping time, they are not immediately coupled to their overt rhythms. Hence, the pineal NAT rhythm in the rat pups studied by Deguchi did not appear until the seventh day after birth (Deguchi, 1979), although the timing of the rhythm once it became detectable indicated that the pups' circadian pacemakers had been functioning throughout the seven days. A newborn animal gradually develops overt circadian rhythms in its various body functions as they become coupled to the pacemakers over the first weeks and months of life. The amplitude of the rhythms steadily increases until the adult rhythm is obtained (Hellbrugge, 1960; Davis, 1981).

Both pacemaker function and the expression of overt circadian rhythms may precede the capability to entrain to environmental time cues. In the newborn human infant no circadian rhythm in waking and sleeping is seen initially. As any parent knows, a baby is just as likely to be awake at night as in the daytime and alternates between naps and short periods of wakefulness. Over the first months of life the episodes of wakefulness gradually become extended at certain times of day, and sleep becomes concentrated at others, hence a sleep–wake circadian rhythm emerges (Meier-Koll et al., 1978). However, the rhythm develops before the capability to entrain to light–dark cycles. With a feeding-on-demand schedule maintained around the clock, Kleitman and Engelmann (1953) found that a free-running rhythm emerged first and that it took some weeks before the infant synchronized to the environmental light–dark cycle (Fig. 2.10). Before this happened, the mother observed with relief that the infant appeared to be sleeping during the night, only to find, a week or so later, that as the free-run proceeded the child gradually reverted to daytime sleep and nocturnal wakefulness. Eventually, by about the twentieth week of life the human infant began to achieve synchrony with the schedule of his or her parents.

CRITERIA FOR ENTRAINMENT

In natural environments, multiple time cues exist. To separate out each possible zeitgeber and test its actions it is necessary to study animals (and humans) in controlled environments. In such conditions potential zeitgebers can be introduced and removed and their ability to convey temporal information determined.

To demonstrate that an oscillation in an environmental variable acts as a time cue to the circadian system, one must show that the following criteria are met:

(1) *Absence of other time cues.* The monitored circadian rhythm must be free-running with an independent period before the time cue is imposed upon the animal and must resume free-running after the time cue is removed.

(2) *Period control.* Once the animal is exposed to the environmental cycle, the period of the circadian rhythm must adjust so as to equal the period of the environmental cycle.

(3) *Stable phase relationship.* A stable and reproducible relationship must emerge and be maintained between the timing of the ob-

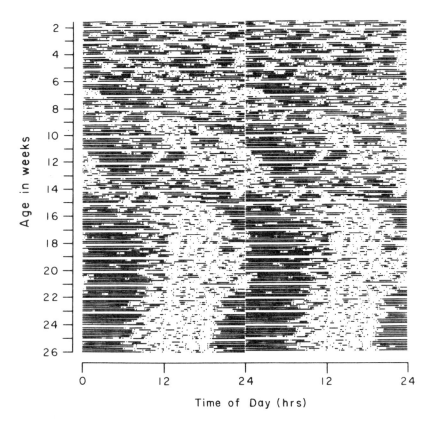

Fig. 2.10 Double-plotted record of the developing sleep–wake cycle in a human infant over the first six months of life. Solid bars represent sleep; dots represent mealtimes. With the infant on demand feeding, a free-running rhythm was seen from the fourth week of life to approximately the seventeenth week, when entrainment with the day–night cycle occurred. The circadian rest–activity cycle appeared to coalesce from multiple ultradian components before entrainment was evident. (From Kleitman and Engelmann, 1953, replotted according to Davis, 1981.)

served rhythm and the timing of the zeitgeber; the rhythm's occurrence must thus be independent of clock time and dependent only on the imposed time cue.

(4) *Phase control.* When the time cue is removed, the rhythm must start to free-run from a phase determined by the environmental cycle and not by the rhythm prior to entrainment.

Figure 2.11 shows that a 24-hour light–dark cycle with 12 hours of 600 lux and 12 hours of darkness fulfills these criteria as an entraining agent for the drinking rhythm of the squirrel monkey. First, the drinking rhythm free-ran both before the light–dark cycle was imposed on the animal and after it was removed. Second, once the light–dark cycle was imposed, the rhythm period adjusted until it equaled the period of the zeitgeber. Third, the phase relationship between the light–dark cycle and the drinking rhythm was constant, in both this experiment and others where the light–dark cycle was applied at different initial phase relationships to the drinking rhythm (and at different clock times). Fourth, after removal of the light–dark cycle the drinking rhythm free-ran from a phase determined by the light–dark cycle and not from the point that would be predicted had the clock been free-running throughout the light–dark cycle exposure (shown by the dashed extrapolation).

The importance of these criteria for identifying zeitgebers will be appreciated if we consider environmental cycles that do not entrain but that may influence the waveform of circadian rhythms. There are three issues to consider: masking, relative coordination, and the range of entrainment.

MASKING

A cycle of hot and cold temperature might suppress a circadian rhythm such as locomotor activity at certain times of day, thus producing an apparent entrainment so that the period of the activity rhythm equaled the period of the temperature cycle and there was a constant phase relationship between the environmental cycle and the rhythm. For example, the animal might move around more during the cold phase to keep warm, thereby masking the expression of the true circadian rest–activity cycle. However, when the environmental cycle was removed, the rhythm would resume its free-run from a phase determined not by the environmental cycle but by the animal's continuously running circadian clock.

An example of such an effect, which has been called *masking*, is shown in Figure 2.12. DeCoursey (1960) subjected flying squirrels (*Glaucomys volans*) to a 24-hour temperature cycle (12 hours at 25° C;

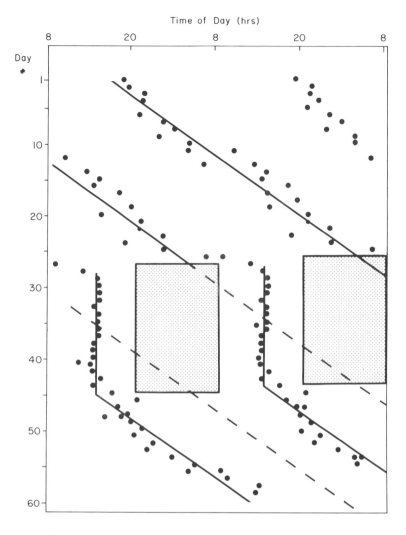

Fig. 2.11 Entrainment of a squirrel monkey's free-running drinking rhythm by a light–dark cycle (LD 12:12). The maximum (*acrophase* determined by least squares fit) of the rhythm on each day is double-plotted against time with a regression line (solid) plotted through the maximum. In constant light (600 lux) the free-running period was approximately 25 hours. With exposure to the LD cycle on days 27 through 44 (shaded area = darkness), the drinking rhythm entrained with a constant phase relationship to the LD cycle (approximately mid-light) so that the period of the rhythm equaled the 24-hour period of the zeitgeber. On release into constant light on day 45, the rhythm again free-ran with a 25-hour period (solid line), but from an initial phase determined by the LD cycle. This was 180° displaced from the phase one would have predicted had no entrainment occurred (dashed line). The criteria for entrainment are therefore met. (Copyright 1982 by Moore-Ede, Sulzman, and Fuller.)

12 hours at 15° C). While the rhythm showed apparent entrainment, with activity confined to the cold phase, there was a little burst of activity each day at a time predicted by the free-running clock. Furthermore, when the animal was released into constant temperature, the phase of the activity rhythm was better predicted by the previous free-running rhythm than by the phase of the applied temperature cycle. Thus, the temperature cycle cannot be considered an effective zeitgeber for this animal.

RELATIVE COORDINATION

In some cases an environmental cycle may not be strong enough to entrain a circadian rhythm but may still exert some phase control. This effect was first studied in depth by von Holst (1939), who was concerned not with circadian rhythms but with the oscillatory fin movements of fishes. He named the process *relative coordination* in which the frequencies of two oscillators, if weakly coupled, would interact at certain phase relationships. This interaction might, for example, speed up one oscillator at one phase relationship and slow it down at another. If the coupling was increased, both oscillators would adopt the same frequency, with what he called *absolute coordination,* or, as far as we are concerned, *entrainment.*

Relative coordination is frequently observed in circadian systems. Swade and Pittendrigh (1967), for example, studied the wheel-running rhythms of hamsters under the influence of a bright-light–dim-light cycle that did not provide enough contrast between the light intensities to entrain the rhythms (Fig. 2.13). The light intensity, varying between 180 lux and 80 lux, did not entrain the rhythm to a 24-hour period but did speed up the rhythm at certain phases and slow it down at others. As Figure 2.13 shows, the wheel-running rhythm slowed down to a period near but not quite 24 hours when activity coincided with the dimmest light. Had the time cue been stronger, with greater contrast between light and dark, entrainment would have occurred at this phase. This phase relationship would then have been appropriate for a nocturnally active animal such as a hamster.

RANGE OF ENTRAINMENT

All endogenous oscillators show a limited range of periods to which they can be entrained. A zeitgeber with a period that is either too short

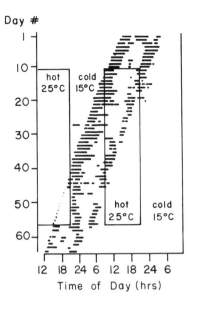

Fig. 2.12 Masking effects of an environmental cycle which does not act as a zeitgeber. The free-running wheel-running rhythm of a flying squirrel appeared to become synchronized to the temperature cycle and adopt a 24-hour period, but when released into constant conditions, the rhythm resumed free-running with a phase predicted by the prior free-running period. (After DeCoursey, 1960b.)

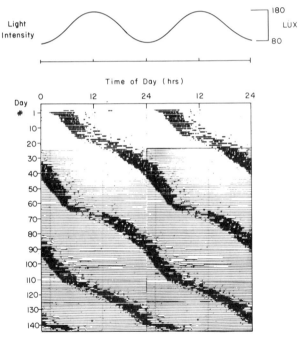

Fig. 2.13 Relative coordination in a hamster subjected to a light–dark cycle of insufficient strength to entrain the wheel-running rhythm. *Upper panel*, timing and waveform of the LD cycle, which varied between a maximum of 180 lux and a minimum of 80 lux each day. *Lower panel*, double-plotted activity rhythm. The rhythm shortened at certain phases of the LD cycle and lengthened at other phases, but the shortening was never enough to entrain the animal to a 24-hour period. (After Enright, 1980; replotted from Swade and Pittendrigh, 1967, by permission of the University of Chicago Press, copyright 1967 The University of Chicago.)

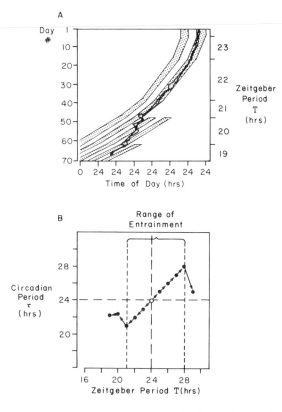

Fig. 2.14 Range of entrainment of activity rhythms of two mice. (A) Plot of the light-dark cycle (shadow area = darkness) and the maxima of activity in the two mice (solid lines) as the zeitgeber period was reduced from 23 to 19 hours. Entrainment was lost when the zeitgeber period was reduced below 21 hours, although relative coordination was still seen between the activity rhythm and the LD cycle. (B) The range of entrainment of 21–28 hours where the period of the rhythm equaled the period of the zeitgeber. Outside the range of entrainment the period reverted toward its original free-running length. (After Aschoff, 1978a; replotted from Tribukait, 1956.)

or too long will not synchronize a rhythm to its period, even though the time cue may be very effective when the period is within the *range of entrainment*. An example of the range of entrainment for the activity rhythm of mice, studied by Tribukait (1956) and reanalyzed by Aschoff (1978a), is shown in Figure 2.14. The timing of the peak of the daily activity rhythm was studied as the period of the light–dark cycle was reduced one hour at a time from 24 hours to 19 hours (Fig. 2.14A). Both mice maintained entrainment down to a zeitgeber period of 21 hours, but once the period was reduced to 20 hours, entrainment was lost and the activity rhythm began to free-run with its own endogenous period. It still showed relative coordination, as would be expected from a free-running rhythm exposed to a time cue not quite strong enough to entrain it.

In this series of experiments the period of the rest–activity cycle synchronized to the zeitgeber period between 21 and 28 hours, which therefore constituted the range of entrainment. Outside that range the rhythm moved toward its previous free-running period, but the rhythms seen after breaking away from 20-hour and from 29-hour zeitgeber periods were not the same. The period remained closer to the period of the zeitgeber which had entrained it for two reasons. First, relative coordination exerted an influence on the mean free-running period, slowing or speeding it up at certain phases of the cycle. Second, there were aftereffects of the prior entrainment period, so that when the animal broke away from a longer zeitgeber period, its free-running period was initially longer (Fig. 2.14B).

Zeitgebers in Mammals

A number of environmental cycles have been reported to entrain circadian rhythms in mammals. Not all experiments, however, have been designed to meet unequivocally all the criteria for entrainment listed above. We will restrict our discussion to those results which show enough evidence to suggest the presence or absence of zeitgeber activity.

LIGHT–DARK CYCLES

The light–dark (LD) cycle is an important time cue in all mammalian species, both nocturnal and diurnal. Figure 2.13 provided an example of entrainment by the LD cycle in the squirrel monkey, showing that each of our criteria was met. All mammalian species that have been studied have proved to be similarly entrained by LD cycles.

Man is the only mammalian species for which the role of LD cycles as a time cue has been disputed, although when isolated in an environment free of time cues, the circadian rhythms of human subjects free-run just like those of other species. The first systematic investigation of effective zeitgebers in humans was the pioneering work of Aschoff and Wever at the Max Planck Institut für Verhaltensphysiologie in West Germany. Aschoff and Wever presumed initially that LD cycles would be important synchronizers in man and therefore designed their facility to allow for the extensive manipulation of illumination conditions (R. Wever, personal communication). As expected, free-running circadian rhythms were observed in both self-selected LD cycles and constant light of various intensities (Wever, 1970a). Initial studies reported that an imposed light–dark cycle was a very effective time cue (Aschoff, Poppel, and Wever, 1969). However, in one experiment, because of equipment malfunction the light–dark cycle was applied without a presumably insignificant system of gong signals that had been used to signal or awaken the subject for periodic urine collections. To his surprise, Wever (1970a) found that the imposed LD cycle without the periodic gong signals was not sufficient to entrain the subjects to a 24-hour day (Fig. 2.15). Entrainment occurred only when the LD cycle and gong signals were used together.

These experiments led Wever to conclude that a simple LD cycle, complete with dawn/dusk transitions, was an insufficiently strong time cue to entrain human circadian rhythms to a period of 24.0 hours. He concluded that the gong signal was interpreted by the subjects as a personal call from the experimenter and, therefore, a social contact; as a result, social cues were concluded to be more important zeitgebers for man than LD cycles (Wever, 1970a, 1974).

A number of features of Aschoff and Wever's experimental technique suggest that LD cycles cannot be dismissed as a time cue in man, however. First, the subjects were permitted to use kitchen, bathroom, bedside, and desk lamps for light during the "dark" phase of each cycle. Illumination was therefore available to the subject, even when a "dark period" was imposed upon him. Thus, the tests of light–dark cycle entrainment in man were not comparable to those carried out with other species, being more similar to experiments with self-selected LD cycles. It has repeatedly been shown that animals (Walstrom, 1965; Aschoff et al., 1971; Cone, 1974; Warden and Sachs, 1974)

Time of Day (hrs)

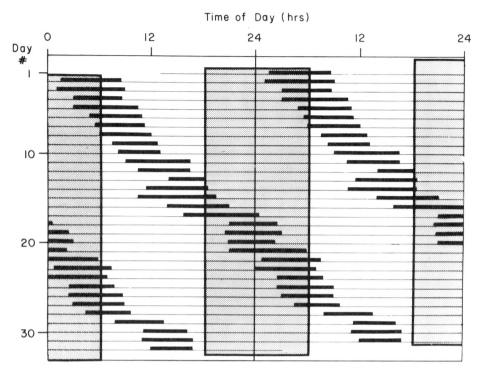

Fig. 2.15 Double-plotted sleep–wake rhythm of a human subject exposed to a weak light–dark cycle. Dark bars = sleep, shaded area = darkness (although subject had access to reading and bathroom lights). The sleep–wake cycle free-ran through this LD cycle, although it showed some evidence of relative coordination. (After Wever, 1970a.)

and humans (Aschoff and Wever, 1962; Siffre, 1964; Colin et al., 1968; Wever, 1969) will free-run when the experimental subject has the ability to switch on and off the LD cycle.

This led Czeisler, Richardson, Zimmerman, and colleagues (1981) to test the role of the LD cycle as a time cue in humans in an experiment with a design similar to that used for most animal studies. Without knowing the purpose or design of the experiment, human subjects with free-running rhythms living in an isolated apartment were exposed to an LD cycle, complete with dawn/dusk transitions (which gave some warning of impending changes in illumination). Figure 2.16 shows the results of an experiment in which the subject was not al-

Time of Day (hrs)

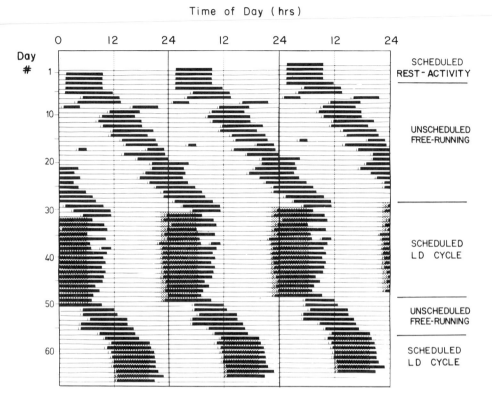

Fig. 2.16 Triple-plotted sleep–wake cycle of a human subject exposed to a strict light–dark cycle. Horizontal black bars = time asleep. On days 1–5 the subject was entrained to normal day–night cues, including scheduled rest–activity, light–dark, and mealtimes. On days 6–31 the subject was allowed to self-select times of sleep, meals, and lighting, and showed a free-running period of approximately 25 hours. On days 32–50 an LD cycle (shaded area = darkness) was imposed with gradual transitions of light intensity at dawn and dusk. The subject's sleep–wake cycle entrained to this zeitgeber. When the subject was released into unscheduled conditions on days 51–58, the rhythm free-ran from a phase not predicted from the prior free-running period, but instead from the phase determined by the prior light–dark schedule. From day 59 on, the LD cycle was again provided. (Reprinted from Czeisler, Richardson, Zimmerman, et al., 1981, by permission of Raven Press.)

lowed control over any light switch. The subject was clearly free-running before the LD cycle was imposed and then became entrained to that LD cycle after some transients. When the subject was again released into the time-cue-free environment, he commenced to free-run from a phase predicted by the environmental LD cycle rather than one predicted by a continuously free-running circadian clock. The distinction between this experiment and the studies of Aschoff and Wever was that an absolute LD cycle was imposed on Czeisler's subjects, comparable to the LD cycles which effectively entrain circadian rhythms in animals, whereas Wever weakened the entraining cue by allowing self-selected light during the subject's night. As a result of these studies, we must conclude that the LD cycle is indeed an effective entraining agent in humans.

FOOD AVAILABILITY CYCLES

Cycles of eating and fasting (EF) might be expected to provide some important temporal restraints on an animal's behavior, especially under natural conditions when food may be available only at certain times of day. Richter (1922) was the first to demonstrate this in a laboratory setting. Using a rodent running wheel (the first documented use of this traditional tool in circadian research), he demonstrated that in constant darkness a single meal once a day could synchronize the rest–activity pattern of a rat. Furthermore, he showed that the activity rhythm persisted over several days of food deprivation, with some indication of a free-running pattern. Since that time, however, there has been relatively little investigation of the capability of EF cycles to entrain circadian rhythms. A number of authors have shown that periodic availability of food can modulate the phases of the expressed rhythms in animals concurrently exposed to LD cycles (Bruce, 1960; Fuller and Snoddy, 1968; Dowse and Palmer, 1969; Krieger, 1974; Lindberg and Hayden, 1974). However, the criteria ordinarily used to demonstrate entrainment cannot be satisfied by this paradigm. There is, for example, no way to distinguish between masking and entrainment effects, nor is it possible to determine if the time cue is strong enough to entrain the circadian system in the absence of other zeitgebers.

Feeding cycles will entrain the circadian system of squirrel monkeys according to the criteria listed above (Sulzman, Fuller, and Moore-Ede,

1977a, b). Monkeys with free-running circadian rhythms of drinking and activity, in constant light with food continuously available, were submitted to a regimen in which food was provided only between 9 A.M. and 12 noon each day. Thus the animals were on a cycle of 3 hours of eating and 21 hours of fasting (EF 3:21). In Figure 2.17A the circadian rhythm of activity showed a transient phase shift after the EF cycle was imposed and then entrained with a 24-hour period and a constant phase relationship to the time of feeding. It is interesting to note that once stable entrainment was achieved, the animal's activity anticipated the daily provision of food. Then, when the animal was released into constant conditions, with food again available on demand, the circadian rhythm of activity free-ran from a phase determined by the prior EF cycle. All other rhythms monitored in the squirrel monkey, including body temperature and urinary electrolyte and water excretion, were similarly entrained by EF 3:21 cycles (Sulzman, Fuller, and Moore-Ede, 1977a, b, 1978c). Figure 2.17B shows the body-temperature rhythm on the last day of entrainment and on release into constant conditions (in this case with no food available). The body temperature rhythm rose in anticipation of the time of feeding both when food was provided and also at about 8–9 A.M. on the days when no food was given. A comparison of the rhythm on the day when food was available versus the subsequent day, when it was not, indicates that food consumption caused very little in the way of a passive, or masking, response.

In other mammalian species the evidence for entrainment by EF cycles is somewhat equivocal. In the rat, Edmonds and Adler (1977) reported that EF cycles in otherwise constant conditions entrained the wheel-running rhythm, while Boulos, Rosenwasser, and Terman (1980) found that only some of the animal's rhythmic drinking behavior entrained, while the rest continued to free-run. Takahashi, Hanada, and Takahashi (1979) found that EF cycles entrained the circadian rhythm of plasma corticosterone in rats and exerted phase control so that, when released from the EF cycle, the rhythm free-ran from a phase determined by the EF cycle, although the rhythm damped out within a few cycles. However, Morimoto and co-workers (1979) found only a masking effect of EF cycles on the plasma corticosterone rhythm, without any persistence of phase control. In mice, Nelson and

A.

Time of Day (hrs)

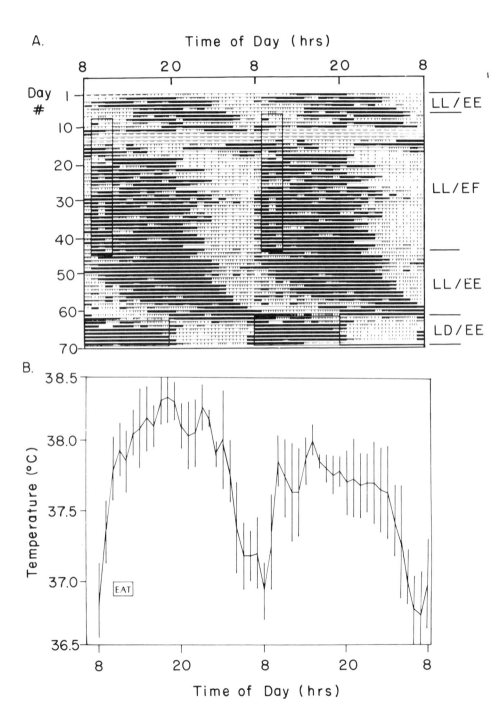

B.

Fig. 2.17   Entrainment of circadian rhythms in the squirrel monkey by food timing (EF) schedules. (*A*) Free-running activity rhythm in constant light (600 lux). On days 8–45 there was an EF 3:21 cycle with food available (indicated by box) from 0900 to 1200 hrs each day. The activity rhythm entrained to this cycle, with activity onset anticipating the provision of food each day. When food was made available constantly (EE), the activity rhythm free-ran from a phase determined by the prior EF cycle. On days 63–71 the animal entrained to an LD cycle with lights on from 0800 to 2000 hrs each day. (*B*) Body temperature rhythm (mean ± SEM) on the last day before and the first day after release into constant conditions (fasting FF) for four monkeys maintained on an EF 3:21 cycle in constant light (600 lux). Body temperature started to rise before the time of feeding on each day whether or not food was actually provided, and the phase of the rhythm was determined by the phase of entrainment. The waveform of the body temperature rhythm was essentially unchanged on the second day, demonstrating that rhythmic feeding does not have a major passive effect on body temperature. (*A* from Sulzman, Fuller, and Moore-Ede, 1977b; *B* copyright 1982 by Sulzman, Fuller, and Moore-Ede.)

colleagues (1973) demonstrated synchronization of the body temperature rhythm with an EF 4:20 cycle of food presentation, but they did not demonstrate phase control in constant conditions.

There is no doubt, however, that rodents show anticipatory behaviors prior to mealtimes, provided the interval between feedings is within the circadian range (Boulos, Rosenwasser, and Terman, 1980). Thus the animal will anticipate meals regularly spaced at 23- or 25-hour intervals but cannot anticipate meals when they are spaced 18 or 30 hours apart. Such studies indicate that food availability cycles do entrain the circadian system. The full implications of these findings will be discussed in Chapters 3 and 5.

TEMPERATURE CYCLES

Since mammals are well regulated against changes in environmental temperature, it might be assumed that hot cold (HC) temperature cycles in the environment would play only a minor role as a time cue for the circadian system. This is in contrast to the effective action of HC cycles in nonthermoregulated organisms (Sweeney and Hastings, 1960; Roberts, 1962; Hoffman, 1969; Rence and Loher, 1975; Gander, 1979). However, environmental temperature is sensed by temperature receptors in the skin, so changes in core body temperature need not be induced for temperature entrainment to occur. There is thus no reason why warm-blooded animals should differ from nonthermoregulated organisms in this regard.

Temperature cycles are not effective in entraining some rodent species (Bruce, 1960; Stewart and Reeder, 1968), including the flying squirrel (DeCoursey, 1960), as was shown in Figure 2.12, and some primate species, including the squirrel monkey (Sulzman, Fuller, and Moore-Ede, 1977b). For example, we have shown that temperature cycles having 12 hours of heat and 12 hours of cold (HC 12:12) with a range as wide as 32° C to 17° C do not entrain this species. Figure 2.18A shows an example of a squirrel monkey subjected to a 28:20° C temperature cycle which produced masking effects but through which its circadian rhythms free-ran.

However, other mammalian species can be entrained by HC cycles. Tokura and Aschoff (1980) have demonstrated that HC cycles of 33:17° C will entrain the pig-tailed macaque monkey. The results from one of their experiments are shown in Figure 2.18B, in which a monkey with free-running rhythms of activity in constant light entrained when the temperature cycle was applied and then resynchronized after a phase shift of the temperature cycle. When finally released into constant conditions at a temperature of 17° C, the animal again free-ran from a phase determined by the previous phase of the entraining cycle. Another mammal for which HC cycles appear to be an effective time cue is the pocket mouse. Lindberg and Hayden (1974) have shown that a temperature cycle of as little as 23:20° C will entrain this species. This sensitivity of pocket mice to temperature cycles may be related to their high surface-area/volume ratio, which means that changes in environmental temperature constitute more of a challenge than in larger mammals.

SOCIAL CUES

Cycles of social contact and isolation (SI) are harder to quantify and isolate than the physical cycles just discussed. However, it does appear that some form of social interaction can entrain certain species such as bats (Marimuthu et al., 1978). Furthermore, blind mice housed in the same room with sighted mice may become synchronized to LD cycles after several months, presumably through social cues from the entrained mice (Halberg et al., 1954). The nature of such social cues is

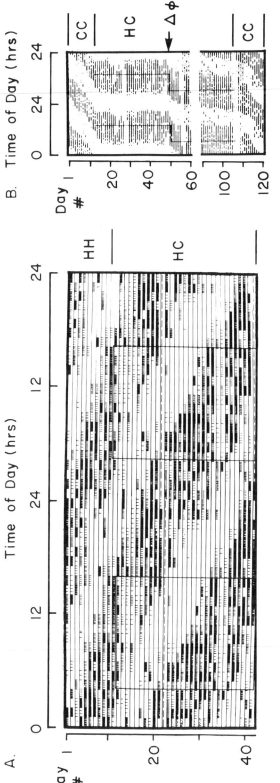

Fig. 2.18 Effect of environmental temperature cycles in two primate species. (A) Temperature cycle (HC) does not entrain the drinking rhythm of a squirrel monkey. In constant light (600 lux) the drinking rhythm (double-plotted) free-ran through the HC cycle, which was provided from day 12 through the end of the experiment. (B) In a pig-tailed macaque monkey (*Macaca nemestrina*) maintained in constant illumination (450 lux), a temperature cycle (12 hours at 17° C and 12 hours at 35° C) on days 12–105 did entrain the activity rhythm. On day 49 the temperature cycle was shifted by 8 hours, and the activity cycle readjusted to this phase shift. After release on day 105 into constant conditions, the rhythm again free-ran from a phase predicted by the prior entraining cycle. (A from Sulzman, Fuller, and Moore-Ede, 1977b; B from Aschoff, 1979a, reprinted by permission of Raven Press.)

unclear, but they may involve sounds or smells emitted by the en-training animal that are meaningful to the recipient.

In some other mammalian species social cues have been demonstrated not to be effective time cues. For example, for a squirrel monkey free-running with constant light and food, having another squirrel monkey visit for 8 hours each day (SI 8:16) did not entrain the first monkey even when the visitor was otherwise synchronized by a strict 24-hour light–dark schedule. Even when a squirrel monkey in a cage covered by a cloth was placed in a room full of monkeys synchronized to a 12:12 LD cycle it did not entrain if kept in constant dim light under the opaque cloth (Sulzman, Fuller, and Moore-Ede, 1977b).

There has been considerable discussion of the role of social cues as a zeitgeber for humans. When several human subjects are isolated together, they tend to remain synchronized with each other even if there are no external time cues (Wever, 1979). There is a problem in interpreting many of these experiments, however, if the subjects can self-select the times of lights-on and lights-off. When they are in self-selected cycles, clearly the dominant individual who switches on the lights in the morning or switches them off at night may have an entraining effect mediated not directly by social cues but indirectly by the light–dark cycle. This may therefore really be a form of light–dark cycle entrainment.

A study suggesting a more direct action of social cues in humans was conducted by Vernikos-Danellis and Winget (1979) at the NASA Ames Research Center. Two groups of volunteers, each comprising four subjects, were isolated in separate facilities. Each individual maintained synchrony with all the others in the same room, but the mean free-running period of the subjects in one room was different from that of the subjects in the other, with one room averaging 24.4 hours and the other averaging 24.1 hours. Each group was maintained in constant illumination, so there was no possibility of a confounding factor being introduced through self-selected light–dark cycles. To confirm whether social cues were playing a role, one of the subjects was moved from one room to the other. He subsequently showed a progressive phase shift and resynchronization with the group in that room. However, the nature of the social cues which appear to be effective entraining agents in humans is unclear. For example, entrainment could be mediated by the rest–activity cycle selected by the individual out of social pressure to conform to the demands of his peer group.

Social cues may be important as auxiliary time cues to strengthen the total time cues perceived by humans. After flying across six time zones, subjects who were required to remain in their hotel rooms entrained more slowly to the new environmental time than those who were allowed to go outdoors and be subjected to other cues from their environment (Klein and Wegmann, 1974). The subjects who went outside, of course, were also more exposed to ambient light–dark cycles, feeding routines, and sounds than those who stayed in the hotel room, where they could close the shades or switch on the lights. Social cues may also be important in blind people, who can often (Sieber, 1976; Pauly et al., 1977) although not always (Miles, Raynal, and Wilson, 1977), maintain entrainment with the 24-hour day.

When attempting to identify zeitgebers for humans, the problem of motivation has to be considered. The subject's motivation to get out of bed and thus cut short his sleep may determine whether a signal acts as an effective time cue. The alarm clock may wake one up in the morning, but that alarm clock signal will have a different effect depending on the day of the week and the consequences of ignoring it. If it is a weekday and one may lose one's job, there is a strong motivation to get out of bed and subject one's body to light–dark, food, noise, and other potential zeitgebers. In contrast, during the weekend when the only consequence of late rising is a lawn not mowed, a person may elect to remain in bed and fall back to sleep.

The problem of Monday morning hangover may be related to the tendency to free-run during the weekend, since in the modern world one's environmental cues are largely self-selected. On Friday, Saturday, and Sunday nights one may go to bed progressively later and wake up later each day. On Monday morning, the circadian system is phase-shifted with respect to environmental time, so the time one must wake up is much earlier in relationship to subjective body time. In contrast, by the end of the week the circadian system has had time to reentrain, and getting up earlier in the morning is easier. In extreme cases this may present a severe problem, resulting in chronic insomnia (Chapter 7).

SOUND CUES

In certain mammalian species, cycles of noise and quiet (NQ) may act as a time cue. It is difficult to separate these signals from social cues, however, since in both instances where sound cues have been shown to act as zeitgebers in mammals, the sounds conveyed some social information. For example, Meyer (1968) has shown that hamster sounds played into the chamber in a cyclic fashion will entrain hamsters. With human subjects, gong signals (Wever, 1970a, 1979) promoted entrainment when coupled with light–dark cycles, but these could also have been interpreted as social cues.

Not all mammalian species entrain to NQ cycles. We have tested squirrel monkeys exposed to 12-hour cycles of squirrel monkey sounds or white noise, alternating with 12 hours of quiet, or NQ 12:12 (Sulzman et al., 1977b). Neither the conspecific nor the white noise cycles were capable of entraining this species. In contrast, birds seem to be much more strongly entrained by NQ cycles. For example, cycles of conspecific sound have been shown to entrain the activity rhythms of house sparrows (Menaker and Eskin, 1966), and a 6-hour intermittent buzz, 25 decibels above ambient noise levels, can also entrain bird circadian rhythms (Enright, 1965).

ELECTROMAGNETIC FIELD STRENGTH

The wheel-running activity of mice can be entrained by cycles of electrostatic field strength (Dowse and Palmer, 1969). In humans Wever has shown, curiously, that the imposition of a periodic electromagnetic field with field-on for 11.75 hours and off for 11.75 hours each day in a shielded chamber will entrain human circadian rhythms to a 23.5-hour period (Wever, 1970b). The mechanisms responsible for these effects are not understood, although there is evidence that animals (Walcott, Gould, and Kirschvink, 1979) and even humans (Baker, 1980) can detect magnetic fields.

ATMOSPHERIC PRESSURE CYCLES

There is a report that cycles of ambient pressure (12 hours at 1.00 atmosphere alternating with 12 hours of 1.09 atmospheres) can entrain the body temperature rhythm in mice (Hayden and Lindberg, 1969). However, barometric pressure changes are effective only at levels much larger than those found in normal environments. The implications of this finding as well as the mechanisms of entrainment are unknown.

WATER AVAILABILITY CYCLES

In contrast to cycles of food availability, cycles of water availability do not appear to entrain animals to a 24-hour period (Sulzman, Fuller, and Moore-Ede, 1977b; Moore, 1980). For example, squirrel monkeys on a water–thirst (WT) cycle in which water was available for only three hours each day show a free-running circadian rest–activity cycle. This result implies, as we will see in Chapter 4, that since EF cycles entrain this species, some constituents in the diet may act as the effective entraining agent. The behavioral act of oral ingestion, or the effect of simply distending the gastrointestinal tract alone, cannot be effective as a time cue.

Phase Relationship to Zeitgeber

One of the most important functions of the circadian system is to ensure that behaviors and internal metabolic adjustments are appropriately timed with respect to the daily events in the environment. Most species adopt a characteristic temporal niche that complements their spatial ecological niche. Mammals may be nocturnal, diurnal, or crepuscular (dusk and dawn active), or they may confine their activity to an even more limited time such as early night. These relationships between circadian rhythms and environmental time cues probably do much to maximize the survival of each species in a world where food supplies and predator activity are themselves cyclic. In short, there is truth in the old saying "The early bird catches the worm." It is the circadian system that wakes up that bird, often before dawn (Aschoff et al., 1971).

The phase relationship (ψ) between a zeitgeber and a rhythm is defined either in terms of the difference in timing (hours) or in terms of the phase-angle difference (degrees) between reference phases in the two oscillations. The reference phases chosen might be, for example, the onset of light in a light–dark cycle and the onset of activity. If an animal's daily bout of activity began an average of 2 hours before light onset, then ψ would equal +2 hours. Alternatively, the value of ψ can be expressed in degrees of phase-angle difference. If the light–dark cycle period was 24 hours, then 24 hours would equal 360° and ψ in this case would be +30°.

RELATIVE PERIODS OF ZEITGEBER AND PACEMAKER

The steady-state phase relationship maintained between an entrained circadian rhythm and its zeitgeber is dependent on the period of the zeitgeber and the natural period of the pacemaker driving the rhythm. In Figure 2.19A this relationship is illustrated. When the natural period of the rhythm is shorter than that of the time cue, the entrained rhythm "phase-leads" as the time cue and hence ψ is positive. In contrast, when the rhythm has a longer free-running period than the entraining time cue, then the entrained rhythm "phase-lags" with respect to the time cue and ψ is negative. A corollary of this is shown in Figure 2.19B. Entrainment by a time cue with a period shorter than the rhythm's natural period results in a phase-delayed rhythm (with ψ negative), whereas when the zeitgeber cycle is longer than the natural free-running period of the rhythm, the entrained rhythm is phase-advanced with respect to the time cue. Whether a phase lag or lead is actually seen between two rhythms also depends on the phase reference points chosen. Thus, with shorter rhythm periods and longer zeitgeber periods the rhythm may phase-lead the entraining cycle more (as in Fig. 2.19) or just phase-lag it less, depending on the initial phase relationship.

These relationships are illustrated in the experiment conducted by Tribukait (1956) and reanalyzed by Aschoff (1978a), shown in Figure 2.14. Aschoff systematically examined the rest–activity rhythms of mice exposed to light–dark cycles of different periods. With a 24-hour light–dark cycle (LD 12:12), the mice demonstrated their peak activity at their characteristic phase just after lights-out. However, when the period of the LD cycle was shortened, the peak of the activity rhythm stabilized at a phase later in the night. As Figure 2.20 shows, ψ became increasingly more negative with each shorter time-cue period, as was predicted from Figure 2.19. The reverse was true when longer-period LD cycles were applied. As the zeitgeber period was altered, a wide range of phase relationships was observed. However, with zeitgeber periods outside the range of entrainment for the circadian pacemaker

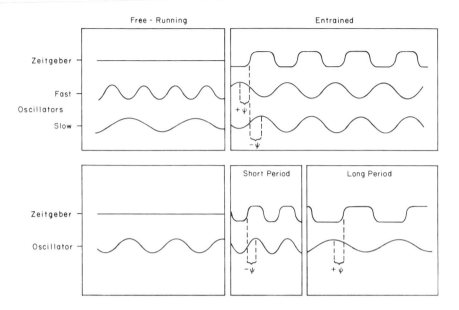

Fig. 2.19 Schematic drawing of the influence of zeitgeber period and rhythm free-running period on the phase relationship (ψ) between rhythm and zeitgeber during entrainment. *Top left,* free-running rhythms of fast and slow oscillators. After entrainment to a zeitgeber cycle (*top right*), both rhythms show a period identical to that of the zeitgeber, but the fast oscillator is phase-advanced relative to the zeitgeber cycle, whereas the slow oscillator is relatively phase-delayed. *Lower panels,* oscillator with an intermediate free-running period entrained to a short-period and a long-period zeitgeber. With a short-period zeitgeber, the oscillator is relatively phase-delayed with reference to the zeitgeber, whereas with the long-period zeitgeber, the oscillator is relatively phase-advanced. Whether an actual phase lag or phase lead is seen depends on the phase reference points used. If a later phase reference point such as the maximum of the zeitgeber cycle were picked, no phase-lag rhythms would be seen. The same relative directional effects on the phase relationship would have occurred, but the oscillator reference phase would always have been phase-advanced with respect to that of the zeitgeber. (Copyright 1982 by Moore-Ede, Sulzman, and Fuller.)

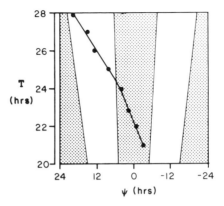

Fig. 2.20 Phase relationships (ψ) between the maximum of the activity rhythm of mice and the light–dark cycle—reference phase (0) is mid-darkness—as the zeitgeber period (T) is changed. The data from Figure 2.14 are replotted to display the relative timing of the maximum of activity and the LD cycle over a range of zeitgeber periods. As was predicted from Figure 2.19, a wide range of phase relationships was observed, depending on the relative periods of the zeitgeber and the rhythm. (From Aschoff, 1978a, replotted from Tribukait, 1956.)

(less than 21 hours or greater than 28 hours), the pacemaker free-ran and hence no consistent value for ψ could be obtained.

Just as changes in the period of the zeitgeber can alter the phase relationship, so can changes in the free-running period of the rhythm. Richter (1977) used deuterium oxide (D_2O), which slows down circadian pacemakers, to demonstrate this. Hamsters were provided with various concentrations of D_2O in their drinking water and studied in an LD 12:12 cycle which was normally capable of entraining them to a 24-hour period with activity onset occurring just after lights-out. Figure 2.21A shows that with increasing concentrations of D_2O (and therefore with a lengthening pacemaker period) the phase of activity onset was entrained progressively later in the night, so that with 10% D_2O it occurred about 3–4 hours after lights-out. It was possible to increase the pacemaker's period using 20% D_2O so that the 24-hour LD cycle was now outside the pacemaker's range of entrainment. The rhythm was no longer entrained but instead displayed relative coordination (Fig. 2.21B). Increasing the amount of D_2O concentration to more than 35% of the drinking water increased the pacemaker's period so far from 24 hours that it free-ran and not even relative coordination was seen.

COUPLING STRENGTH

In addition to being dependent on their relative periods, the phase relationship between a zeitgeber and a circadian pacemaker is dependent on the coupling strength from the zeitgeber to the pacemaker. Coupling strength depends on both the strength of the zeitgeber and the sensitivity of the pacemaker to that zeitgeber, although in practice these two factors are hard to separate. A wide variation in coupling strengths is seen in nature. With some zeitgebers in certain species the coupling is very strong, and tight phase control is exerted over the animal's circadian timing system. In other cases the coupling strength is very weak, so that the zeitgeber can modulate the period of the free-running rhythm only at certain phases, thus producing a pattern of relative coordination.

There are three criteria by which one can judge the strength of entrainment exerted by any given time cue: rate of phase shift, stability of phase control, and range of entrainment. Each may be useful in evaluating the coupling strength between a zeitgeber and the circadian system.

Rate of phase shift If a zeitgeber that is entraining a circadian rhythm is abruptly phase-shifted by several hours, there will be one or more cycles of transients while the circadian rhythm resynchronizes with the new phase of the time cue. The rate of resynchronization can be a useful indicator of the effectiveness of the time cue; when the coupling is strong, resynchronization is rapid, whereas it takes much more time when the coupling is weaker.

However, time cues, besides exerting phase control, also exert masking effects on the rhythmic variable, which means that the rate of resynchronization is not an exact means of determining coupling strength. Thus, while light–dark cycles entrain body temperature rhythms (Aschoff, 1970; Fuller, Sulzman, and Moore-Ede, 1979a), light itself has a direct masking effect on body temperature, causing elevations at whatever phase it impinges (Fuller, Sulzman, and Moore-Ede,

A.

Time of Day (hrs)

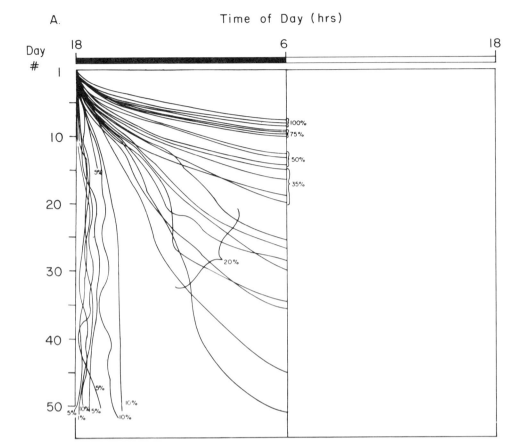

B.

Time of Day (hrs)

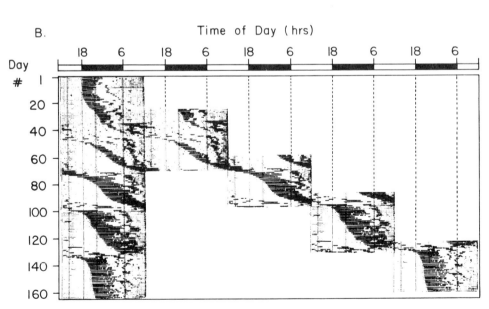

Fig. 2.21 The effect of lengthening the period of a circadian clock by adding heavy water (D_2O) in varying concentrations (1–100%) to the drinking water. (A) Lines are drawn through the succession of activity onsets in 30 individual hamsters provided with D_2O over a 50-day experiment. The animals were kept in an LD 12:12 cycle with lights off from 1800 to 0600 hrs (6 P.M. to 6 A.M.) daily. In an LD 12:12 cycle, hamster activity normally begins just after the onset of darkness. With increasing amounts of D_2O (1–10%), activity began later in the night and then failed to be synchronized by the LD cycle when the D_2O concentration was increased to 20% or more. (B) At a 20% concentration of D_2O, the activity rhythm shows relative coordination with the LD cycle, slowing as it passes through the period of darkness (shaded area) each day. (From Richter, 1977.)

1978a). This can be observed, for example, when an animal is subjected to high-frequency light–dark cycles which do not entrain the circadian system (see Fig. 5.14).

This masking effect may make the rate of resynchronization to the new phase of the zeitgeber appear faster or slower than it really is. One way to obviate this problem is to examine the waveforms of the rhythms before and after reentrainment; when the previous waveform is replicated, it is reasonable to assume that reentrainment is complete. An alternative approach is to remove all environmental and behavioral variations for one cycle during the course of phase shifting a rhythm so as to evaluate the phase of the rhythm independently of the masking effects of the environment (Mills, Minors, and Waterhouse, 1978b). Provided the problem of masking is overcome, the rate of resynchronization can be a valuable means of assessing the entrainment strength.

Stability of phase control Another indication of coupling strength is the stability of the rhythm's phase during entrainment to the time cue. There are two convenient ways of assessing this. One way is to determine the timing of a phase reference point for each cycle, for example, the *acrophase* (the maximum of the best least-squares fit sine wave) could be used. The standard deviation of the phase reference point over a number of cycles will provide an indication of how tightly a circadian rhythm is coupled to the zeitgeber.

Another way of assessing stability of phase control is to examine the stability of the rhythm's waveform. A mean waveform may be educed by computing the mean value at each phase point of the cycle, using the period of the time cue as the averaging interval. The stability of the phase control is assessed by determining the mean variance of all the educed phase points. If one finds that during entrainment one rhythm has a much greater mean variance than another, this may indicate less precise phase control. Again, in this particular measure, problems may creep in for the unsuspecting as a result of the masking effects from the time cue. Similar steps to those used in evaluating phase shifts may be necessary to minimize such problems.

Range of entrainment Determining the range of entrainment of the rhythm by the zeitgeber is another way to indicate coupling strength. When the coupling is strong, the rhythm will entrain to a much wider range of zeitgeber periods than when the coupling is weak. In evaluating the range of entrainment, however, some attention should be paid to the sequence of trial zeitgeber periods used. Because of the aftereffects of prior zeitgeber periods, a steady increase (or decrease) in zeitgeber period may not be the most appropriate assay. It is better to return the animals to the same standard period (such as 24 hours) for sufficient time to reentrain before testing with the next trial zeitgeber period.

EFFECTIVENESS OF A ZEITGEBER

Using these criteria, we can determine that the following characteristics of an environmental time cue contribute to its coupling strength.

Range The coupling strength of a zeitgeber is a function of the range of its oscillation. Thus, for example, the contrast in intensities between the light and dark portions of a light–dark cycle will influence coupling strength. Increasing the illumination during the dark phase of an LD cycle has been shown by Swade and Pittendrigh (1967) to result in loss of entrainment to a 24-hour light–dark cycle (Fig. 2.13). Similarly, Erkert (1976) found a reduction in the rate of resynchronization after a shift of the LD cycle when the contrast between the light and dark fractions of the cycle was decreased. In both situations the coupling strength of the zeitgeber was decreased.

Waveform   When the ratio of duration of light to duration of darkness is reduced in a light–dark cycle, the range of entrainment in hamsters has been shown to be reduced (Aschoff and Pohl, 1978) and the rate of resynchronization in bats after a phase shift of the LD cycle is similarly slowed (Erkert, 1981), indicating that coupling strength is reduced.

An important ingredient of the waveform of light–dark cycles in nature is the gradual transition between light and darkness that occurs at dawn and dusk. Light intensity roughly halves (or doubles) every three minutes during twilight. Most light–dark cycles in the laboratory have abrupt transitions between light and darkness, because it is simplest just to switch an electric light on and off. However, Kavanau (1962) has shown that adding simulated dawn–dusk transitions to LD cycles in the laboratory increases the strength of entrainment. Mice (*Peromyscus maniculatus*) will entrain readily to a 16-hour light–dark cycle (LD 8:8 or 10:6) when there are dawn and dusk transitions, but without these transitions the animals free-run because the zeitgeber period is outside the range of entrainment.

Quality   The coupling strength exerted by a zeitgeber may also depend on the quality of the signal. For example, it has been shown that only light within a certain range of wavelengths is effective in entraining rats (McGuire, Rand, and Wurtman, 1973). The effective spectrum of light correlates with that of the retinal pigment rhodopsin, suggesting that the light information conveyed to the circadian pacemakers is detected by the rhodopsin-containing rods of the retina (the story is actually more complicated; see Chapter 4). The extent to which the zeitgeber can stimulate the animal's receptors thus is important in determining the coupling strength. Sound frequencies outside the audible range of that species or dietary ingredients that do not stimulate the putative chemoreceptors linked to the circadian system will obviously not be effective at conveying temporal cues.

Differences in the period of a zeitgeber and pacemaker, or the coupling strength between them, however, do not adequately account for the differences in phase relationship between species. The circadian systems of diurnal and nocturnal species must be organized differently to account for the dramatic differences in the phase relationships of their rhythms to the light–dark cycle. It is possible that the differences lie in the coupling between zeitgeber and pacemaker. However, as we will show later in this chapter, the similarities between nocturnal and diurnal species in the way that light resets circadian pacemakers make it more likely that the difference in the phase relationships of the rhythms of nocturnal and diurnal animals actually depends on differences in the coupling mechanisms between the circadian pacemaker and the rhythms it drives.

The age of an animal also determines the strength of entrainment by a zeitgeber. In newborn animals it may take some weeks before the neural pathways responsible for entrainment mature (Campbell and Ramally, 1974; see Chapter 4). In the human infant entrainment to local time after a flight from California to Holland took much longer at 3.5 months of age than at 16.5 months of age (van den Hoed and Boukamp, 1980), presumably because the entrainment pathways had not fully developed.

With old age there is again a weakening of the coupling to environmental time cues. Rats resynchronize more slowly after a phase shift of the light–dark cycle as they age (Quay, 1972; Ehret, Groh, and Meinert, 1978). The evidence in humans is less complete, but the sleep–wake cycle appears to be less precisely synchronized to environmental time with age (Kahn and Fisher, 1969). The amplitude of circadian rhythms also diminishes with age in rodents (Yunis et al., 1974; Sacher and Duffy, 1978) and humans (Lobban and Tredre, 1967; Finkelstein et al., 1972; Kapen et al., 1975), and immediately before death circadian rhythmicity may disappear altogether (Albers, Gerall, and Axelson, 1981b).

Mechanism of Entrainment

For a zeitgeber to entrain the circadian timing system, it must in each cycle reset the phase of an otherwise free-running circadian pacemaker by an amount that corrects for the difference between the period of the time cue and that of the pacemaker. This resetting is itself achieved through a circadian rhythm in the sensitivity of an organism to a time cue. Light, for example, will induce a phase delay, a phase advance, or no phase shift at all, depending on when in its *subjective day* or *night* an animal is exposed to light (that is, the time scale defined by the animal's own circadian system). The largest phase delays occur in early subjective night, when a diurnal animal has just gone to sleep or a nocturnal animal has just woken up, and the largest phase advances occur in late subjective night. The circadian system is relatively unresponsive to light during the subjective day.

This key property of entrainable circadian systems—a periodically changing sensitivity to light—was first observed in plants by Klein-hoonte (1932). In 1956 Rawson demonstrated a similar differential sensitivity to light in mammals. He observed phase shifts in the activity rhythm of mice free-running in constant darkness when he switched on the room light temporarily to feed his animals at certain phases of their subjective night, but no phase shift when he switched on the light at other times. Although he did not precisely measure the responses he observed, he did recognize their significance for entrainment of circadian rhythms. Burchard (1958) later extended these findings by exposing hamsters to light for several hours at different circadian phases.

The phase shifts induced by light pulses were first precisely quantified by Pittendrigh and Bruce (1957; Pittendrigh, 1958) in *Drosophila* and by Hastings and Sweeney (1958) in the unicellular alga *Gonyaulax polyedra*. The observed relationship between the time in the animal's subjective day when a light pulse is given and the phase shift obtained can be conveniently plotted as a *phase-response curve* (PRC) (De Coursey, 1959, 1960a, b; Pittendrigh, 1960). The main features of the phase-response curve, with phase delays in early subjective night and phase advances in late subjective night, are similar in all species whether they are single-celled algae or primates and whether they are nocturnally or diurnally active.

The first detailed examination of the phase-response curve of a mammal to short (ten-minute) light pulses was undertaken by De Coursey (1959, 1960a, b). In a series of elegant studies using the flying squirrel (*Glaucomys volans*), she determined the phase shift in the wheel running rhythm at hourly intervals throughout subjective day and night. DeCoursey was aided by the high uniformity of the flying squirrel's activity rhythm (Fig. 2.22A), so that phase shifts of a few minutes could be reliably detected. One of her first phase-response curves to ten-minute light pulses (Fig. 2.22B) shows the general features of all phase-response curves to light.

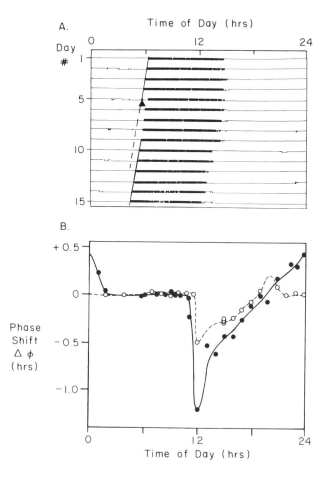

Fig. 2.22 Phase-response curve to light in the nocturnal flying squirrel (*Glaucomys volans*). (A) Wheel-running activity is a highly precise rhythm, free-running on days 1–4 with a period slightly shorter than 24.0 hours. A light pulse (solid triangle) at the normal time of activity onset in late subjective night phase-delays the rhythm, which then free-runs with the same period but delayed by about 50 minutes. (B) Phase shifts obtained by applying identical light pulses at different phases of the rhythm in two animals (solid line, dashed line). Each shows no phase shift in the middle of subjective day, maximum phase delays in early subjective night, and maximum phase advances in late subjective night. However, the phase-response curve (PRC) differs in the two animals although they are the same species. (After DeCoursey, 1960a; copyright 1960 by the American Association for the Advancement of Science.)

The discovery of the phase-response characteristics of the circadian system has contributed significantly to our understanding of how light–dark cycles (and other zeitgebers) entrain an animal's circadian pacemakers. Because brief pulses of light are so effective, the daily light–dark and dark–light transitions have become recognized as major entraining features. Furthermore, the PRC has become important in the analysis of circadian systems because it enables us to study the properties of pacemakers independently of an animal's overt rhythms.

OBTAINING A PHASE-RESPONSE CURVE

In Figure 2.23 we provide a schematic diagram of the most straightforward (although not always the easiest) technique for obtaining a phase-response curve. We use as an example the rest–activity cycle of an animal in constant darkness with a free-running period of 25 hours being reset by 1-hour pulses of light. The horizontal bars represent wheel-running in a nocturnal animal (or sleep in a diurnal animal) plotted as a standard circadian actogram (as in Fig. 1.9). Before the pulse was given, the free-running period was stable at 25 hours, with the rhythm phase-delaying by 1 hour every day as compared to 24-hour clock time. Panels A through E in Figure 2.23 illustrate the consequence of applying a single 1-hour light pulse at different times of subjective day or night. Not all light pulses produced phase shifts of the rest–activity cycle. In mid-subjective day (panel A) or mid-subjective night (not shown), no phase shift was obtained.

In panel B the light pulse fell at late subjective day, just before a nocturnal animal would commence running. On the next cycle the animal's activity began later than would have been predicted from the free-running rhythm. By the second day after the pulse, the free-running rhythm of activity was beginning a full hour later than it would have without the light pulse (dotted line). Thus a net phase delay of 1 hour was obtained, with the rhythm regaining its original 25-hour free running period after transients were over.

When an identical light pulse was applied in early subjective night, just after a nocturnal animal would start its activity, the onset of activity was more delayed (panel C). By the second day the rhythm had stabilized at a net phase delay of −3 hours. In contrast, similar light pulses in late subjective night (panel D) and early subjective day

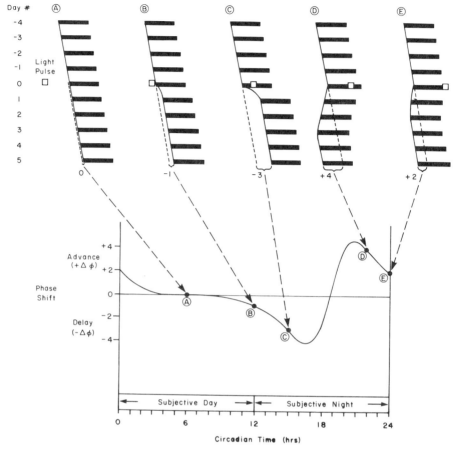

Fig. 2.23 Derivation of a phase-response curve. (A–E) Five experiments with one individual nocturnal animal. A free-running activity rhythm with a period of 25.0 hours is seen on days −4 to −1. On day zero, a light pulse is given at mid-subjective day (A), at late subjective day (B), at early subjective night (C), at late subjective night (D), and at early subjective day (E). The light pulse in mid-subjective day (A) has no effect, whereas the light pulses in late subjective day and early subjective night (B and C) produce phase delays of the activity rhythm that are complete within one cycle. The light pulse in late subjective night and early subjective day (D and E) produce phase advances with several cycles of transients before reaching a steady-state shift by day 5. *Lower panel,* direction and amount of phase shifts plotted against the time of light pulses, to obtain a phase-response curve. When light pulses are given at frequent intervals throughout subjective day and night, the waveform for the phase-response curve follows the solid line. In mammals there is normally a gradual transition between maximum phase delay and maximum phase advance, with a point in mid-subjective night (like the one in mid-subjective day, A) where there is no phase shift. (Copyright 1982 by Moore-Ede, Sulzman, and Fuller.)

(panel E) produced phase advances of +4 hours and +2 hours, respectively, instead of phase delays.

With the data obtained from such experiments, we can quantify the response of the animal's circadian system to light. In the lower panel of Figure 2.23 is shown the resultant phase-response curve, which has a waveform characteristic of the PRCs for mammalian species.

Several additional points can be made about this schematic PRC. First, responsiveness to light is confined mostly to subjective night in both nocturnal and diurnal species (although there are little data from diurnal mammals). Thus, the PRC can explain how an animal phase-shifts its rhythms when placed in a suddenly shifted light–dark cycle, since light will then fall on a different portion of the PRC and produce a corresponding phase shift of the circadian system.

Second, at the time when the animal is normally exposed to light (in mid-subjective day), light has no effect on the circadian system. Most of the daytime illumination falls on the inactive portion of the PRC and thus has little or no influence on the entrainment process.

Third, the PRC documents how the sun rising in the morning (during the animal's subjective day) tends to produce a phase advance (so a diurnal animal would start activity earlier), whereas light falling at dusk (late subjective day) causes a phase delay (so that the animal would continue activity longer). Thus a natural daily light–dark cycle is constantly nudging the circadian clocks of animals forward in the morning and backward in the evening with each 24-hour day. Notice that the PRC in Figure 2.23 indicates that the dawn pulse (E) causes a phase advance of 2 hours each day, whereas the dusk pulse (B) causes a phase delay of only 1 hour. The net effect is a 1-hour phase advance per day, just sufficient to reset the 25-hour free-running period to 24 hours and therefore to ensure entrainment to the 24-hour day–night cycle of the environment. Although this schematic diagram oversimplifies the process, it demonstrates the general principle of how the fine tuning of the circadian system's phase normally occurs each day at dawn and dusk.

Transient and steady-state measurements   After an animal receives a pulse from a time cue such as those used in deriving the phase-response curve in Figure 2.23, there may be several circadian cycles of transients before a steady-state phase shift is achieved. Some authors have derived PRCs using the phase shifts observed on the first day of

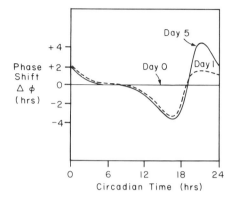

Fig. 2.24 The phase-response curve on the first day after the light pulses were given (day 1, dashed line), and on day 5 (solid line) after all transients were complete from the experiment in Figure 2.23. Because phase delays occur much faster than phase advances, the waveform of the delay portion of the curve was essentially complete by day 1. The full waveform for the advance portion of the curve was not attained until the transients were complete on day 5. (Copyright 1982 by Moore-Ede, Sulzman, and Fuller.)

study, but for most purposes it is better to use the ultimate phase shifts achieved after several cycles of transients. The shape of the PRC will differ depending on the experimental technique. Figure 2.24 shows the progressive change in the PRC in Figure 2.23 using data from each successive day of study. In this example the phase delays occurred more rapidly than the phase advances and such a difference between delays and advances in the length of transient behavior is commonly found. Although the final-state PRC is usually more useful, it is not always possible to measure the PRC after several cycles; for example, long-term changes in the period of the rhythm may make reliable estimates of the steady-state phase difficult.

Advances and delays With large phase shifts, especially when they approximate 180° changes in phase angle, it may be difficult to distinguish between an advance and a delay. Simply counting the number of maxima in the time series may not help if there is transient behavior for several cycles with disturbances of the waveform after the time cue is provided. One of the most effective ways to distinguish between advances and delays is to vary the strength of the time cue. A less intense light pulse given at the same time of the subjective day–night cycle as the original pulse will result in a smaller phase shift, and it can then be determined whether the original phase shift was an advance or a delay.

In the literature there are inconsistencies in the plotting of phase-response curves. The first reported PRCs were plotted with delays (indicated by +) above the axis and advances (−) below it (DeCoursey, 1959, 1960a, b; Pittendrigh, 1960). By 1965 most plots were still oriented with delays above the axis, but phase delays were indicated by a minus sign and advances by a plus (Aschoff, 1965a). According to normal scientific convention, this method of plotting was inverted; to correct this, phase advances are now plotted above the line and designated +, and delays are plotted below the line and designated −. For consistency, all the phase-response curves presented in this book are in accordance with the current convention, which has necessitated replotting much of the earlier data.

Other experimental approaches   Aschoff (1965a) has pointed out that there are at least six different ways in which a phase-response curve can be obtained for a light–dark cycle. Not all of them are feasible for other time cues, however. Aschoff's six strategies are summarized in Figure 2.25.

The first strategy is the classic form of the phase-response curve used by DeCoursey (1960a), shown in Figure 2.22 and schematically derived in Figure 2.23. As we have discussed, animals free-running in constant darkness are given pulses of light at predetermined circadian phases, and the phase shift is then assayed. There are, however, a number of limitations to this approach. In order to know at what phase the light pulse is being applied, one should ideally be able to assay the rhythm day by day as the experiment proceeds. Moreover, problems arise in analyzing the phase shift if there are changes in the free-running period before or after the light pulse. The necessary extrapolations of the phase may become unreliable, especially if the rhythm is obscured by noise or somewhat irregular. As a result, there may be difficulty in deciding at what subjective phase the light pulse was delivered.

A second strategy to obtain a PRC involves releasing the animal

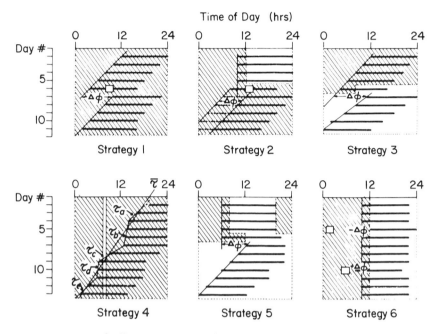

Fig. 2.25 Aschoff's six strategies for obtaining phase-response curves elicited by single stimuli. Horizontal bars = activity of a hypothetical light-active animal; hatched area = darkness; white area = bright illumination. Onsets of activity are connected by sloped lines. (From Aschoff, 1965a.)

from a light–dark cycle into constant conditions and then giving a light pulse during the first few subjective days of the free-run. This approach has several advantages: Prior to the light pulse, the animals have been in a fixed regimen so that long-term effects, such as aftereffects, are predictable. Furthermore, the phase of the light–dark cycle prior to release into constant conditions enables the experimenter to know precisely the phase of the circadian rhythm when the time cue was applied. Finally, this method allows the experimenter to choose a convenient time to deliver the time cue, since the light–dark cycle used prior to release into constant conditions can be used to shift the rhythm to any local clock time. However, there are problems with this approach because in computing the subjective time when the pulse is given, allowance must be made for the transient behavior of the circadian system after release from the light–dark cycle into constant conditions. One way to approach this problem is to compare the resultant phase with that from control studies in which the animal was released from the same light–dark cycle into the same constant conditions but with no light pulse given.

The third strategy to obtain a PRC uses a step-down or step-up in light intensity. Thus, in the example in Figure 2.25, the animal had been free-running in constant darkness and then was abruptly placed in constant light at a particular phase of the subjective circadian day. One problem with this approach is that the free-running period may be different in constant darkness and in constant light, and thus it may be hard to determine the phase shift reliably from extrapolating the free-running phases before and after the step change. In addition, as Aschoff (1965a) pointed out, the relative amounts of activity and rest tend to change with light intensity (following Aschoff's rules), and therefore the apparent phase of the free-running rhythm may be influenced, since the phase relationship between the onset and cessation of activity will be changing.

The fourth strategy uses the phenomenon of relative coordination. If the light pulse is given at the same time each day with an intensity not quite strong enough to entrain the animal, then relative coordination may be observed, with the rhythm showing increases in period at certain phases of the subjective day and decreases at others. The shift in the rhythm's phase relative to the timing of the light signal can be used to determine a PRC. Obviously the strength of the light pulse that can be used with this technique is limited, because it must be strong enough to induce relative coordination but not strong enough to cause entrainment.

The fifth strategy is essentially a combination of the second and third strategies, so that the animal is released from a light–dark cycle into constant conditions just after a phase shift of the last light–dark cycle.

In strategy six the animal is maintained throughout the experiment in a regular light–dark cycle, for example, and pulses of light are given during the dark phase. With this approach, only certain phases of the circadian rhythm can be assayed, and the complicating effects of the light–dark transition that follows the phase shift must also be taken into account.

Experimenters may choose one of these strategies depending on what is most convenient or most relevant to the issue under study. PRCs obtained by each of these various methods have been reported in the literature.

CHARACTERISTICS OF THE PHASE-RESPONSE CURVE

Most of the PRCs derived for mammalian species have been for short (10- or 15-minute) light pulses applied to an animal in constant darkness. The vast majority of experiments has used free-running animals in constant conditions (strategy 1) or animals just released into constant conditions after being entrained by a light–dark cycle (strategy 2).

Nocturnal and diurnal species appear to show the same general phase-response curves to light pulses, although the data for diurnal mammals are very sparse. All species show responses similar to those diagramed in Figure 2.23, with phase delays in early subjective night, phase advances in late subjective night and a breakpoint in the middle of the night. During the animal's subjective day there is little or no sensitivity to light pulses. Thus, the pacemaker(s) of the circadian system appear to be similarly phase-related to environmental time in nocturnal and diurnal species, although many of the overt rhythms of the animals are 180° out of phase.

The size of the phase shifts, and therefore the waveform of the PRC, depends on the nature of the zeitgeber signal. The following features are important:

Intensity. The intensity of the light stimulus influences the magnitude of the phase shifts and therefore the amplitude of the PRC.

Duration. The duration of the light pulse may determine the magnitude of the resultant phase shift in two ways. First, a prolonged light stimulus may have a *tonic* effect (a continuous action of light on the circadian system such as the changes in τ described by Aschoff's rules—also referred to as *parametric* effects). The longer the light pulse, the more light the animal is exposed to and thus the larger may be the phase-resetting effect. Second, a lengthening of the light pulse means that the on and off transitions (both *phasic* influences of light) are further separated in time and may fall on quite different parts of the PRC.

Sign of signal. Since phasic signals such as the transitions between light and dark are usually effective phase-resetting influences, then it might be expected that transitions in opposite directions would yield opposite results. This prediction has been tested by Subbaraj and Chandrashekaran (1978) on bats maintained in constant light and given dark pulses. In this species transitions in opposite directions do, indeed, produce mirror-image inversions of the PRC. For example, instead of phase delays occurring in early subjective night, phase advances occur. However, Ellis, McKlveen and Turek (1981) have not found such a mirror-image relationship between the PRCs produced by light pulses and dark pulses in hamsters. Dark pulses did not have any effect in late subjective night, while in early subjective night phase advances were obtained in some animals and rhythm splitting in others.

Wavelength. As discussed earlier, a time cue must be detected by the animal's receptors if it is to have any effect, and within the detectable range the wavelength will determine the strength of the stimulus. Therefore the wavelength of the light pulse is another determinant of the waveform of the PRC.

STRONG AND WEAK PHASE RESETTING

In most mammals the transition, in mid-subjective night, between the phase-delay portion of the PRC and the phase-advance portion (the breakpoint) is gradual (Fig. 2.26A). However, when much larger phase shifts are achieved, for example, by using long light pulses with *Rattus exulans* (Gander, 1980), the breakpoint becomes an asymptotic function with an instantaneous switch from extreme phase delays to extreme phase advances of the PRC (Fig. 2.26B).

Winfree (1971) has distinguished between the strong phase-resetting PRC with an instantaneous transition (his "type 0") and the weak phase-resetting PRC with a gradual transition ("type 1"). This nomenclature is derived from Winfree's plot of final phase on the y axis and initial phase on the x axis so that the typical weak PRC has an average slope of 1, and the typical strong PRC has an average slope of 0. The disadvantage of this nomenclature is that it is not simple to remember, and, while there are exceptions, in most cases it is easier to talk of weak and strong PRCs.

A PRC with an abrupt transition can be converted into one with a gradual transition by reducing the strength of the phase-resetting stimulus, for example, by reducing the light intensity of the pulse. That

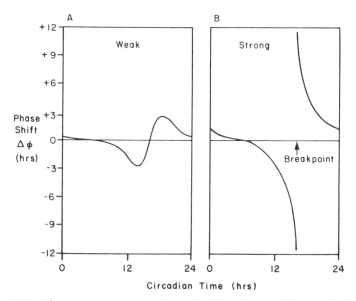

Fig. 2.26 Phase-response curves to weak (*A*) and strong (*B*) stimuli plotted against circadian time. As the maximum phase delay in early subjective night and the maximum phase advance in late subjective night increase with increasing stimulus strength, the transition in mid-subjective night between delays and advances gets more abrupt. With the strong PRC there is a breakpoint in the curve so that stimuli a few minutes apart may produce either a very large phase delay or a very large phase advance. (Copyright 1982 by Moore-Ede, Sulzman, and Fuller.)

strong PRCs are typically seen in lower organisms rather than mammals (although they have been obtained in mammals; Gander, 1980) suggests that the mammalian system may be more highly buffered against phase-resetting perturbations.

INTERSPECIES AND INTERINDIVIDUAL DIFFERENCES

Although the general characteristics of PRCs are similar in all species, there are some consistent differences in the waveform. To examine this, one must take care that all variables which influence the waveform of the PRC, such as intensity, duration, sign, and wavelength, are standardized. Daan and Pittendrigh (1976b) have done this, undertaking interspecies comparisons with several animals of each species subjected to light pulses at each circadian phase. While each of the four rodent species they studied shows the same general pattern of minimal phase shifting in subjective day, phase delays in early subjective night, and phase advances in late subjective night, the amplitude and the relative balance between advances and delays differ between species. In *Mus musculus* and *Peromyscus maniculatus* the delay portions of the phase-response curve are much more prominent than the advance portions, whereas in the golden hamster, *Mesocricetus auratus*, the advances are larger than the delays (Fig. 2.27). With 15-minute light pulses, the maximum possible phase shifts also differ. A

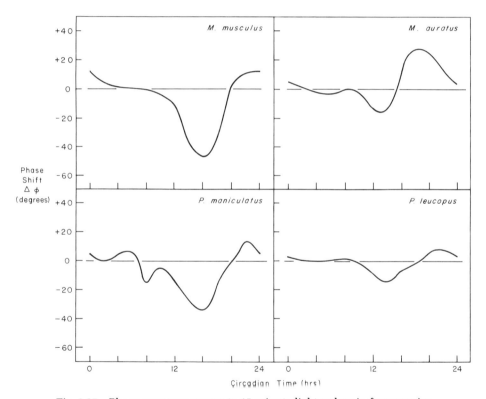

Fig. 2.27 Phase-response curves to 15-minute light pulses in four species of rodents. For comparison, the phase shifts are expressed in degrees to allow for the different free-running periods typical of each species. For an animal with a 24.0-hour free-running period, 30° = 2 hours of phase shift. The species vary in the relative proportion of delays to advances and the amplitude of PRC. (From Pittendrigh and Daan, 1976b.)

shift of approximately 50 minutes can be achieved in *Mus musculus,* whereas in the white-footed mouse, *Peromyscus leucopus,* the maximum phase shift with a similar light pulse is less than 20 minutes.

Deriving a phase-response curve for an individual animal rather than groups of animals is a time-consuming process. As we have already discussed, many days must be allowed for an initial steady state to be obtained, then all transients have to cease before a measurement can be made. Studying the effect of a light pulse once every month, an experimenter would need the best part of a year before completing the PRC of each animal. For animals whose lifespans are only one or two years, this clearly can cause problems, particularly since the free-running period of the circadian system changes with age in some species (Pittendrigh and Daan, 1974).

However, DeCoursey (1960a) was able to document carefully the phase-response curves for 12 individual flying squirrels and showed some clear differences in the waveform although they all conformed to the same general pattern. Figure 2.22B shows the phase-response curves of two animals with clear differences in amplitude and waveform.

Comparisons between this species and the four species studied by Pittendrigh cannot be readily made because Pittendrigh used 15-minute light pulses, whereas DeCoursey used 10-minute light pulses. Furthermore, Pittendrigh studied the final steady-state phase shift, whereas DeCoursey studied the phase shift after the first day of transients (see Fig. 2.24), which would tend to underestimate phase advances, which take several cycles to be completed, although not phase delays, which occur usually within one cycle.

As we have discussed, all zeitgebers act in a similar way, relying on the periodically changing sensitivity of the animal's pacemaker to the zeitgeber signal. Thus, phase-response curves could in theory be derived for each zeitgeber that has been reported, although so far most have been for light. Phase-response curves have now been obtained for a wide range of species, and these have been compiled in an atlas of PRCs edited by Pittendrigh (1981). The atlas also includes the PRCs for a number of agents that do not act as zeitgebers in the natural environment, such as certain chemicals. These substances are of increasing interest because they provide possibilities for the pharmacological manipulation of circadian clocks and thus for the probing of clock mechanisms. This subject we deal with in greater detail in Chapter 7.

What the Phase-Response Curve Tells about the Pacemaker

A circadian pacemaker is in some ways analogous to a pendulum. A tap with a hammer in the plane of the pendulum's swing will phase-delay the pendulum if the pendulum is at that instant displaced on the far side of the vertical, or will phase-advance the pendulum if it is displaced on the near side of the vertical. No change in phase (although there will be a change in amplitude) will occur if the pendulum is vertical at the time of impact. We can, therefore, produce a phase-response curve to hammer taps which will be a function of the properties of the pendulum-pacemaker and its initial conditions.

The circadian clockwatcher is, however, in the dark about the mechanism of the clock and cannot directly measure its phase. Imagine then that the room containing the pendulum in our analogy is in complete darkness. However, the pendulum is connected to a clock face which is lit so that one can see the hands of the clock but not the position of the pendulum. We could still gain information indirectly on the phase of the pendulum-pacemaker by obtaining a phase-response curve from the advances and delays in the hands of the clock produced by lever taps on the pendulum at different clock hours. If a hammer tap caused the hands of the clock to be set forward (a phase advance) then the pendulum must have been displaced on the far side of the vertical. Conversely, if a phase delay were obtained, the pendulum must have been on the near side of the vertical. If no phase shift was seen, the pendulum must have been vertical at the time of impact. Thus the PRC of an overt rhythm can be used to gain information about the circadian pacemaker, even if we do not know its structure and cannot observe its behavior directly.

MEASUREMENT OF PACEMAKER PHASE

Once a phase-response curve has been documented for short light pulses, these light pulses can be used to probe the state of the pacemaker in other experiments. For example, one can examine whether an overt rhythm is an accurate reflection of the behavior of the pacemaker that drives it. After a phase shift induced by a single pulse of a zeitgeber such as light (especially after a phase advance), there may be several cycles of transient behavior before the final phase shift is achieved. We saw such transient behavior during a phase advance of the rest–activity cycle in Figure 2.23. This could be because the circadian pacemaker shifts slowly after the light pulse. Alternatively, the pacemaker could shift very rapidly but there might be a delay in the

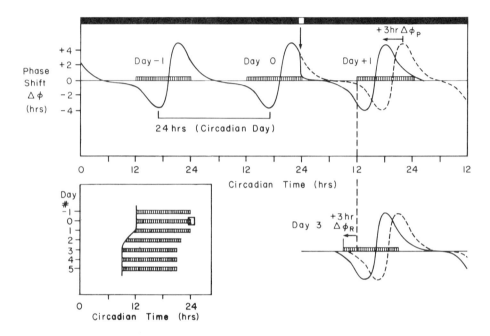

Fig. 2.28 Resetting of a circadian pacemaker and rhythm by a single light pulse. *Top panel,* solid line = PRC of the pacemaker (derived in Fig. 2.23); hatched bar = the animal's nocturnal activity bout. On day 0, a 1-hour light pulse in late subjective night hits a point in the PRC that produces a 3-hour phase advance. No shift in the activity rhythm is seen on day 1, but by day 3 *(lower right)* the complete 3-hour phase advance of activity has occurred. The solid line of the PRC represents what would happen if the pacemaker (and PRC) phase shifted instantaneously (the actual case, see text). The dashed line on day 1 shows a PRC which, like the activity rhythm, does not immediately shift. *Lower left,* plot of the activity rhythm, indicating the timing of the light pulse on day 0 and the gradual phase advances that were observed. (Copyright 1982 by Moore-Ede, Sulzman, and Fuller.)

resynchronization of the secondary oscillators in the circadian timing system (see Chapter 3 for definition) which generate the rest–activity cycle. Of course, some combination of slow pacemaker and slow secondary oscillator synchronization is also possible.

These explanations are illustrated in Figure 2.28, which uses the PRC developed in Figure 2.23 for the effects of 1-hour light pulses. On day −1 the free-running animal in constant darkness exhibits a stable phase relationship between the PRC of the pacemaker and the activity rhythm. On day 0 a single 1-hour pulse of light is given at the time marked by the arrow in late subjective night when the PRC predicts a 3-hour phase advance will be obtained. However, on the next cycle (day 1) the locomotor activity rhythm has not yet started to shift, and it will not complete its phase shift until after several more cycles on day 3 (see inset in the figure). Is this because the pacemaker, and therefore its PRC, has not yet started to shift (shown by the dotted continuation of the PRC), or has the pacemaker instantaneously shifted (shown by the solid line) but the secondary oscillators that are driving the activity rhythm have not yet started to resynchronize?

This question can be answered by using a second pulse of light given shortly after the initial stimulus to probe the pacemaker's new phase. By comparing the additional phase shift obtained after all transients are complete with the phase shifts obtained in control animals receiving only the initial stimulus, the phase of the pacemaker can be determined. This is done (Fig. 2.29) using the documented PRC and comparing the final steady-state phase shifts with those predicted. As Figure 2.29 shows, it is possible to determine from the combined response to the two light pulses if there was no immediate phase shift of the pacemaker (dotted line), or if the pacemaker had completed its phase shift by the time of the second pulse (solid line). This procedure is analogous, then, to our clockwatcher tapping the pendulum with his hammer a second time after the initial tap and examining the total combined displacement of the hands on the clock face.

In Figure 2.29 we can see that in addition to the phase advance of 3 hours produced by the initial light pulse, there is either a superimposed phase delay of 1 hour, if the pacemaker has phase shifted (solid line) with a net phase shift of $(+3) + (−1) = +2$ hours, or no additional phase shift, if the pacemaker has not shifted by the time of the second

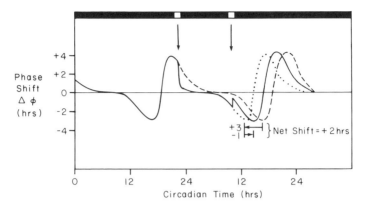

Fig. 2.29 Double-pulse experiment to probe the phase of a pacemaker after the single light pulse in Figure 2.28. The first light pulse, in late subjective night, produced a phase advance of 3 hours (as in Fig. 2.28). To determine whether the phase-response curve had been immediately shifted by the light pulse (solid line) or had not begun to shift on the first day (dashed line), a second light pulse was given in the next late subjective day, which would produce no additional phase shift if it hit an inactive portion of the unshifted PRC (dashed line) or would produce an additional phase delay of 1 hour if the PRC had already shifted (solid line) so that the original shift in the PRC (dotted line) was corrected by 1 hour. (Copyright 1982 by Moore-Ede, Sulzman, and Fuller.)

pulse (dashed line: +3 + 0 = +3 hours). Thus the two predictions suggest different final phases after all transients are complete and the steady-state phase of the rest–activity cycle can be measured.

Such double-pulse experiments were first conducted by Pittendrigh (1967b) in the fruit fly, *Drosophila,* and showed that the resetting of pacemaker phase is almost instantaneous, even though the overt rhythm may take several days to phase-shift. A second light pulse as close as 10 minutes after the first produced a combined phase shift that matched the prediction for an instantaneous and complete phase shift of the pacemaker within that 10-minute interval. Light pulses less than 10 minutes apart do not show this effect (Pittendrigh, personal communication). However, this does not necessarily mean that the pacemaker has not phase shifted. It may be that the photoreceptors of the retina have not fully recovered to their previous state before the second pulse. For example, the retina might not have had time to fully dark-adapt.

It is now important to extend these double-light-pulse studies to mammals. At the time of this writing, at least two groups of investigators were undertaking these laborious studies, but the results had not yet been published. When these studies are complete, it should be possible to determine whether it is the coupling of the pacemaker to the secondary oscillators of the circadian timing system or the coupling of the pacemaker to the zeitgeber that is largely responsible for the transient behavior of overt rhythms after phase shifts.

ENTRAINMENT BY ONE LIGHT PULSE PER CYCLE

We are now in a position to examine what the PRC tells us about the process of entrainment by zeitgeber cycles. A circadian clock is like a cheap wristwatch (at least before the days of the electronic quartz crystal) that runs consistently fast or slow and so must frequently be reset. A daily resetting can be accomplished by a short light pulse every 24 hours, provided the pulse falls on the portion of the PRC that will produce a phase shift of the required magnitude and direction. Thus, if the animal has a free-running period of 22 hours, the daily light pulse must fall on the portion of the PRC that will induce a 2-hour phase delay each cycle, so that the entrained rhythm will have a period of 24 hours.

Figure 2.30 illustrates this process of entrainment by short light pulses for a pacemaker with an endogenous 22-hour period. The solid line is the PRC for light pulses (of this particular intensity, duration, and wavelength) on this pacemaker, with a free-running period of 22 hours. The 1-hour light pulses are applied at 24-hour intervals. In our diagram, the first pulse falls on the part of the PRC that instantaneously phase-delays the pacemaker by 2 hours, so that the whole PRC becomes shifted by −2 hours. The next light pulse, applied 24 hours later, falls at an equivalent phase of the PRC on the animal's new subjective time scale. Again a 2-hour phase delay is obtained. This process is repeated with each light pulse so that the pacemaker's period is effectively lengthened to 22 + 2 = 24 hours and is therefore entrained to the 24-hour period of the light pulses. The fact that a single light pulse each day can entrain a rhythm provides further evidence that pacemaker phase resetting is completed rapidly.

The events depicted in Figure 2.30 represent the simplest example of entrainment. The first light pulse after the free-run happened to hit the

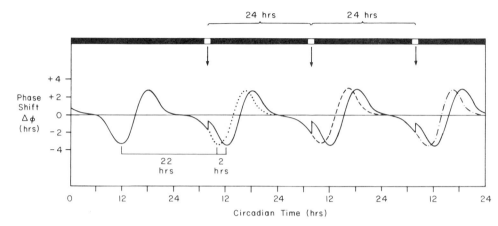

Fig. 2.30 Entrainment by a single 1-hour light pulse per day of a pacemaker with a free-running period of 22 hours. By hitting the phase-response curve in early subjective night, the 1-hour light pulse produces an instantaneous 2-hour phase delay (solid line) of the pacemaker (and PRC). The pacemaker period is thus lengthened to 24 hours and becomes entrained to the 24-hour period. On each subsequent cycle the 1-hour pulse again falls on that portion of the PRC which provides the requisite 2-hour phase delay for entrainment. (Copyright 1982 by Moore-Ede, Sulzman, and Fuller.)

PRC at the phase required to induce the exact phase delay needed for entrainment to a 24-hour period. What would happen in the more likely case where the daily light pulse commences at an arbitrary phase relationship to the pacemaker? In that case there would be a succession of transient cycles while the circadian system becomes entrained to the zeitgeber. Let us consider a pacemaker with a 26-hour period being entrained by a succession of light pulses 24 hours apart. The only steady-state phase relationship that can be achieved is one that phase-advances the pacemaker by 2 hours each cycle so that the periods are equal. The reason for this can be appreciated if one works through the sequence of events. The initial light pulse might produce a phase advance or a phase delay of various magnitudes or no phase shift at all. Let us consider the example in Figure 2.31, in which the first light pulse (on day 1) hits the PRC in the dead zone so that it induces no phase shift. Since the pacemaker period is 26 hours and the zeitgeber period is 24 hours, by the next cycle (day 2) the pacemaker

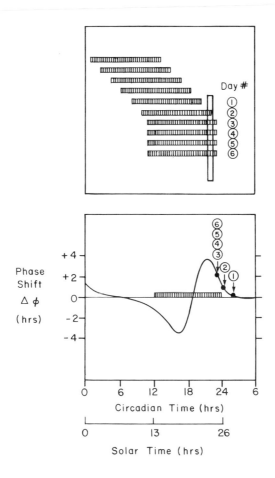

Fig. 2.31 Initial transients during entrainment to a single 1-hour light pulse per day. *Top panel,* a free-running activity rhythm with a 26-hour period. On day 1 a light pulse is applied, and over the next 2 days the rhythm gradually reaches a stable phase relationship with the 1-hour light pulse so that it occurs in late subjective night. *Lower panel,* the phase-response curve of the pacemaker plotted against circadian and solar time, showing the points on the PRC hit by the 1-hour light pulse each day (day numbers in circles). After day 3 a stable phase relationship is achieved, with the light pulses producing the requisite 2-hour phase advance to entrain a 26-hour pacemaker to a 24-hour zeitgeber. (Copyright 1982 by Moore-Ede, Sulzman, and Fuller.)

will be 2 hours later with respect to the time cue. The second light pulse will therefore hit the PRC at a point 2 hours earlier, which this time produces a 1-hour phase advance. The next light pulse, on day 3, consequently will fall on a PRC shifted by an hour and will therefore fall at a point 1 hour phase delayed from the previous cycle (instead of 2). This would then produce a phase advance of 2 hours so that entrainment would be achieved. Because the period of the pacemaker and zeitgeber for that cycle are now the same, this 2-hour phase advance will occur with each subsequent cycle. By a series of successive approximations, therefore, a steady state is gradually reached.

Determination of phase relationships We can now appreciate how the specific phase relationship between a circadian pacemaker and zeitgeber is determined. For a pacemaker with a particular free-running period and a zeitgeber with a given period, there is a specific phase relationship between the zeitgeber and the PRC (and therefore the pacemaker) where entrainment can occur. This is determined by where on the PRC the zeitgeber signal must fall to adjust the pacemaker to its own period. If the zeitgeber signal falls on any other part of the PRC, the phase relationship is automatically shifted back toward the specific phase relationship for stable entrainment (see Fig. 2.31).

With this model in mind, we can understand how the phase relationship between a pacemaker and a zeitgeber is influenced by the natural periods of both and by the coupling strength between them. If the period of either the pacemaker or zeitgeber is changed, then the daily phase shift necessary to achieve entrainment is altered. This means the zeitgeber signal must fall at another point on the PRC to achieve the required phase shift, and the phase relationship between the PRC (and pacemaker) and the zeitgeber cycle is thus altered.

Different coupling strengths change the waveform of the PRC. Thus an increase in zeitgeber strength, by changing the intensity, duration, or wavelength of a light pulse, will produce larger phase shifts at each phase of subjective day and night and therefore result in a PRC with a larger amplitude. This may mean that the particular phase shift required for entrainment will occur at a different point on the PRC. Consider the two PRCs to light pulses of different intensities in Figure 2.32. To obtain the required 2-hour phase delay to entrain a 22-hour free-running period to a 24-hour zeitgeber cycle, the light pulses must fall at different points on the animal's subjective time scale. The steady-state phase relationship between the light stimulus and the pacemaker at which entrainment will occur is thereby altered by the change in light intensity.

The phase relationships between the zeitgeber and the overt circadian rhythms of an animal are not determined entirely by the PRC of the pacemaker. While the pacemaker adjusts instantaneously to each zeitgeber signal, the animal's overt rhythms show several intermediate cycles of transients before they adjust. Because the process of entrainment requires a phase shift every cycle, the steady-state phase relationship observed between the overt rhythm and the pacemaker when the circadian system is free-running may never be achieved. The secondary oscillators (and rhythms) will tend to either lag behind, if entrainment is reached by phase advances, or lead the pacemaker, if entrainment involves phase delays. The phase relationship between zeitgeber and overt rhythm is therefore a product not only of pacemaker phase but also of the coupling strength between the pacemaker and the rhythm. The stronger the coupling, the fewer transients after a pacemaker shift and the more closely the phase relationship between pacemaker and rhythm will approximate that observed during free-running conditions.

Unstable phase relationships   Not all phase relationships between a zeitgeber and a pacemaker are compatible with steady-state entrainment. Consider the PRC waveform in Figure 2.33. There appear to be three areas of instability where stable entrainment cannot be achieved. First, pulses falling anywhere within the dead zone (zone A) will have no phase-shifting effect. Stable entrainment cannot be achieved with light pulses falling in this area, unless the zeitgeber period happens to equal the period of the pacemaker.

Second, in the portion of the PRC between the maximum phase delay and the maximum phase advance (zone B), single pulses of light usually cannot produce stable entrainment. At first sight it might not seem to matter whether a light pulse induces a 1-hour phase delay by falling on point x or point y in Figure 2.33. However, if one thinks through what happens as the steady state is approached, the reason for the instability of point y becomes apparent. If the light pulse falls

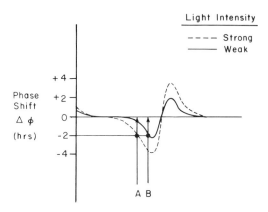

Fig. 2.32 Dependency of the phase relationship (ψ) on the strength of the zeitgeber signal, showing PRCs obtained for the same pacemaker with strong (dashed lines) and weak (solid lines) zeitgeber stimuli. As the amplitude of the PRC increases, its waveform broadens. A phase delay of 2 hours (such as might be necessary to entrain a 22-hour-period pacemaker to a 24-hour light–dark cycle) requires that the light pulse fall on a different phase of each PRC. With a strong pulse (A), the phase relationship between pacemaker and zeitgeber is shifted by several hours as compared to that with a weak pulse (B). (Copyright 1982 by Moore-Ede, Sulzman, and Fuller.)

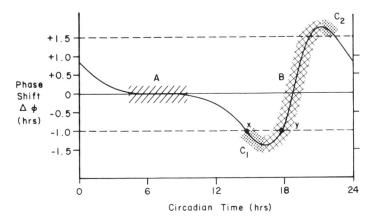

Fig. 2.33 The unstable zones for entrainment in a phase-response curve. Areas A, B, C_1, and C_2 are not compatible with stable entrainment, which requires that light pulses fall on the unhatched portions of the phase-response curve. The maximum range of stable entrainment is shown by the dashed lines. (Copyright 1982 by Moore-Ede, Sulzman, and Fuller.)

later in the PRC than the target steady-state point, the appropriate response is to lengthen the next pacemaker cycle by inducing a larger phase delay so as to correct the error and minimize the discrepancy in the next cycle. This happens at point x because a pulse that falls later will induce a larger phase delay. At point y, however, the opposite occurs, and a later timed pulse will induce a smaller phase delay, thus increasing the error on the subsequent cycle and causing instability in the phase relationship. A similar phenomenon occurs in the advance portions of the curve. In PRCs that show an abrupt transition between phase delay and phase advance, there is no zone B of instability, and a greater fraction of the cycle may be available for stable entrainment.

There may be a third zone of instability where the value of the slope of the PRC is highly positive or negative, such as zones C_1 and C_2. In this situation a small error in the timing of the light pulse with respect to the PRC of the pacemaker may induce a correction on the next cycle that is much bigger than the original error—thus leading to an unstable oscillation in the phase relationship.

Range of entrainment   As we have discussed, circadian oscillators can entrain to only a limited range of zeitgeber periods. Usually this range is within a few hours, plus or minus, of their natural free-running period. An inspection of the PRC, together with a consideration of the forbidden phase relationships shown in Figure 2.33, will demonstrate why the range of entrainment is limited. The maximum phase shifts that can be obtained from the PRC in Figure 2.33 are a 1.5-hour phase advance and a 1-hour phase delay. Thus, if the free-running period is 25 hours, then the pacemaker can entrain to time cue periods only between 23.5 and 26 hours, a range of 2.5 hours.

Relative coordination   An appreciation of the phase-response curve also helps us understand the unstable entrainment condition termed relative coordination, discussed earlier. In this condition, the period of a free-running rhythm is modulated as it passes through certain phase relationships with a zeitgeber cycle that is not strong enough for entrainment. The signal causes phase shifts and modifies the free-running period, but the maximum phase shift that can be obtained at any phase relationship is not enough to correct for the entire difference between the free-running period and that of the time cue. With each successive cycle the zeitgeber signal falls at a different point on the phase-response curve, sometimes retarding the period of the pacemaker, sometimes accelerating it. The observed result on the overt rhythm is a speeding-up and then slowing-down of period coordinated with the timing of the zeitgeber.

ENTRAINMENT BY NATURAL LIGHT–DARK CYCLES

It is somewhat remarkable that many of the entraining properties of natural light–dark cycles can be mimicked by a single short light pulse per cycle. For example, as little as 1 second of light per 24 hours can entrain the diurnal squirrel monkey; see Figure 2.34 (Sulzman, Fuller, and Moore-Ede, 1979). It might be expected that the natural light–dark cycle would be a much richer source of temporal information. However, the full light–dark cycle mostly serves to modulate the entrainment process so as to stabilize entrainment and to conserve the phase relationship of the animal's rhythms to its environment.

It is useful to start our analysis of the role of the natural light–dark cycle by breaking it down into its components. First, there is a gradual transition from dark to light at dawn, followed by a maintained level of light with a seasonal variation in duration (from 9 to 15 hours in temperate zones). Finally, there is a transition, again gradual, from light to dark at dusk. There may be a low level of moonlight or starlight illumination during the dark phase. There may also be variations of incident light intensity during both day and night depending on the thickness of the cloud cover and whether the animal is in a heavily leafed forest or on an open sandy beach.

Skeleton photoperiods   We will start by considering the influence of the light–dark transitions that occur at dawn and dusk. These may be mimicked by two light pulses per circadian cycle, one of which is interpreted as the lights-on signal at dawn and the other as the lights-off signal at dusk. Such two-pulse-per-cycle waveforms are usually referred to as skeleton photoperiods because they provide the temporal framework of a full photoperiod. It is interesting that such skeleton photoperiods may be a natural phenomenon for nocturnal animals that spend most of the day in a darkened burrow; they may be exposed to light only at dawn and dusk, and thus are exposed to two short pulses of light per cycle.

Time of Day (hrs)

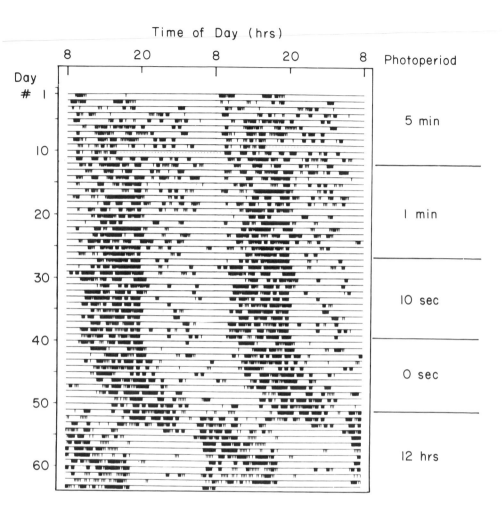

Fig. 2.34 Entrainment of drinking rhythm in the squirrel monkey by a single brief light pulse (600 lux) per day. From day 1 to day 39 the daily light pulse, starting at 0800 hrs each day, was gradually reduced from 5 minutes to 10 seconds. On days 40–52, no light pulse was given and the rhythm free-ran. On day 52, a light–dark cycle (LD 12:12) was applied, again providing entrainment. Other experiments have shown that a 1-second light pulse is an effective zeitgeber in this species. (Copyright 1982 by Moore-Ede, Sulzman, and Fuller.)

Two pulses per cycle provide a more stable entrainment than a single pulse, and they mimic well the entrainment to full photoperiods. Each pulse instantaneously resets the pacemaker, and the net phase shift of the two pulses determines the total phase shift per cycle. In entrained conditions, of course, this shift must equal the difference between the free-running pacemaker period and the period of the zeitgeber. Thus in Figure 2.35 the "dusk" pulse induces a 3-hour phase delay, and the "dawn" pulse induces a 1-hour phase advance, resulting in a net phase delay of 2 hours, thereby adjusting a 22-hour period pacemaker to 24 hours. With two pulses and thus two corrections of period per cycle, the pacemaker reaches its steady-state phase relationship with the zeitgeber more quickly.

Skeleton photoperiods, however, can be ambiguous. Consider two light pulses, 8 hours apart, per 24-hour cycle. The animal may interpret this either as an 8-hour day and a 16-hour night, or a 16-hour day and an 8-hour night, depending on which is taken as the dawn pulse and which as the dusk pulse. Pittendrigh and Daan (1976b) have conducted a series of experiments with hamsters in which the interval between pulses has been manipulated. With two pulses mimicking a short-day photoperiod of less than 10 hours, there is good agreement between the phase relationship of the activity rhythm in a full light–dark cycle and that seen with the skeleton photoperiod (Fig. 2.36). However, as the photoperiod is increased by moving the dawn pulse earlier, so that it follows the dusk pulse more closely, there is an increasing deviation between the phase relationships obtained with the skeleton photoperiod and with the complete photoperiod. Despite this, the hamster activity rhythm will entrain with skeleton photoperiods of up to 15 hours. When the photoperiod is increased to 16 or 17 hours by moving the dawn and dusk pulses still closer together, the animals' activity jumps from the short to the long interval between the two light pulses. It thus reaches a new phase relationship with the two pulses, with dawn and dusk reversed. In other words, the animals convert their interpretation of the skeleton photoperiod from a 16:8-hour to an 8:16-hour LD cycle and begin their daily activity just after what had been the dawn pulse.

The full light–dark cycle, therefore, ensures that day and night are unambiguous. While the transitions from dark to light and from light to dark convey much of the information needed for stable entrain-

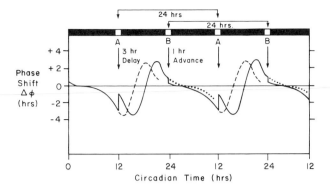

Fig. 2.35 Entrainment by two light pulses per circadian day. The PRC of a pacemaker with a 22-hour free-running period in constant darkness, subjected to a light pulse in early subjective night (*A*), which produces a 3-hour delay, and a light pulse in late subjective night (*B*), which produces a 1-hour phase advance. The net effect is a 2-hour phase delay, thus lengthening the pacemaker's period to 24 hours. (Copyright 1982 by Moore-Ede, Sulzman, and Fuller.)

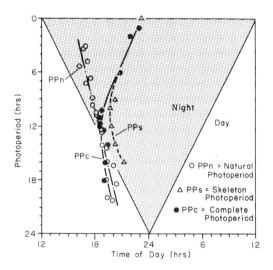

Fig. 2.36 The phase of activity onset in hamsters subjected to light–dark cycles with different photoperiods. The phase of activity onset in a skeleton photoperiod (PP$_s$) with two light pulses per day corresponds well to that of an equivalent complete photoperiod (PP$_c$) up to a photoperiod of 16 hours. However, activity onset coincided most closely to the onset of darkness in the natural photoperiod (PP$_n$), which provided a complete light–dark cycle with dawn and dusk transitions. (From Pittendrigh and Daan, 1976b.)

ment, light falling during the day enables the animal to become entrained to long photoperiods. The tendency of the activity rhythm to phase-jump from evening to morning is counteracted because light would then fall on the portion of the PRC that produces large phase shifts. Computer simulations of entrainment by Pittendrigh and Daan (1976b) further suggest that the aftereffects of long photoperiods on the pacemaker period may also help to prevent phase jumps.

To complete full circle in our analysis of entrainment by natural light–dark cycles, we have to consider the influence of the gradual changes in light intensity that occur at dawn and dusk. We have already discussed how entrainment is strengthened by the use of light–dark cycles with gradual dawn and dusk transitions (Kavanau, 1962). As Figure 2.36 shows, the phase relationship that is significant for the animal—the time when it starts its nightly bout of activity—adheres much more closely to the timing of dusk if the animal is in a natural light–dark cycle rather than in a cycle that is abruptly switched on and off. As with so many other physiological adaptations, species appear to have utilized this subtle feature of the environment to enhance their survival.

Organization of the
Circadian Timing System

In Chapter 2 we conducted what engineers might refer to as a black-box analysis of the circadian timing system (Fig. 3.1). We treated the whole animal as a system whose contents are unknown but whose overall function can be studied by recording certain rhythmic outputs (such as wheel-running activity) while manipulating environmental inputs (such as the light–dark cycle). Using this approach, we established that mammals have an endogenous ability to measure approximately 24-hour intervals of time and that their circadian timing systems have some clearly defined properties that hint at the internal structure.

If one looks at several different circadian rhythms within an individual subject (Fig. 3.2), the highly specific temporal order of the circadian system becomes apparent (Moore-Ede and Sulzman, 1981). The various rhythms play an intricate counterpoint, reaching their peaks and troughs at different phases of the circadian day. These few rhythms provide just a glimpse of the elaborate timekeeping that underlies all physiological processes.

The waveforms of the rhythms in Figure 3.2 are dramatically different. Some are sinusoidal, others approach square-wave patterns, still others more resemble a single short-lived pulse once every 24 hours. We are just beginning to appreciate the significance of the details of rhythm waveform. Often, for analytical convenience, rhythms are described by the timing (or *phase*) of a single definable point, for exam-

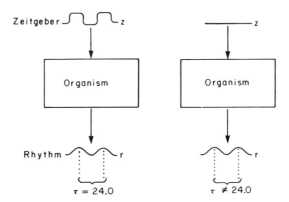

Fig. 3.1 Black-box analysis of the circadian timing system, in which the contents of the box are unknown and can be probed only by altering the input and observing the output. *Left,* the system in nature with normal 24-hour periodic environmental inputs; the circadian timing system produces rhythmic outputs with a 24-hour period. *Right,* when the timing system is placed in constant environmental conditions, the output is still rhythmic but free-runs with a non-24-hour period. Such experiments lead to the conclusion that there is a self-sustained oscillator within the timing system. (Copyright 1982 by Moore-Ede, Sulzman, and Fuller.)

ple, the maximum. Indeed, for many species, phase maps (Fig. 3.3) have been compiled which demonstrate the multiplicity of circadian rhythms within an animal and the specific phase relationships of each rhythm (Szabó, Kovats, and Halberg, 1978).

While different species may have quite different phase relationships to the outside world, the internal temporal order is usually quite similar from one animal to another. Thus, the activity rhythm of a nocturnal species such as the rat is approximately 180° out of phase with that of a diurnally active species such as the squirrel monkey, but in both species the maximum of the plasma corticosteroid rhythm coincides with the beginning of activity (Fig. 3.4). Hence, the internal structure of the circadian timing systems appears to be similarly organized in these species.

It is that internal structure, or at least the general principles of its organization, that will concern us in this chapter. We will focus on such questions as: Is there a single clock or are there multiple oscillators? If

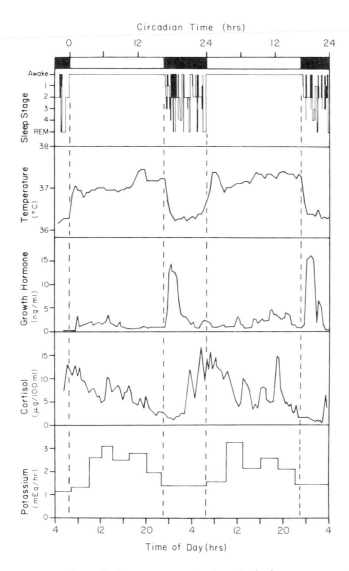

Fig. 0.0 Circadian rhythms in several physiological parameters in a human subject monitored in a light–dark cycle (LD 16:8), with lights-on from 0700 to 2300 hrs EST. From top to bottom, the rhythms plotted are sleep, colonic temperature, plasma levels of growth hormone and of cortisol, and the rate of urinary potassium excretion. Each rhythm has a characteristic phase relationship and waveform. (From Moore-Ede and Sulzman, 1981; data from Czeisler, 1978.)

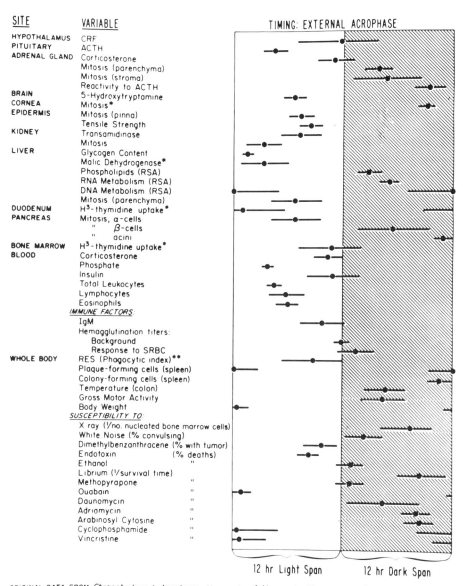

Fig. 3.3 A phase map of the circadian system of the mouse (*Mus musculus*). This figure shows the acrophase (time of occurrence of the best-fit maximum value) of numerous rhythms. The mean values of the acrophases are plotted ± SEM. (From Szabó, Kovats, and Halberg, 1978.)

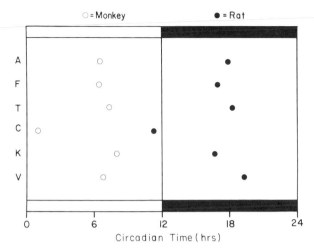

Fig. 3.4 Internal phase relationships of various circadian rhythms for a diurnal and a nocturnal mammal. The acrophase of the rhythms of activity (A), feeding (F), body temperature (T), plasma corticosteroids (C), urinary potassium excretion (K), and urinary volume (V) are plotted for the squirrel monkey (*Saimiri sciureus*) and the rat (*Rattus rattus*) maintained in an LD 12:12 cycle. (From Moore-Ede and Sulzman, 1981; rat data from Halberg, 1969.)

there are multiple oscillators, are they hierarchically or nonhierarchically arranged? How are the oscillators which generate circadian rhythms coupled to each other and to environmental time cues? And how do circadian oscillators drive the animal's overt rhythms?

Evidence for Multiple Oscillators

Even in lower organisms the circadian system can best be described as having more than one self-sustained oscillator (Pittendrigh, 1960). In mammals there is now little doubt that this is the case; only a multioscillator arrangement could account for all the various behaviors of the mammalian circadian timing system (Pittendrigh, 1960, 1974; Aschoff, 1965; Wever, 1975; Moore-Ede et al., 1976, 1979).

SPLITTING OF RHYTHMS

Splitting of the activity patterns of rodents (Pittendrigh, 1960) was one of the earliest indications of multiple, potentially independent oscillators in mammals. On occasion the activity portion of the daily rest–activity cycle, which is normally consolidated, splits so that two or more separate components can be distinguished. Each split component can on occasion free-run with a different period. Three different types of splitting in hamster activity are shown in Figure 3.5. The split pattern in the left side of the figure, in which the two components show similar periods and an approximately 180° phase difference, is probably the most common form of splitting. The middle section shows a case in which the two components have different periods and never achieve a stable phase relationship, then coalesce after many days. The right panel shows how split components can emerge out of an arrhythmic pattern and finally join to form a cohesive free-running rhythm.

The activity patterns of both nocturnal and diurnal mammals have been observed to split, especially when there is a change in the intensity of continuous illumination. A step-up in light intensity has induced splitting of activity rhythms in the nocturnal golden hamster (Pittendrigh, 1967a, 1974). Diurnal mammals have been less consistent, with a decrease in light intensity inducing splitting in the prosimian tree shrew (*Tupaia belangeri*; Hoffman, 1971) and the palm squirrel (*Funambulus palmarum*; Pohl, 1972), and an increase in intensity promoting splitting in the Arctic ground squirrel (*Spermophilus undulatus*; Pittendrigh, 1960).

Various physiological interventions may also induce splitting. After lesions which destroy the suprachiasmatic nuclei, a circadian pacemaker in mammals (see Chapter 4), the normally consolidated wheel-running patterns in hamsters may split into two or three discrete activity bouts within the 24-hour day under constant conditions (Rusak, 1977). Changes in hormonal levels can also induce splitting. Under the influence of testosterone, the free-running rhythm of locomotor activity of the starling (*Sturnus vulgaris*) tends to split into two components that temporarily run with different circadian periods (Gwinner, 1974).

Pittendrigh and Daan (1976c) have noted that hamsters in light–dark cycles often exhibit two distinct bouts of activity, one at lights-off and the other before lights-on. At the top of Figure 3.5 (left panel), the hamster was exposed to LD 12:12 before being placed in constant darkness. During the time of LD entrainment, the separate morning

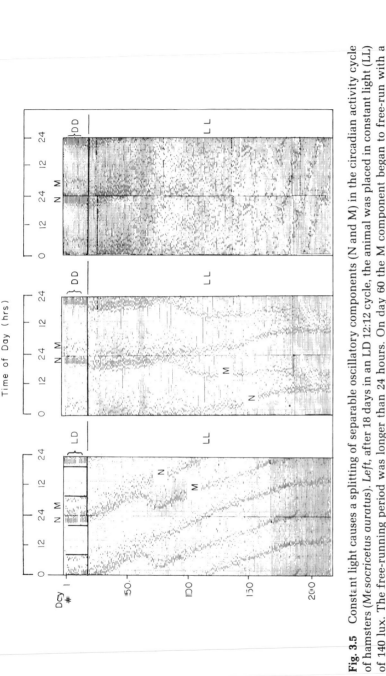

Fig. 3.5 Constant light causes a splitting of separable oscillatory components (N and M) in the circadian activity cycle of hamsters (*Mesocricetus auratus*). *Left*, after 18 days in an LD 12:12 cycle, the animal was placed in constant light (LL) of 140 lux. The free-running period was longer than 24 hours. On day 60 the M component began to free-run with a shorter period than N; it continued to free-run for 17 days, when it assumed a new steady-state phase (actually 180° out of phase) relative to M. The two components, in their 180° phase relationship, free-ran thereafter. *Middle*, this hamster was released from DD into LL on day 18. On day 90, M started an independent free-run but failed to lock onto N in the 180° anti-phase relationship, and by day 200 ultimately resumed its normal phase relative to N. *Right*, a third hamster released from DD into LL became totally aperiodic after 30 days. Subsequently a very long period rhythm coalesced out of the arrhythmic pattern. (After Pittendrigh, 1974.)

(M) and night (N) bouts of activity can be seen. These components may be the source of the two split components seen under constant conditions. In fact, Pittendrigh and Daan have proposed that the circadian pacemaker is composed of two oscillators (M and N) which respond differently to changes in light intensity.

In addition to the locomotor rhythms, splitting has also been observed in the body temperature rhythm of bats (Menaker, 1959) and squirrel monkeys (Fuller, Sulzman, and Moore-Ede, 1979a). Furthermore, there is often a prominent circa-12-hour component in the urinary potassium excretion rhythm of monkeys, which may represent splitting of kidney oscillators (Kass et al., 1980b). These several examples suggest that splitting behavior may be widespread.

TRANSIENT INTERNAL DESYNCHRONIZATION

With the advent of long-range commercial jet airplanes, attention has focused on a second type of evidence for multiple oscillators. Before commercial jet travel became possible, people crossed time zones relatively slowly, but now a person may have dinner in New York, travel across six time zones in as many hours, then have breakfast in Paris. The endogenous circadian timing system requires several days to resynchronize to the new local time, but what is more, different physiological rhythms show different rates of resynchronization. The state of temporal disarray after a change in time zone, before all rhythms return to their original internal phase-angle relationships, has been termed *transient internal desynchronization* (Moore-Ede, Kass, and Herd, 1977).

That different rhythms have characteristic rates of resynchronization was first documented by Sharp and co-workers in experiments with human subjects living in isolation in Spitsbergen, Norway (Sharp, 1960; Sharp, Slorach, and Vipond, 1961; Martel et al., 1962). Rhythms of water, sodium, potassium, creatinine, and ketogenic steroid excretion were studied during the constant light of the Arctic summer. First the individuals lived on an LD 16.5:7.5 cycle with darkness achieved by blindfolding the subjects during each "night." After the subjects demonstrated stable rhythms, the timing of day and night was reversed. In the days following the reversal, each rhythm shifted at a different rate. For example, the sodium excretion rhythm required four days to phase shift, whereas the ketogenic steroid excretion rhythm took eight days.

Transient internal desynchronization has also been demonstrated in human subjects exposed to shifts in artificial light–dark cycles while they were living in an isolation laboratory (Wever, 1979). Figure 3.6 shows the activity, body temperature, and urinary sodium rhythms of individuals subjected to 6-hour phase advances (attained by shortening the dark portion of one light–dark cycle) and 6-hour phase delays (by lengthening darkness for one night). A plot of the rate of phase shift (Fig. 3.7) shows that the activity rhythm shifted the most quickly, followed by the temperature and sodium excretion rhythms. Furthermore, in this case each rhythm appeared to shift more quickly after a phase advance than after a phase delay. Five days after the phase advance, the rhythm of sodium excretion, for example, showed reentrainment, while even seven days after an equivalent phase delay the rhythm was only about 75% resynchronized.

Other studies have measured the rates of resynchronization of human rhythms while the subjects were being flown across several time zones. For example, Klein and co-workers have flown subjects between Germany and the United States and have demonstrated transient internal desynchronization between the body temperature and performance rhythms (Klein, Wegmann, and Hunt, 1972). As measured by simple psychomotor tests, the performance rhythm shifted more quickly than the temperature rhythm, but if more complex tasks were used to measure performance, the rhythm shifted more slowly than the temperature rhythm. Phase shifts were also faster in one direction than in the other, but this directional asymmetry was the reverse of that shown in Wever's laboratory studies. In keeping with the common experience of transatlantic passengers, eastbound travelers (who were subject to a phase advance) required more time to adjust than westbound travelers (subjected to a phase delay).

This discrepancy between real and simulated flights is probably due to the masking effects produced by passive responses to zeitgebers in the environment, which tend to obscure the rate and even the direction of rhythm phase shifts, as discussed in Chapter 2. Mills, Minors, and Waterhouse (1978b) have separated the rate of resynchronization of the endogenous rhythms of human subjects from their passive responses to the environment by interspersing 24 hours of constant environmental conditions during phase-shift experiments conducted in an isolation facility. When the light–dark (and scheduled activity-

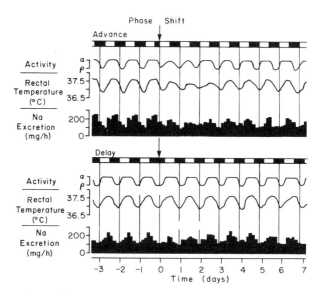

Fig. 3.6 Phase shift of circadian rhythms of rest (ρ)–activity (α), rectal temperature, and urinary sodium excretion in human subjects living in temporal isolation. *Top*, the results of a phase advance (accomplished by shortening the dark period on day 0); *bottom*, the result of a phase delay (by lengthening the dark period on day 0). (After Wever, 1979.)

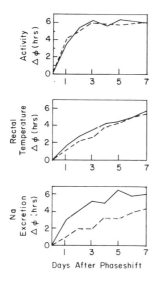

Days After Phaseshift

Fig. 3.7 Phase shift of the rhythms of activity, rectal temperature, and urinary sodium excretion in human subjects in isolation, subjected to either a 6-hour phase advance (solid line) or a 6-hour phase delay (dashed line) of environmental time cues. Each rhythm phase-shifted more rapidly after a phase advance of the zeitgeber than after a phase delay, and some rhythms shifted more rapidly than others. Thus, the activity rhythm resynchronized before the temperature rhythm which, in turn, resynchronized before the sodium excretion pattern. (After Wever, 1979.)

bed rest) cycles were phase-advanced by 8 hours, the endogenous rhythms of most subjects resynchronized by phase-delaying 16 hours. Furthermore, after a 12-hour phase shift, 90% of the adaptations to the new environment occurred as phase delays. Thus, many apparent phase advances of observed rhythms were really phase delays of the endogenous rhythm masked by passive responses to the environment.

Transient internal desynchronization is less well documented in other mammalian species, probably not because it does not occur but because it is difficult to measure multiple rhythms simultaneously in the laboratory. Where this has been done, in the chair-acclimatized squirrel monkey (Moore-Ede, Kass, and Herd, 1977), the circadian rhythms of activity, feeding, drinking, body temperature, and excretion of urinary potassium, sodium, and water each resynchronized at its own characteristic rate after a phase shift of the light–dark cycle (Fig. 3.8). Thus, the activity rhythm resynchronized in one to two days after an 8-hour phase delay of the light–dark cycle, but the body temperature rhythm took three days, and the rhythm of urinary potassium excretion, seven days.

FORCED INTERNAL DESYNCHRONIZATION

It is also possible to separate circadian rhythms by exposing a subject to day–night cycles of abnormal length. Lewis and Lobban (1957) studied two groups of students living in isolated communities in Spitsbergen. During the seven weeks of the experiment one group of subjects lived on a 21-hour day and the other group lived on a 27-hour day. This was accomplished by giving the members of each group specially adjusted wristwatches that showed 12 clock hours during 10.5 solar hours (for 21-hour days) or 12 clock hours during 13.5 solar hours (for 27-hour days). Because Spitsbergen is in the Arctic Circle (latitude 79°N), there was relatively constant illumination from the sun during the 24-hour midsummer day. In fact, during the summer of 1955, when these experiments were conducted, the sky was overcast much of the time, so the sun gave no indication of the real time of day. Outside the living quarters the average temperature difference between day and night was only 4° C. Thus, there were no natural cues to indicate the real time of day, and the special watches facilitated strict adherence to the experimental routine, even when subjects were away from their

Time Elapsed (hrs)

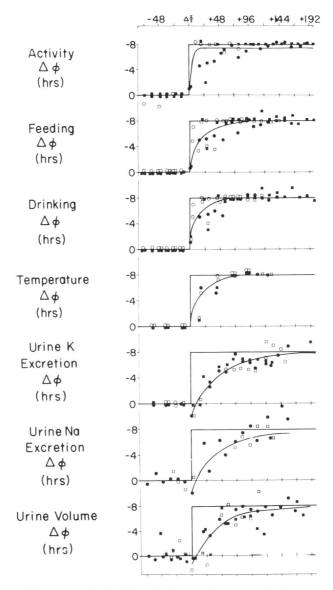

Fig. 3.8 Response of circadian rhythms of activity, feeding, drinking, body temperature, and excretion of urinary potassium, sodium, and water in four squirrel monkeys after an 8-hour phase delay of the LD 12:12 cycle. The time of the phase reference point for each cycle in the control days before the phase shift of the LD cycle is plotted as 0 phase. After two control days there was an 8-hour phase delay in the LD cycle, and the times of the phase reference points (as compared to the control values) are plotted for eight days. The average rate of phase shift is shown by the exponential functions that were fitted to the phase reference points of each rhythm after the LD phase shift. (From Moore-Ede, Kass, and Herd, 1977.)

camps. Recordings were made of body temperature and of the urinary excretion of potassium, chloride, and water. While the body temperature and urine volume rhythms could adapt to the non-24-hour days, the potassium excretion rhythm became internally desynchronized and maintained an independent, near-24-hour period.

Wever (1979) has similarly produced internal desynchronization between the activity–rest patterns and colonic temperature rhythms in subjects living in isolation and exposed to a zeitgeber comprising a light–dark cycle with gong signals. By varying the period of the cycles, Wever observed that the range of entrainment of the body temperature rhythm was narrower than that of the activity rhythm. For example, although both rhythms could be entrained to a 24-hour cycle, the activity rhythm would entrain to a 30-hour period but the temperature rhythm would not, instead exhibiting a period of 25 hours (Fig. 3.9).

Analogous experiments have been undertaken with other species. Sulzman, Fuller, and Moore-Ede (1981) have simultaneously monitored the feeding, colonic temperature, and urinary potassium excretion rhythms of squirrel monkeys while exposing them to long and short LD cycles in which the light and dark periods were of equal length (Fig. 3.10). In a 28-hour cycle (LD 14:14), for example, the feeding and temperature rhythms remained synchronized to the LD cycle, while the urinary potassium excretion rhythm free-ran. Eastman (1980) has also been able to produce a separation between the sleep–wake and body temperature rhythms in rats exposed to light–dark cycles with non-24-hour periods.

Another way to force internal desynchronization is to provide the animal with two zeitgebers conveying different temporal information. For example, Sulzman and co-workers (1978) have simultaneously provided squirrel monkeys with a 23-hour light–dark cycle (LD 11.5:11.5) and a 24-hour food-availability cycle (EF 1:23). The rhythms of each animal separated into two groups (Fig. 3.11). Body temperature, for example, followed the LD cycle, whereas urinary potassium excretion followed the EF cycle. A similar separation can be obtained by shifting the relative phases of LD and EF cycles and then maintaining both at a 24-hour period.

SPONTANEOUS INTERNAL DESYNCHRONIZATION

Both transient and forced internal desynchronization have provided highly suggestive evidence for the existence of multiple oscillators in

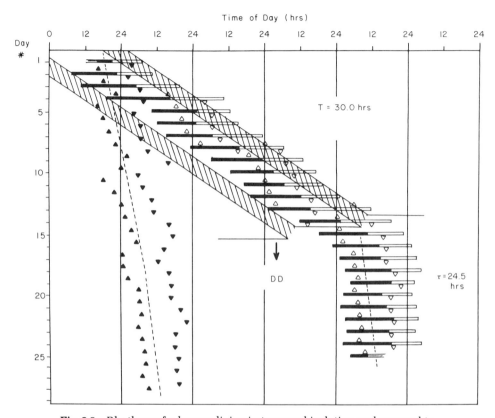

Fig. 3.9 Rhythms of a human living in temporal isolation and exposed to a 30-hr light–dark (and noise–quiet) cycle for 15 days, then constant darkness (DD). Hatched areas represent the daily dark period in the LD cycle; dark bars represent the active portion of the day; light bars indicate the rest portion. Upright triangles show the time of the maximum of the rectal temperature rhythm; inverted triangles show the time of the temperature minimum. The temperature rhythm is double-plotted with closed and open triangles. The activity rhythm was entrained by the zeitgeber cycle, but the temperature rhythm was not. (From Wever, 1979.)

the circadian timing system. However, because the animal must be exposed to zeitgebers to achieve these forms of desynchronization, there must always be concern whether the phenomena represent only a separation of the masking responses to the zeitgeber from the animal's endogenous rhythms. Furthermore, transient internal phase shifts be-

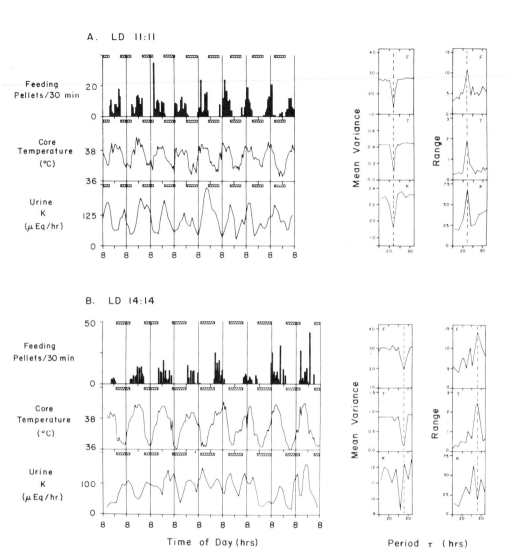

Fig. 3.10 The rhythms of squirrel monkeys exposed to an LD 11:11 cycle (*top*) and an LD 14:14 cycle (*bottom*). The dark period of the cycle is represented by the hatched boxes above each panel. On the left are plotted the rhythms of feeding (pellets/30 min), colonic temperature (° C), and urinary potassium excretion (μEq/hr). On the right are shown the period analyses for the two parameters, mean variance and circadian range, from the data on the left. The dashed line is drawn at 22 hours (top) and 28 hours (bottom). *Top panel,* all of the rhythms were synchronized to the 22-hour period, but only the feeding and temperature rhythms were synchronized to the 28-hour period (*bottom panel*). (From Sulzman, Fuller, and Moore-Ede, 1981.)

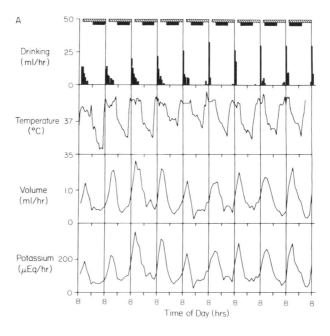

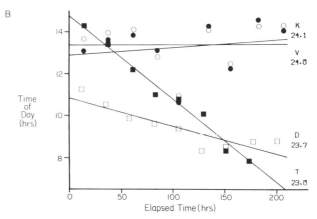

Fig. 3.11 Circadian rhythms of a squirrel monkey exposed to concurrent 23-hour LD and 24-hour EF cycles. (A) Rhythms of drinking (ml/hr), colonic temperature (° C), urinary volume (ml/hr), and urinary potassium excretion (μEq/hr). Vertical lines indicate 24-hour intervals. The fasting portions of the EF cycle are indicated by the hatched boxes at the top of the panel, and the dark periods of the LD cycle are represented by the dark boxes. (B) Phase plots of the rhythms of drinking (D; open squares), colonic temperature (T; closed squares), urinary volume (V; open circles), and urinary potassium excretion (K; closed circles) of the data from A. Times (in EST) of the phase reference points are plotted as a function of elapsed experimental time. (From Sulzman, Fuller, Hiles, and Moore-Ede, 1978.)

tween rhythms could occur with only one circadian clock if the various overt rhythms were coupled to it with different strengths.

The most convincing evidence for multiple circadian oscillators has been provided by Aschoff (1965b) and Wever (see his review: 1979). Aschoff reported that human circadian rhythms would on occasion spontaneously desynchronize, as illustrated in Figure 3.12, which shows the rhythms of a human subject in isolation. The rest–activity and urinary calcium excretion rhythms spontaneously began to oscillate with a period of 32.6 hours, while the rhythms of body temperature, urinary potassium, and water excretion continued to oscillate with a period of 24.7 hours. What is more, these two groups of rhythms continued to free-run with independent periods for a sufficient number of cycles so that one group "lapped" the other. In other words, by the end of the experiment some rhythms had completed more circadian days than the others, and all 360° of internal phase relationships were observed between the two groups of rhythms. This evidence strongly

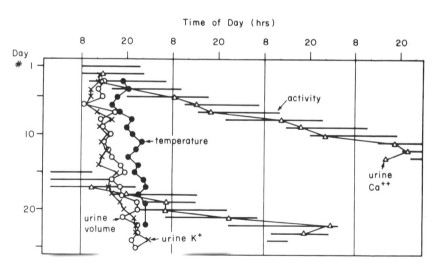

Fig. 3.12 Desynchronization of circadian rhythms of a human subject living in isolation without time cues. The times of wakefulness (black bars), the maxima of body temperature (closed circles), and the maxima of urinary excretion of calcium (open triangles), water (open circles), and potassium (Xs) are plotted. (From Aschoff, 1965b; copyright 1965 by the American Association for the Advancement of Science.)

indicates the presence of two or more circadian oscillators, since one circadian clock could not drive two quite independent rhythms.

The experiments of Aschoff and Wever demonstrated spontaneous internal desynchronization in less than 25% of their subjects, raising the possibility that only certain individuals might be susceptible. Indeed, it has been suggested that internal desynchronization is a result of aging (Wever, 1979) or of the person's psychological state (Lund, 1974). However, the original experiments rarely continued longer than a month. More recent studies by Czeisler, Weitzman et al. (1980) have shown that most individuals studied in an environment free of time cues for more than two months will eventually show internal desynchronization, although in some subjects it occurs much sooner than in others.

For over ten years after Aschoff's (1965b) report of spontaneous internal desynchronization in human subjects, no evidence was found of this phenomenon in other species. There remained one possible explanation for internal desynchronization that did not require more than one circadian clock. This explanation was based on the ability of humans to control consciously whether they will go to sleep or stay awake. If a person had a lot of work to do each day, he might delay the time of going to bed on successive days, thus prolonging his activity–rest cycle. His other rhythms, however, might not be able to entrain to the lengthened sleep–wake cycle but would still follow his single, endogenous circadian clock.

However, in the last few years we have shown that spontaneous internal desynchronization can also occur in the squirrel monkey, a species in which the above criticism does not apply (Sulzman, Fuller, and Moore-Ede, 1977c, 1979a). We observed spontaneous internal desynchronization in about one-quarter of the experiments conducted with chair-restrained monkeys in constant illumination. In the usual pattern, the feeding, drinking, and colonic temperature rhythms desynchronized from the rhythms of urinary potassium and water excretion (Fig. 3.13). A similar incidence of desynchronization occurred at three different light intensities (1 lux, 60 lux, and 600 lux) and was not always evident in the same monkey (Sulzman, Fuller, and Moore-Ede, 1979a). For example, the monkey shown in Figure 3.13 was internally synchronized at 1 lux and 600 lux but internally desynchronized at 60 lux. The distributions of free-running periods for feeding, colonic

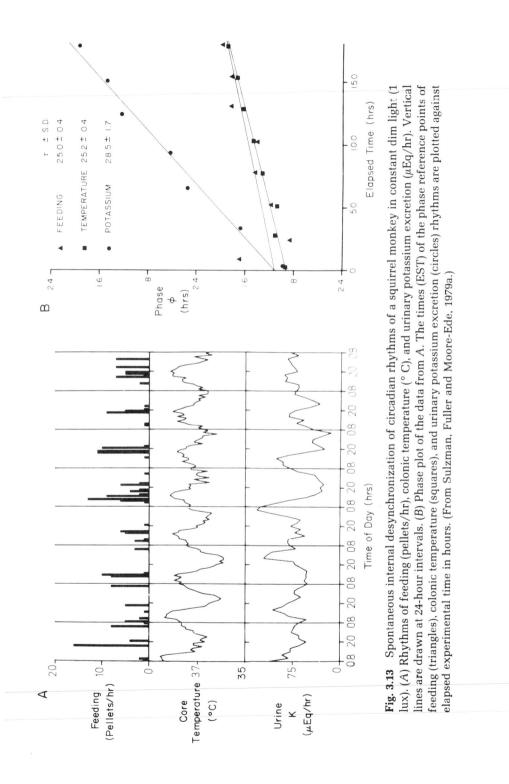

Fig. 3.13 Spontaneous internal desynchronization of circadian rhythms of a squirrel monkey in constant dim light (1 lux). (A) Rhythms of feeding (pellets/hr), colonic temperature (°C), and urinary potassium excretion (μEq/hr). Vertical lines are drawn at 24-hour intervals. (B) Phase plot of the data from A. The times (EST) of the phase reference points of feeding (triangles), colonic temperature (squares), and urinary potassium excretion (circles) rhythms are plotted against elapsed experimental time in hours. (From Sulzman, Fuller and Moore-Ede, 1979a.)

temperature, and urinary potassium excretion are shown in Figure 3.14. The feeding and temperature rhythms show relatively narrow distributions, with the feeding period ranging from 24.0 to 26.0 hours and the temperature period varying from 23.0 to 25.5 hours. However, there was a wide range of periods for the rhythm of urinary potassium excretion (20.5 to 29.5 hours). In those experiments with urinary potassium periodicities outside the range seen for feeding or temperature cycles, internal desynchronization was observed.

Although spontaneous desynchronization between the rest–activity and the colonic temperature rhythms has been observed in man, we have no unambiguous evidence at the present time that those two rhythms desynchronize spontaneously in the squirrel monkey. This could indicate some important difference in the structure or function of the circadian timing systems of human and nonhuman primates. However, neurophysiological evidence, which we will discuss more fully in Chapter 4, indicates that the rest–activity and body-temperature rhythms are generated by separate oscillators in the nonhuman primate, just as they are in man. Thus, although these observations of spontaneous internal desynchronization indicate the presence of at least two groups of potentially autonomous oscillators in both human and nonhuman primates, the interoscillator coupling strengths may differ between species, thus determining the incidence of desynchronization between particular rhythms.

While internal desynchronization provides useful information on the organization of the circadian timing system, it is not its normal state. Under both entrained and free-running conditions, phase relationships among the animal's various circadian rhythms are usually stable. In blinded rats, for example, the free-running rhythms of plasma corticosterone, pineal N-acetyltransferase activity, and serotonin remained synchronized with the rest–activity cycle over at least 60 days of observation (Gibbs and Van Brunt, 1975; Pohl and Gibbs, 1978). Similarly in the squirrel monkey, internal synchronization between the free-running rhythms of feeding, temperature, and urinary potassium excretion is usually seen in animals maintained in constant illumination (Sulzman, Fuller, and Moore-Ede, 1979a). Studies of human subjects in isolation for a month indicate that such rhythms as sleep–wakefulness, body temperature, and plasma cortisol concentra-

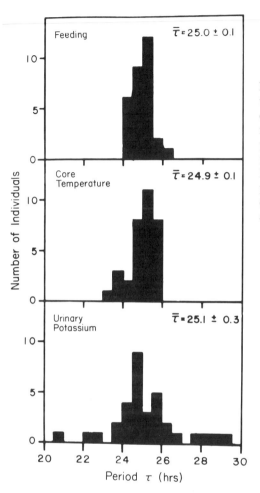

Fig. 3.14 Plot of the distribution of the free-running periods for three circadian rhythms of the squirrel monkey: feeding (*top*), colonic temperature (*middle*), and urinary potassium excretion (*bottom*). In the upper right corner of each panel is shown the mean period ($\bar{\tau} \pm$ SEM). (After Sulzman, Fuller, and Moore-Ede, 1979a.)

tion are usually free-running with a similar periodicity and stable phase relationships throughout the period of study in most individuals examined (Aschoff, 1965b; Czeisler, 1978; Weitzman, Czeisler, and Moore-Ede, 1979). Such studies provide evidence that there is a reasonably strong internal coupling mechanism among the various circadian rhythms and that internal entrainment does not depend on the presence of strong environmental time cues. It is important, consequently, to stress that splitting or internal desynchronization, despite their important implications for multioscillator theory, are not the usual state of affairs. Most circadian systems remain internally coupled for many cycles in the absence of environmental time cues.

PERSISTENCE OF RHYTHMICITY *IN VITRO*

Additional compelling evidence for multiple oscillators is the persistence of circadian rhythms in tissues maintained in culture outside the body. Isolating tissues *in vitro* is more easily said than done, however. Although maintaining an organ culture for a few hours is relatively routine, it is much more difficult to maintain for the several days necessary to conclusively determine whether rhythmicity is present or not. Furthermore, the absence of the exactly appropriate hormonal milieu and ionic conditions may preclude the demonstration of spontaneous rhythmicity. On occasion, however, it has been possible to successfully demonstrate *in vitro* that mammalian tissues contain circadian oscillators.

Adrenal glands The first demonstration that mammalian adrenal glands could show a continuing circadian rhythm of corticosteroid secretion *in vitro* was provided by Andrews and Folk (1964). Adrenal oxygen consumption and steroid synthesis were shown to be rhythmic for two to five days. A free-running rhythm was apparent from their data, with a period shorter than 24 hours. Shiotsuka, Jovonovich, and Jovonovich (1974) have also shown persisting circadian rhythms in corticosterone production by cultured hamster adrenal glands, although they observed a wider range of periods than that reported by Andrews and Folk.

The phase of the corticosteroid secretion rhythm of adrenal glands cultured *in vitro* appeared to be determined by the phase of the light–dark cycle in which the donor animal was maintained (Andrews and Folk, 1964). This finding, although not surprising, documents that the *in vitro* circadian rhythms are probably the continuation of the same function seen *in vivo*. Further, the period of the rhythm was found to be temperature–compensated, as would be expected of a circadian clock (see Chapter 2): when adrenal glands were maintained at 37° C, 25° C, and 15° C, no significant differences in period were observed.

In living animals the adrenal circadian rhythms are probably synchronized by the rhythm of plasma adrenocorticotropic hormone (ACTH) released from the pituitary, although other pathways may play a role (Dallman, Engeland, and McBride, 1977; Dallman et al., 1978). Andrews and Folk (1964) have shown that the addition of ACTH to the adrenal culture is not necessary for the demonstration of adrenal rhythms. ACTH does, however, influence the amplitude of the circadian oscillation; higher concentrations of ACTH in the culture medium produce larger-amplitude oscillations in corticosteroid production. Although ACTH may not be necessary to maintain the rhythm, there is evidence to suggest that it can synchronize the adrenals by a phase-resetting mechanism (Fig. 3.15). Andrews (1968) administered a pulse of ACTH to adrenal gland cultures at the maximum and minimum times of the circadian cycle. The pulse given at the peak of the cycle resulted in no phase shift of the rhythm as compared to the control groups of glands. However, a pulse given at the minimum point resulted in an apparent phase advance; the first four cycles of the free-running period after the ACTH pulse were shortened, then the period stabilized to 24 hours, with the peak advanced 9 hours as compared to the control glands. It is not entirely clear from the data, however, whether the new steady-state rhythm had a similar periodicity to that of the control gland.

Liver Suspension cultures of liver cells from young rats show circadian rhythms in oxygen consumption (Langner and Rensing, 1972), tyrosine aminotransferase activity, protein synthesis (Hardeland, 1973), and nuclear size (Rensing et al., 1974). All of these rhythms are observed when cells are maintained in an LD cycle but are probably not passively induced by it. For example, Langner and Rensing (1972) demonstrated a circadian rhythm in oxygen consumption in cultures of rat liver cells kept in an LD 12:12 cycle; the rhythm was observable for at least 48 hours (Fig. 3.16) and could be shifted by a 4-hour phase delay of the LD cycle. Langner and Rensing argued that the rise in oxygen consumption was not directly induced by the LD cycle, because the daily increase occurred before the dark light transition. Only the tyrosine aminotransferase (TAT) and nuclear size rhythms have been shown to persist in constant illumination. The activity of TAT showed a two- to threefold oscillation which is synchronized to the LD cycle, with the peak activity occurring during the light period (Hardeland, 1973a). In constant light the rhythm persisted with a near-24-hour pe-

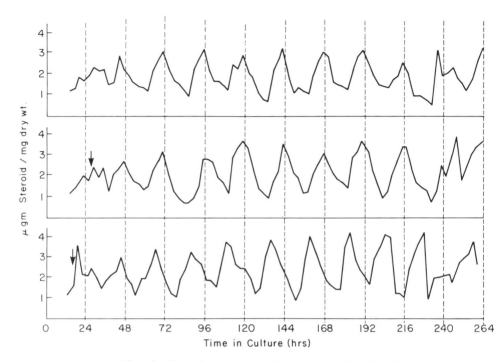

Fig. 3.15 The rhythm of corticosteroid secretion of isolated adrenal glands maintained in culture for 10 days. *Top panel,* untreated (control) mean corticosteroid secretory rates for 20 paired adrenal cultures; *middle and lower panels,* data for glands treated with 1.0 i.u. pulses of ACTH at the times indicated by the arrows. The pulses were given at presumptive high and low points of the secretory cycle. Note the displacement of the rhythm in the lower plot as compared to the other two. (Reprinted from Andrews, 1968 by permission of Pergamon Press.)

riod. The TAT rhythm would gradually damp out after two weeks, but the oscillation could be immediately induced by a single 1-hour pulse of dark and then would persist for successive cycles in constant light (Fig. 3.17; Hardeland, 1973b).

This action of LD cycles directly on *in vitro* cultures is surprising, since *in vivo* very little light reaches an animal's liver. Normally, mammalian tissues receive light–dark cycle information only indirectly, via specialized receptor cells of the retina. However, these experiments indicate that liver cells are sensitive directly to light, so at least *in vitro* light can synchronize the circadian TAT activity rhythm.

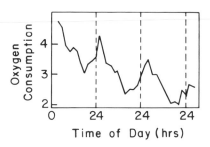

Fig. 3.16 Circadian rhythm of oxygen consumption in rat-liver suspension cultures. The volume of oxygen consumed (in arbitrary units) is plotted against time of day. (After Langner and Rensing, 1972.)

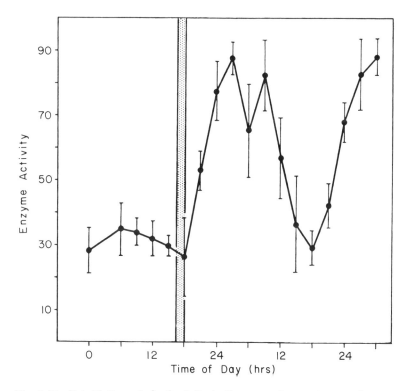

Fig. 3.17 Reinitiation of rhythmicity in liver tyrosine aminotransferase (TAT) activity by a dark pulse. After three weeks in constant light, the culture was exposed to 1 hour of darkness (shaded bar); the rhythm resumed at the end of the pulse. The mean values of TAT activity are plotted with their threefold standard errors. (After Hardeland, 1973, reprinted by permission of Pergamon Press.)

Rensing and colleagues (1974) demonstrated a circadian rhythmicity in the nuclear size of liver cells in rats ranging in age from three to seven days. The one-day-old culture suspensions reached a maximum size just after lights-off, with a slight second peak occurring at lights-on. Under constant illumination a rhythm with two clear peaks per cycle was observed, with a period of approximately 25 hours. This evidence suggests the presence of a second component that did not express itself until environmental time cues were removed, allowing the apparent uncoupling of two oscillators in the liver cells.

Other tissues   It has been known since the time of Galen (129 A.D.) that a heart can beat after being removed from an animal. Less well known is that an isolated perfused heart will show persisting ultradian or circadian rhythms in heart rate. Tharp and Folk (1965) demonstrated that isolated perfused rat hearts or heart-cell networks can show periods ranging from 1 to 24 hours. Most hearts had the strongest period between 7.5 and 14 hours. A similar range of periods was observed in isolated heart cell networks, which had periodicities predominantly between 11 and 17 hours. Even embryonic hearts can display rhythms in heart rate. Lunell, Cunningham, and Rylander (1961) demonstrated rhythms in electrical activity in a seven-day-old chick embryo heart maintained in tissue culture, finding a 20-hour period in one study and a 17-hour period in another.

Some years ago Bünning (1958) reported that isolated sections of intestine from a golden hamster would show a persisting circadian rhythm of spontaneous contraction when placed in organ cultures. The rhythm persisted for two to three days with a period that was close to 24 hours. Between 20° and 39° C the incubation temperature did not have much effect on the period length, so this rhythm also appeared to be generated by a temperature-compensated circadian oscillator.

Red blood cells   The development of *in vitro* preparations of mammalian tissue which will show persisting circadian rhythms in culture offers some prospect for learning more about the mechanism of circadian clocks. One key issue is whether the nucleus of the cell is required for the generation of rhythmicity. Thus, initial reports of persisting circadian rhythms in human red blood cell suspensions were greeted with considerable excitement. Not only did these cells have no nucleus, but they were also an easily obtainable human tissue.

Ashkenazi and co-workers (1975) reported that several enzymes in suspensions of human red blood cells showed circadian rhythms either with one peak (acid phosphatase, acetylcholine esterase) or two peaks (glucose-6-phosphate dehydrogenase, glutamate oxaloacetate transaminase) per day. Additionally, Hartman, Ashkenazi, and Epel (1976) reported an *in vitro* circadian rhythm in erythrocyte membrane properties, and they correlated enzyme activity with changes in membrane electrical potential (indicated by a fluorescent chemical probe).

However, it has become apparent that there is a great deal of variability in the *in vitro* rhythms of red blood cells described by Ashkenazi and co-workers (1975), and attempts by several laboratories (Dunlap et al., 1976; Cornelius, 1980; Palmer, personal communication) to reproduce their findings have been unsuccessful. At the present time there has been no confirmation that statistically significant rhythms occur in suspensions of human red blood cells, and it needs to be determined whether this system is valid for circadian rhythm research.

Structural Elements

The three types of evidence described in the previous section—splitting, internal desynchronization, and *in vitro* circadian rhythms—provide firm confirmation for Pittendrigh's (1960) early proposition that "there must be many distinct oscillatory physiological systems in the individual." We can reach a number of important conclusions from this evidence.

First, the circadian timing system has two or more oscillators, each capable of generating self-sustained circadian rhythms. Second, these potentially independent oscillators are normally synchronized whether or not the animal is in a periodic environment. This means that the oscillators must be internally coupled by mechanisms of sufficient strength to maintain synchrony. Third, each oscillator is normally synchronized to certain specific environmental cycles, so there must be transducers that convert the environmental signal into biological signals that can entrain each oscillator.

We will now discuss the general structural elements of the circadian

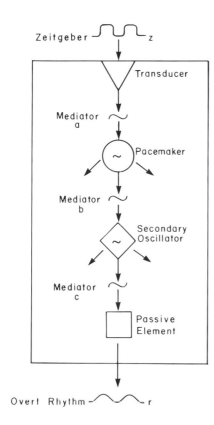

Fig. 3.18 A simplified diagram of the major functional elements of the circadian timing system. Temporal information from the zeitgeber is received by the transducer and flows to the pacemaker via mediator *a*, then to a secondary oscillator via mediator *b*, and then to a passive element via mediator *c*, producing an overt circadian rhythm *r*. (Copyright 1982 by Moore-Ede, Sulzman, and Fuller.)

timing system. It seems safe to say that all mammals, and indeed many simpler life forms, show combinations of the same structural elements (see Fig. 3.18).

TRANSDUCERS

The *transducers* of the circadian timing system are the receptors that detect periodic changes in the environment and transform this temporal information into signals that are meaningful for the time-keeping elements. Not all of an animal's receptors play a significant role in detecting environmental time cues. For example, only light–dark cycles, which impinge on the retina, and eating–fasting cycles, which impinge on some as-yet-unknown receptor, are normally capable of entraining circadian rhythms in the squirrel monkey. Thus, thermoreceptors, auditory receptors, and many other types of receptor do not normally play a major role in timekeeping. However, recent work suggests that auditory cues may play a role when the pacemaker is damaged, and that multiple subliminal transducers may exist which are not usually used (Fuller, Sulzman, and Moore-Ede, 1979c; Fuller, Lydic, et al., 1981b).

Through various mechanisms, a transducer may alter the waveform of the timing signal from the environment; for example, a transducer might transmit information only about changes in input and not about the steady-state level. A light–dark cycle might thus be converted into a two-pulse-per-day (lights-on and lights-off) signal from the retinal transducer to the circadian pacemakers. Thus, transducers may be able not only to receive information but also to alter the waveform of the rhythmic signal.

PACEMAKERS

Pacemakers are the primary oscillators that measure time in the absence of external periodic cues. Like other inherently rhythmic physiological systems (such as the cardiovascular and respiratory systems), the circadian timekeeping system has one or more pacemakers which provide timing signals that synchronize the system as a whole. The pacemaker has two main functions: receiving information about daily changes in the environment and providing daily timing signals to the body, whether or not it receives temporal cues from the environment via the transducers.

Over the last 10 to 15 years there has been some debate over whether circadian pacemakers are single entities that can themselves measure circa-24-hour intervals. Alternatively, they could consist of a network of elements, none of which is individually capable of measuring such intervals but which when coupled through mutual feedback interactions can measure circa-24-hour intervals. An analogous example of the first type of pacemaker organization is the sinoatrial node in the right atrium of the heart, which determines the contraction frequency (Noble, 1975). Other areas of the heart are capable of producing rhythmic contractions when removed from the influence of the sinoatrial node, but in the normal, intact heart all cardiac cells are synchronized by the pacemaker. The other alternative, the network type of pacemaker, is analogous to the respiratory control centers in the brainstem.

There are multiple oscillatory neurons in the medulla and the pons but no clear evidence for a discrete pacemaking center. Instead, it has been proposed that the respiratory cycle is the product of multiple network interactions (Cohen and Feldman, 1977).

At the present time it is not clear whether the pacemakers of the circadian timekeeping system are of the distinct or network types. This is a difficult issue to resolve experimentally, because ablation of one element of the network may produce the same disruption in rhythmicity as ablation of an entire pacemaker. Indeed, it is not even clear that the fundamental period of circadian pacemakers is approximately 24 hours. It is quite possible that each pacemaker is composed of a population of high-frequency oscillators which by feedback interaction produce a lower-frequency summed output. Elegant analyses of such model systems have been conducted by Pavlidis (1973).

Another problem in clarifying the structure of pacemakers is in the experimental techniques involved. If a discrete nucleus in the brain is identified, the question of pacemaker type shifts to a lower level of organization; maybe that discrete nucleus is itself a network of individual oscillatory cells. Obviously, similar reasoning may apply at the subcellular level, and there comes a point where the distinctions between a discrete and a network pacemaker may be semantic.

Currently we do not know how many circadian pacemakers there are in mammals. Substantial evidence accumulated from studies of rodents and primates indicates that the suprachiasmatic nuclei (SCN) of the hypothalamus act as a circadian pacemaker. We will review in Chapter 4 the structure and function of the SCN as well as the evidence which suggests that there is at least one other pacemaker in the circadian timing system.

SECONDARY OSCILLATORS

Secondary or peripheral oscillators are capable of self-sustained oscillations, are synchronized by mediators (see below), and are responsible for driving the overt rhythms in physiological variables. Not all tissues having rhythmic function are capable of self-sustained oscillations, and it is often difficult to ascertain whether or not this capability is present. Some secondary oscillators may be capable only of damped oscillations; the greater the degree of damping, the more difficult it is to document inherent rhythmicity. A minimum criterion is the persistence of rhythmicity for one cycle in the absence of time cues from the environment or input from rhythm mediators in the body. Because the elements of the circadian timing system are normally coupled, secondary oscillators need not have the same characteristics as pacemakers. When isolated from the rest of the timing system, peripheral oscillators may show a different free-running period and less phase stability.

An example of a secondary oscillator is the adrenal cortex; prominent circadian rhythms occur in the plasma levels of adrenal steroids. Since the adrenal cortex is stimulated by ACTH from the pituitary, the rhythmic patterns of adrenal steroids could be forced responses of the adrenal to rhythms in plasma ACTH. However, as we have discussed, persisting rhythmicity has been demonstrated in adrenals maintained *in vitro,* so the gland apparently has an inherent oscillatory ability. The adrenal gland appears to be a secondary oscillator rather than a pacemaker, because its free-running period often is much shorter than 24 hours and the rhythm is less stable than would be expected for a pacemaker, as discussed earlier.

PASSIVE ELEMENTS

Tissues or physiological subsystems that are not capable of self-sustained circadian rhythmicity may be considered the target tissues of the circadian system. These *passive elements* show circadian rhythms in their function when they are driven by a circadian rhythm in a mediator but become arrhythmic as soon as the mediator is maintained at a constant level. In short, passive elements behave as all physiological systems have been assumed to act in most classic textbooks of physiology, where the existence of an endogenous circadian timing system has been either overlooked or ignored.

One example of a passive element may be the rat pineal gland. There are several pronounced circadian rhythms in pineal function, including melatonin synthesis and N-acetyltransferase activity. Temporal information is transmitted to the pineal via neural mediators, including the sympathetic nervous system and its superior cervical ganglion. Stimulation of these ganglia produces changes in pineal biochemistry that mimic changes in the gland produced by environmental light–dark cycles. However, when the neural input into the pineal is disrupted, rhythmicity ceases. Thus, it appears that the rat pineal gland is not capable of self-sustained oscillations, but rather that its rhythms are driven by a pacemaker elsewhere in the body via the sympathetic nervous system (Klein and Moore, 1979).

MEDIATORS

Mediators transmit temporal information from one site in the body to another. Secondary oscillators that produce overt rhythms are normally coupled to the pacemaker through mediators, and the pacemaker is coupled to the transducer by other mediators. The most obvious way to transmit such period and phase information in mammals is through the rhythmic activity of neural and endocrine systems.

There is extensive evidence for the rhythmic activity of neural systems. Circadian rhythms in neuronal firing rate (Schmitt, 1973; Koizumi and Nishino, 1976), in the concentration of the neurotransmitters 5-hydroxy-tryptamine (Hery et al., 1977), norepinephrine (Manshardt and Wurtman, 1968; Reis, Weinbren, and Corvelli, 1968; Bobillier and Mouret, 1971), and dopamine (Bobillier and Mouret, 1971), and in synaptic excitability (Barnes et al., 1977) have been reported. Furthermore, there is evidence for the transmission of circadian information by neural pathways from the hypothalamus to the pineal (Axelrod, 1974), liver (Black and Reis, 1971), and other hypothalamic and brainstem centers (Moore, 1978).

Similarly, there are prominent circadian rhythms in the plasma concentration of a wide variety of hormones, including growth hormone (Weitzman, 1976), prolactin (Sassin et al., 1972), cortisol (Weitzman et al., 1971), and testosterone (Lincoln, Rowe, and Racey, 1974). Of course, the effective rhythm in plasma hormone concentration which provides the mediating time cue to the target oscillator or to the passive target tissue is a combination of rhythms in endocrine secretion; in the volume of the body fluid compartment (Cranston and Brown, 1963); in degradation, for example by liver enzymes (Marotta et al., 1975); and in renal excretion (Kobberling and von zur Muhlen, 1974). However, despite these factors which modify the waveform of the signal, hormonal circadian rhythms appear to act as effective mediators. For example, the circadian rhythm of cortisol concentration acts as a synchronizing mediator between the adrenal cortex and the renal tubular sites responsible for controlling the rhythm of urinary potassium excretion (Moore-Ede, Schmelzer, et al., 1977), as we will discuss in Chapter 5.

Mediators play a critical role at all levels of the system, as diagrammatically depicted in Figure 3.18. For example, mediator *a* conveys information from the transducer of environmental time cues to the pacemaker. Mediator *b* in turn conveys the output of the pacemaker to the secondary oscillator, and mediator *c* conducts information from this secondary oscillator to a passive element in the physiological system that generates the overt rhythm.

However, an oscillator's response to a mediator is quite different from the response of a passive element, and also there are some subtle differences in how secondary oscillators and pacemakers respond. Using the same nomenclature as in Figure 3.18, in Figure 3.19 we compare diagrammatically the responses of a pacemaker, a secondary oscillator, and a passive element to manipulations of the mediator. First consider the passive element with output *r* responding to oscillations in mediator *c*:

(1) Phase shifting *c* will produce an equal and immediate phase shift in *r*.

(2) Rhythm *r* will immediately cease to oscillate if *c* ceases to oscillate (that is, if *c* is maintained at a constant level).

(3) Rhythm *r* will follow any period oscillation in *c* without any limitation in the range of entrainment.

(4) A change in the level of *c* may induce an equivalent change in *r* at any time in the 24-hour day.

In contrast, a pacemaker with output *b* responding to oscillations in mediator *a* will show quite different responses:

(1) After a phase shift in *a*, the output of the pacemaker *b* may take several cycles to resynchronize with the new phase of *a*.

(2) If mediator *a* ceases to oscillate, *b* will continue to oscillate but with its endogenous rhythm now no longer entrained to the period of *a*.

(3) The pacemaker output *b* will only follow oscillations in *a* which are close to the natural period of *b*; if *a* oscillates with a period outside the pacemaker's range of entrainment, *b* will return to its endogenous period (perhaps modulated by relative coordination; see Chapter 2).

(4) Changes in the level of *a* will not necessarily induce changes in
b. At some phases of *b* a pulse of *a* may produce a phase shift in
b, but at other phases there will be no effect. Thus, the pacemaker will show a varying phase responsiveness to *a*.

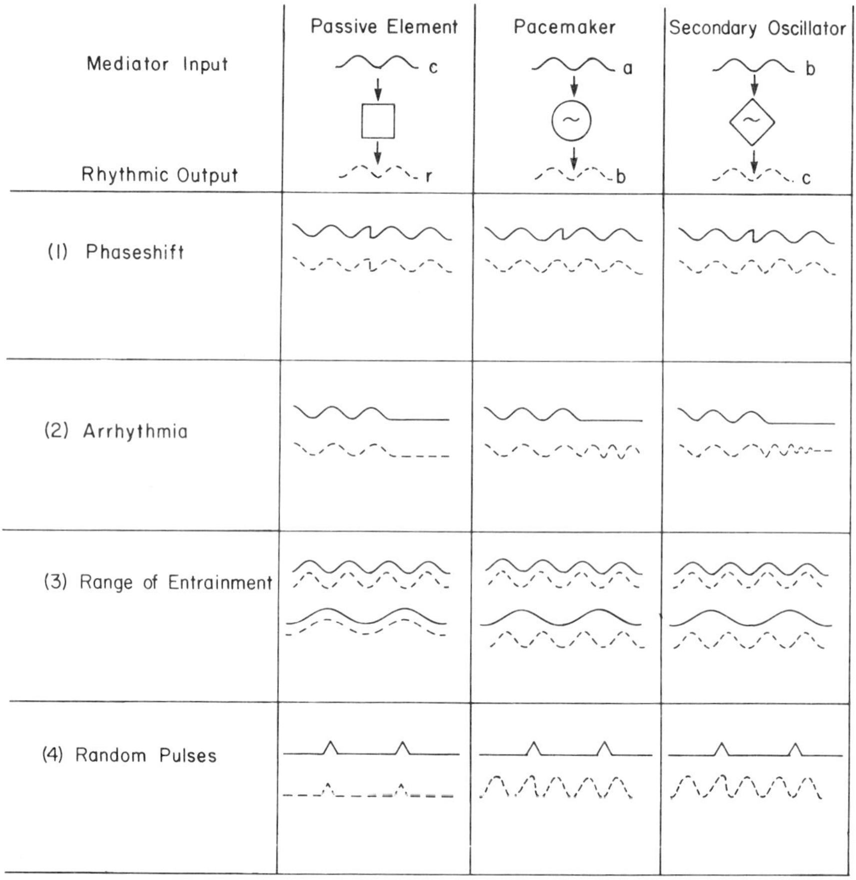

Fig. 3.19 A comparison of the responses of a passive element, a pacemaker, and a secondary oscillator to manipulations of a mediator. The relationships between a mediator (solid line) and the rhythmic output of the
different parts of the system (dashed lines) are shown after the mediator
is (1) phase shifted, (2) becomes arrhythmic, (3) exhibits different periods
as in a range of entrainment study, and (4) after random pulses of the mediator. (Copyright 1982 by Moore-Ede, Sulzman, and Fuller.)

All these responses clearly separate pacemakers from passive elements. Let's finally turn to consider secondary oscillators, which are self-sustained but have less stable properties than pacemakers. In Figure 3.19 the secondary oscillator with output c responds to mediator input b as follows:

(1) After a phase shift in b, the output of the oscillator c may take several cycles to resynchronize with the new phase of mediator b, although there may be fewer transient cycles than with a pacemaker.

(2) If mediator b ceases to oscillate, c will normally continue to oscillate independently, although the period may be very different from the prior circadian entrained period, and the oscillation may damp out after a few cycles.

(3) The secondary oscillator output c will follow oscillations in mediator b only within a limited range of entrainment, although the range may be broader than that of a pacemaker.

(4) Changes in the level of b will not necessarily produce changes in c, although, as with a pacemaker, the output c will normally have a phase-response curve to the timing of pulses of b.

Thus, mediators can be used to differentiate between passive elements, pacemakers, and secondary oscillators. The response of a rhythm to manipulations of the mediator input will depend on whether the rhythm is the product of an oscillator or of a passive element. We will discuss the use of such techniques in Chapter 5.

OVERT RHYTHMS

Overt rhythms are the measurable periodic outputs of the circadian timing system. These can be anything from rhythms in overall behavior to changes in amounts of circulating hormones in the bloodstream. The distinction between overt rhythms and daily changes in mediator activity is a matter of perspective: for example, ACTH levels may be viewed as a mediator between anterior pituitary and adrenal cortex in one context and as an overt rhythm of the pituitary in another.

The rhythmicity of most tissues is probably a combination of endogenous rhythmicity and responses to exogenous factors. For example, the presence or absence of environmental light can modulate feeding and drinking behavior in many species. Take as an example the squirrel monkey, which normally feeds during the day and not during the night (Fig. 3.20A). A strictly exogenous interpretation of the feeding rhythm might propose that feeding behavior is inhibited by dark and that this determines the regular day–night pattern. This view is untenable, because these animals will eat with a circadian periodicity even when maintained in constant darkness. In contrast, a strictly endogenous view of the light–dark feeding pattern might be that the extremely regular pattern is the result of very tight phase control of the circadian pacemakers which time the feeding rhythm. The real situation is a combination of these two extreme views, as can be seen if a monkey is exposed to a higher-frequency light–dark cycle. In Figure 3.20C the feeding rhythm of an animal in an LD 1:1 cycle is free-running with an approximately 25-hour period. Feeding is not continuous during the subjective day; discrete bouts of feeding coincide with the light portions of the LD 1:1 cycle. Thus, it is clear that dark can inhibit feeding behavior, and this dark inhibition probably contributes to the normal sharp on–off transitions in the waveform of the LD 12:12 feeding rhythm. These exogenous effects were superimposed on the endogenous rhythm of feeding as shown in Figure 3.20B. Other exogenous factors which cause stress or additional metabolic demands can similarly contribute to the observed rhythm waveform (Czeisler et al., 1976).

Sometimes the distinction between endogenous and exogenous rhythmicity is not clear-cut. Take for example a rhythm in the plasma concentration of a food substance in the diet. In a constant environment, with food available *ad libitum,* the animal will have an endogenous circadian rhythm of eating and hence the dietary constituent will have a rhythm in the plasma with all the characteristics of an endogenous circadian rhythm. However, if the animal is deprived of that dietary constituent, or is forced to eat equal amounts spaced throughout day and night, there will be no rhythm in the plasma variable. Thus, this overt plasma rhythm, while being a passive response to the time of food intake, is normally endogenously controlled by the circadian timing system. The question that must be asked in examining such a variable is, what is controlling its timing?

Abstract Models

What internal organization of the circadian timing system is indicated by the data we have reviewed? The system could be organized in

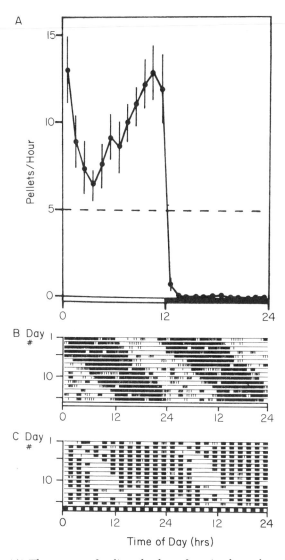

Fig. 3.20 (A) The average feeding rhythm of squirrel monkeys in an LD 12:12 cycle. The mean hourly value of food intake (±SEM) is plotted against circadian time. The horizontal dashed line represents the average 24-hour consumption. (B) The double-plotted feeding rhythm of a squirrel monkey in constant light, showing the endogenous free-running rhythm. (C) The feeding rhythm of a squirrel monkey exposed to an LD 1:1 cycle, demonstrating the passive responses to light superimposed on the endogenous circadian rhythm. The LD cycle is shown at the bottom. Black bars = dark periods. (A from Sulzman et al., 1978; B, C copyright 1982 by Moore-Ede, Sulzman, and Fuller.)

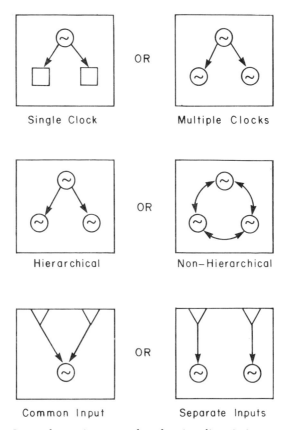

Fig. 3.21 Some alternative ways that the circadian timing system could be organized: *Top,* single and multioscillator organization. *Middle,* hierarchical and nonhierarchical organization. *Bottom,* two transducers connected to one pacemaker or two pacemakers each receiving their inputs from separate transducers. (Copyright 1982 by Moore-Ede, Sulzman, and Fuller.)

any one of numerous ways; some of the fundamental alternatives are illustrated in Figure 3.21. First, all the various rhythms in the organism could be driven by a single pacemaker, or they could be the product of multiple, potentially independent oscillators. However, as we have discussed, there is considerable evidence that mammals have a multioscillator circadian timing system, because separate rhythms can show independent free-running periods simultaneously, and tissues isolated *in vitro* demonstrate circadian rhythmicity.

A more difficult issue to untangle is whether we are dealing with a hierarchical or a nonhierarchical organization of oscillators. With a hierarchical organization, one oscillator, serving as a pacemaker, provides unidirectional phase and period information to the secondary oscillators in the system. These other oscillators, however, cannot significantly influence the phase or period of the pacemaker. With a nonhierarchical organization, each oscillator in the system may both receive and transmit phase information, so that no one oscillator can consistently act as a pacemaker for the whole system.

A third question is how do the transducers of environmental zeitgebers provide information to the circadian system. For example, it is possible that all zeitgebers impinge on a single pacemaker. In this case, the pacemaker acts as the arbitrator of all phase and period information from the environment so that all the other oscillators in the system respond to a unified signal. Alternatively, the zeitgebers might impinge on separate oscillators; in that case, if there is conflicting temporal information from the environment the different rhythms within the organism will be driven by the different zeitgebers. Whether or not the different oscillators uncouple from one another depends on the strength of internal coupling between them relative to the coupling strength exerted by the external environment.

We outline these considerations in Figure 3.21 because, as we start to construct the jigsaw puzzle pieces of the mammalian circadian timing system, we must be concerned with the various possible ways in which the components could be organized. In Chapters 5 and 6, we will refer to these schemes of organization as we start to put together the big picture of how the system works.

The Neural Basis
of Circadian Rhythmicity

The major pacemakers of the circadian timing system in mammals appear to be located in the brain. This is not to say that other tissues in the body are incapable of generating circadian rhythms. Indeed they can, as has been shown by the studies of isolated mammalian tissues maintained *in vitro*. However, from the available evidence it seems that peripheral tissues contain mainly secondary circadian oscillators and that the primary pacemakers reside within the central nervous system.

Identifying the Pacemakers

After several decades of study, Curt Richter identified the hypothalamus as the location of a biological clock. Using the free-running activity rhythm of blinded rats as his marker of the circadian system, he subjected his animals to "almost every conceivable kind of metabolic, endocrinologic and neurologic interference" (Richter, 1965). This included removal of adrenals, gonads, pituitary, thyroid, pineal, or pancreas; electroshock therapy; induced convulsions; prolonged anesthesia; and alcoholic stupor. Each of these procedures, as the examples in Figure 4.1 show, failed to disrupt the rat's free-running activity rhythm.

Richter then placed hundreds of lesions in various parts of the rat brain. The only location where lesions were found to influence the

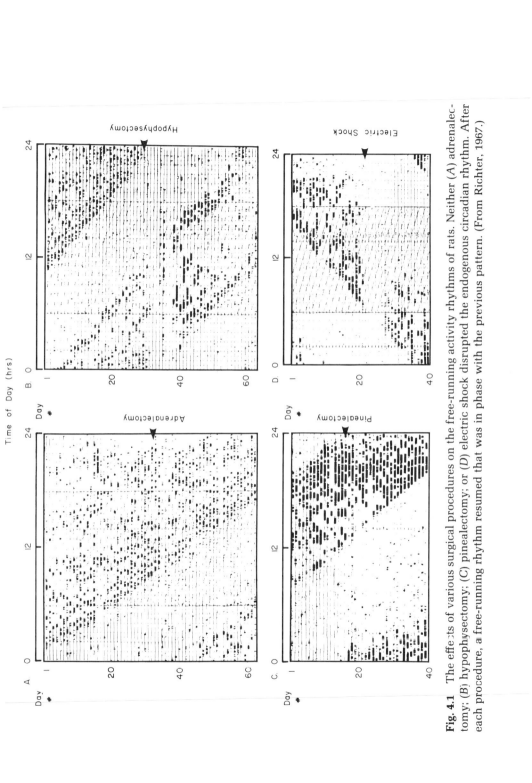

Fig. 4.1 The effects of various surgical procedures on the free-running activity rhythms of rats. Neither (A) adrenalectomy; (B) hypophysectomy; (C) pinealectomy; or (D) electric shock disrupted the endogenous circadian rhythm. After each procedure, a free-running rhythm resumed that was in phase with the previous pattern. (From Richter, 1967.)

free-running rhythm was the hypothalamus (Richter, 1967). From his series of 200 hypothalamic lesions, he was able to identify a general area of the ventral hypothalamus where lesions resulted in complete loss of circadian rhythmicity in activity, feeding, and drinking behaviors (Fig. 4.2). However, he did not report the precise location responsible for the generation of these circadian rhythms.

It has long been recognized that the anterior hypothalamus (Fig. 4.3) participates in the regulation of sleep and wakefulness. In humans, tumors involving the walls of the third ventricle and the optic chiasm (Righetti, 1903; Fulton and Bailey, 1929; Gillespie, 1930), as well as damage to the anterior hypothalamus in lethargic encephalitis victims (von Economo, 1929), have been correlated with disruptions of the sleep–wake cycle. As a result von Economo postulated a center in the anterior hypothalamus that was an "effector of sleep." Subsequently Nauta (1946), by placing an extensive series of knife-cuts in the rat hypothalamus, identified a rostral area of the hypothalamus "roughly conforming to the suprachiasmatic and pre-optic areas (as) the site of a nervous structure which was of specific importance for the capacity of sleeping." This he referred to as a "sleep center" in agreement with von Economo's conclusions from neuropathological studies in man. However, at the time none of the investigators considered the possibility that this area might contain a biological clock that *timed* sleep and wakefulness.

The stage was set for identifying the circadian pacemaker responsible for timing the behavioral rhythms of sleep–wake, drinking, and feeding. Various observations and experimental investigations all pointed to the anterior hypothalamus as a probable site, but one more step was needed before the precise location could be found.

As is so often the case in science, approaching the problem from a different direction and with a new technique provided the key. Since circadian rhythms are synchronized by the environmental light–dark cycle, this suggested that there were neural pathways from the retinae (where light is detected) to the circadian pacemaker(s). Thus, by tracing those visual pathways that originate in the retinae and terminate in the anterior hypothalamus, it should be possible to locate the pacemakers.

Here the technique of autoradiography proved to be useful. Tritiated amino acids injected into the vitreous humor of the eye are taken up by

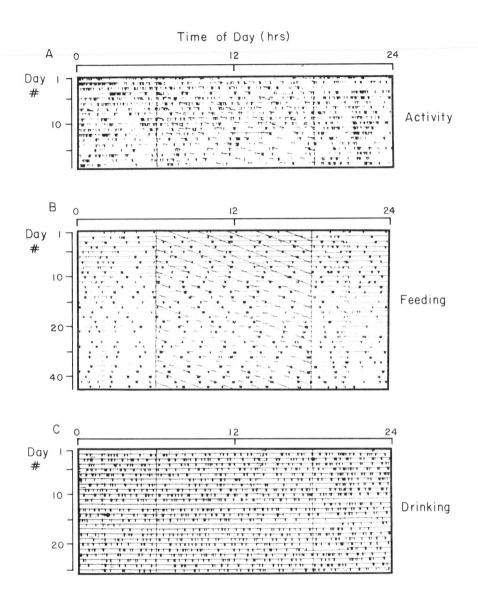

Fig. 4.2 The effects of anterior hypothalamic lesions on circadian behavioral rhythms in blinded rats. The rhythms of wheel-running activity (*A*), feeding (*B*), and drinking (*C*) each became arrhythmic following the lesions. (From Richter, 1967.)

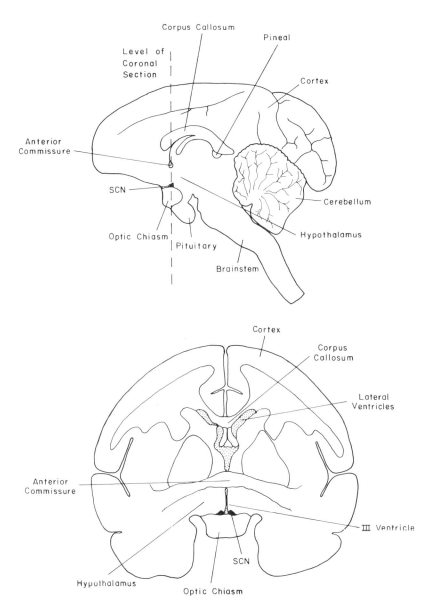

Fig. 4.3 Anatomy of the squirrel monkey brain shown in (top) midline sagittal and (bottom) coronal sections which intersect the optic chiasm. The suprachiasmatic nuclei are in the hypothalamus above the optic chiasm and lateral to the anterior tip of the third ventricle. (Copyright 1982 by Moore-Ede, Sulzman, and Fuller.)

retinal ganglion cells and transported to the synaptic terminals. Some of the tritiated material can also be picked up by the next cell in the neural network and the process repeated. The final location of this material can be recorded by exposing radiation-sensitive film to histologic sections of the brain. Such techniques led to the demonstration of a visual pathway, long suspected (Pate, 1937; Haymaker, Anderson, and Nauta, 1969) but previously unconfirmed. In addition to the previously identified pathways of the visual system (Fig. 4.4), a retinohypothalamic tract was identified (Fig. 4.5), which terminates specifically in the suprachiasmatic nuclei of the hypothalamus (Moore and Lenn, 1972; Hendrickson, Wagoner, and Cowan, 1972).

Moore and Eichler (1972) shortly thereafter showed that lesions which destroyed the suprachiasmatic nuclei resulted in a loss of the circadian rhythm of adrenal corticosterone, and Stephan and Zucker (1972b), using the same method, showed a loss of circadian rhythmicity in drinking and locomotor activity. These initial reports were quickly substantiated by a number of other investigators, and today there is a considerable body of evidence to show that the suprachiasmatic nuclei are integrally involved in the generation of circadian rhythmicity in a wide variety of physiological functions (Rusak and Zucker, 1979; Moore, 1979).

MULTIPLE PACEMAKERS

With the demonstration that lesions of the suprachiasmatic nuclei (SCN) cause the loss of multiple rhythmic functions, it was suggested that the SCN are "the biological clock." The evidence reviewed in Chapter 3 shows that this is clearly not the case—there are multiple circadian oscillators in many body tissues. What is more, the SCN may not be the only circadian pacemaker; there is mounting evidence of another pacemaker in the body which is capable of coordinating circadian rhythms in multiple physiological functions. For example, studies of humans in isolation have repeatedly shown the desynchronization of two groups of rhythmic functions, one group which follows the sleep–wake cycle and the other group coupled with the circadian rhythm of core body temperature (Aschoff, 1965b; Wever, 1979; Czeisler, Weitzman, et al., 1980).

The other pacemaker(s) are not located within the SCN. In monkeys, total bilateral SCN lesions, which result in circadian arrhythmicity of

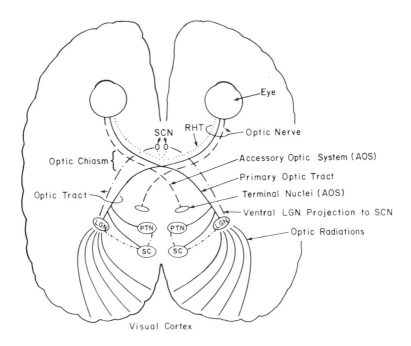

Fig. 4.4 The key visual pathways projecting from the retinae to the brain. The optic nerves arise from each retina and traverse back to the optic chiasm at the base of the hypothalamus. Three major components separate at this point. The first, the retinohypothalamic tract (RHT), terminates in the suprachiasmatic nuclei (SCN) of the hypothalamus. The second, the primary optic tract, provides the primary visual input to the visual cortex via the lateral geniculate nuclei (LGN) and optic radiations. The superior colliculus (SC) and the pretectal nuclei (PTN) also receive inputs from the primary optic tract and LGN. The primary optic tract also contains components of the third major system, the accessory optic system, which has diffuse projections to a variety of areas of the brainstem. This simplified schematic drawing is, however, not to scale and does not show the relative strengths of contralateral and ipsilateral projections. (Copyright 1982 by Moore-Ede, Sulzman, and Fuller.)

the rest–activity cycle, do not eliminate the circadian rhythm of body temperature (Fuller et al., 1981a). The location of this other pacemaker has not yet been defined; however, there are several suggestive pieces of information. First, the SCN must have significant connections with the other pacemaker, because there is considerable evidence of a mutual coupling between them, which ensures the normal internal syn-

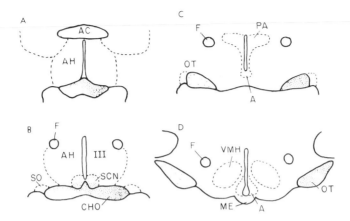

Fig. 4.5 Projection of the RHT to the SCN. ^{3}H-leucine was injected into one eye of a rat, and autoradiographs developed of serial sections from front to back (*A, B, C, D*) of the hypothalamus. Silver grains can be seen in *A* in the optic chiasm (CHO); in *B* in the SCN; and in *C* and *D* in the primary optic tracts (OT). No other areas of the hypothalamus are ^{3}H labeled. AC = anterior commissure; AH = anterior hypothalamus; III = third ventricle; F = fornix; SO = supraoptic nuclei, PA = paraventricular nuclei; VMH = ventromedial hypothalamus; ME = median eminence; A = arcuate nucleus. (From Moore and Lenn, 1972.)

chronization of the circadian system. Furthermore, there is evidence in rodents that this pacemaker can be entrained by food since food entrainment persists after the SCN have been destroyed (Krieger, Hauser, and Krey, 1977; Phillips and Mikulka, 1979; Stephan, Swann, and Sisk, 1979a, b; Boulos, Rosenwasser, and Terman, 1980).

These considerations point to the ventromedial nucleus of the hypothalamus (VMH) or the lateral hypothalamic area (LH) as possible sites of the second pacemaker. Neural pathways have been documented between the SCN and both the VMH and the LH, which could couple the pacemakers (Oomura et al., 1979). Furthermore, both the LH and the VMH are influenced by nutrient intake. Distension of the stomach results in increased vagal nerve activity (Paintal, 1954) and this in turn increases neural activity in the VMH (Sharma et al., 1961). Changes in the plasma levels of glucose, free fatty acids, and insulin also affect both VMH and LH activity (Oomura et al., 1979). According to an initial report, lesions which destroy the ventromedial nuclei result in the loss of the circadian rhythms that persist after SCN destruction (Krieger, 1980). However, the role of the VMH as a circadian pacemaker has yet to be confirmed, because the lesion studies were conducted in animals maintained in a light–dark cycle rather than in constant conditions.

In certain vertebrate species the pineal gland appears to act as a circadian pacemaker. Menaker and his co-workers have shown that if sparrows maintained in constant darkness are pinealectomized, their free-running circadian rhythms of activity and body temperature are abolished (Menaker, 1974). Rhythmicity is restored, however, when the pineal of another sparrow is transplanted into the anterior chamber of the eye, with the rhythm of the recipient assuming the phase of the donor (Zimmerman and Menaker, 1979). Furthermore, Binkley, Riebman, and Reilly (1978), Kasal, Menaker, and Perez-Polo (1979), and Deguchi (1979) have shown that the avian pineal is capable of persisting rhythmicity when maintained *in vitro*.

In other vertebrate species, however, and even in other birds (Rutledge and Angle, 1977; Gwinner, 1978), pinealectomy does not result in a loss of rhythmicity, although there is sometimes a reduction in the stability and uniformity of the rhythm. Richter (1965) showed that pinealectomized rats exhibit persisting free-running activity rhythms, although Quay (1970) demonstrated that some features of the rhythms of pinealectomized rats are altered. The most dramatic changes were seen after phase shifts in the LD cycle. Pinealectomized rats phase-shifted very rapidly following reversal of an LD 12:12 cycle, whereas intact rats required about a week to phase-shift. This indicates that the pineal or some product of it may be involved in the rat's circadian timing system. It may be that in mammals the pineal is normally dominated by the SCN, and that the SCN must be removed for the role of the pineal to be apparent. To our knowledge, such studies of the effects of SCN lesions in pinealectomized mammals have not yet been performed.

Anatomy of the Suprachiasmatic Nuclei

The suprachiasmatic nuclei (SCN) are a pair of small structures, each consisting of a cluster of nerve cells, located in the anterior ventral hypothalamus (Fig. 4.3). They sit on either side of the optic recess of the third ventricle, a fluid-filled space in the midline of the brain. The suprachiasmatic nuclei derive their name from their position just above and usually contiguous with the optic chiasm, which lies under the anterior hypothalamus and third ventricle. The optic chiasm is a large bundle of axons formed by the convergence of the two optic nerves as they pass back from the retinae on their way to the visual cortices of the brain.

The SCN can be visualized by cutting and staining histological sections through the hypothalamus. Figure 4.6 shows such a section in the coronal plane through the hypothalamus of a hamster, a rat, a cat, a squirrel monkey, a rhesus monkey, and man. The SCN are clearly visible on each side of the third ventricle and dorsal to the optic chiasm.

A histological section through the brain provides only a two-dimensional perspective of SCN anatomy. By cutting and staining a series of consecutive sections and then laying them out like slices of bread, one can begin to appreciate the dimensions of an anatomical structure. However, a much more effective way to visualize a three-dimensional structure is to reconstitute the "loaf," using computer graphics techniques (Lydic and Moore-Ede, 1980; Lydic et al., 1981). Figure 4.7 shows such a reconstruction from a squirrel monkey brain, expanded in the anterior-posterior plane to aid visualization. The suprachiasmatic nuclei are immediately above the optic chiasm, and their relationship to the tip of the optic recess of the third ventricle is now readily apparent.

Once the tracings of the histological sections have been entered into the computer, it is possible to rotate the image and examine the SCN on a video screen. Figure 4.8 shows the reconstituted SCN of a rat viewed from various perspectives, showing, for example, how the rat SCN dips down into the optic chiasm posteriorly (Fig. 4.8C). The final step is to collapse the image to get a three-dimensionally correct view of the structure. Figure 4.9 shows the SCN of rat, hamster, cat, squirrel monkey, and rhesus monkey, each within three-dimensional scales to facilitate comparisons. The complex shapes of these nuclei are now readily apparent.

COMPARATIVE ANATOMY

There are some significant phylogenetic differences in the anatomy of the SCN in different mammalian species (Lydic et al., 1981). These

Hamster

Rat

Cat

Squirrel
Monkey

Rhesus
Monkey

Human

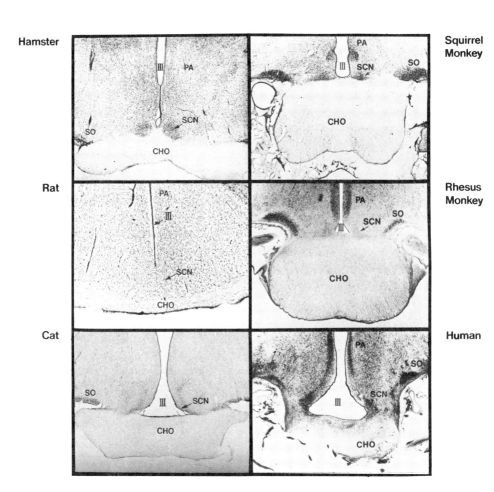

Fig. 4.6 Comparable coronal sections through the anterior hypothalami of six mammalian species. The sections of (A) hamster, (B) rat, (C) cat, (D) squirrel monkey, (E) rhesus monkey, and (F) human brains show the SCN located bilaterally to the third ventricle (III) and dorsal to the optic chiasm (CHO) in each species. The neighboring supraoptic (SO) and paraventricular (PA) nuclei are also visible. (A, B, C copyright 1981 by Lydic, Albers, Tepper, and Moore-Ede; D, E, F after Lydic, Schoene, et al., 1980, reprinted by permission of Raven Press.)

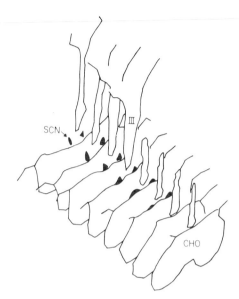

Fig. 4.7 Sequential coronal section tracings at 100-micron intervals of the squirrel monkey hypothalamus. This view, expanded in the anterior-posterior plane, shows the relationship of the SCN to the optic chiasm (CHO) and the anterior tip (optic recess) of the third ventricle (III). (Copyright 1981 by Lydic, Albers, Tepper, and Moore-Ede.)

differences are apparent in both the histological sections in Figure 4.6 and in the three-dimensional reconstructions in Figure 4.9. While the SCN in rat, hamster, cat, squirrel monkey, rhesus monkey, and man are located in the same region of the hypothalamus, the relative proportions and positions of the third ventricle, optic chiasm, and SCN vary.

In rodents the third ventricle is effectively a narrow vertical slit in the brain with the SCN located laterally and ventrally to the ventricle. In cats and primates the ventricle is broader, particularly at its base, and as a result, the SCN are displaced further from the midline. As one moves from New World to Old World primates and then to humans, the SCN are placed progressively more laterally. Thus in humans, with their expanded third ventricle, the SCN are positioned on the lateral walls of the ventricle.

The density of packing of neurons in the SCN also varies between

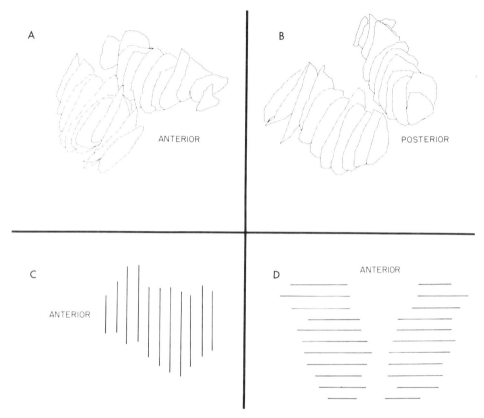

Fig. 4.8 Four different computer graphic orientations of the rat SCN. The outline of each SCN is shown at 50-micron intervals in correct three-dimensional scale, from (*A*) anterior-lateral; (*B*) posterior-lateral; (*C*) lateral; and (*D*) superior viewpoints. (Copyright 1981 by Lydic, Albers, Tepper, and Moore-Ede.)

species (Fig. 4.6). In the rat, hamster, and squirrel monkey the cells are densely packed, and the SCN are easily seen in light microscopy. In the cat and the rhesus monkey the cells are less tightly packed and in humans the cells are even more diffusely organized (Lydic, Schoene, et al., 1980).

The three-dimensional shapes of the SCN in the rodent species correspond most closely to the ovoid shape suggested by one of the earlier names for the SCN, "nucleus ovoideus." In the rat (Fig. 4.9A) the SCN

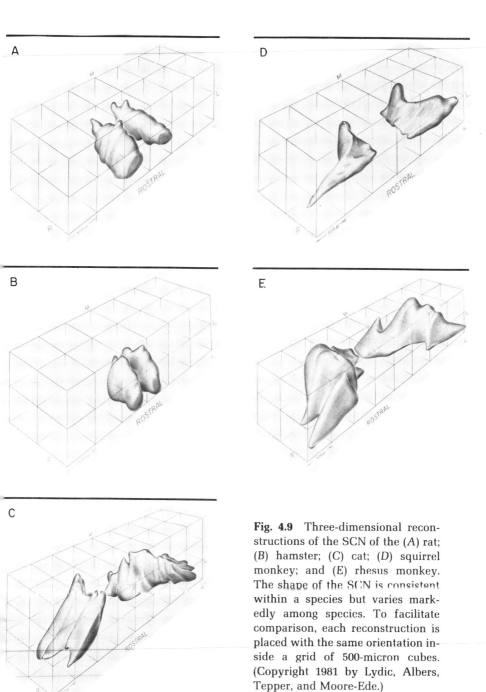

Fig. 4.9 Three-dimensional reconstructions of the SCN of the (A) rat; (B) hamster; (C) cat; (D) squirrel monkey; and (E) rhesus monkey. The shape of the SCN is consistent within a species but varies markedly among species. To facilitate comparison, each reconstruction is placed with the same orientation inside a grid of 500-micron cubes. (Copyright 1981 by Lydic, Albers, Tepper, and Moore-Ede.)

are elongated structures that converge posteriorly but do not fuse. In the hamster (Fig. 4.9B) the SCN actually fuse posteriorly. The SCN are less elongated, extending only 0.6 mm on the rostral caudal axis as compared to 0.7 mm in the rat. In both rats and hamsters the SCN dip into the optic chiasm, particularly posteriorly.

Despite the weak expression of circadian rhythmicity in cats (Hawking et al., 1971; Moore-Ede et al., 1981), this species also has a clearly identifiable SCN (Fig. 4.9C), with a more complex shape than that of the rodent SCN. Because of the broad ventral expanse of the third ventricle, the ventral portions of the SCN are separated by 2–3 mm, as opposed to less than 0.5 mm in rodents. The cat SCN converge dorsally and posteriorly. Only in the cat do the SCN become directly contiguous with the paraventricular nuclei at their most dorsal tip; whether there is any functional significance to this is quite unknown. In the cat the SCN do not dip down into the optic chiasm as they do in rodent species.

The shapes of the squirrel monkey SCN (Fig. 4.9D) change throughout their length and breadth (Lydic and Moore-Ede, 1980). The anterior poles of the SCN have a broad medial to lateral expanse, ranging up to 0.8 mm in width. They become more triangular toward the middle and then have particularly prominent posterior poles oriented at 90° with respect to the anterior portion of the nuclei. The entire structure of each SCN is approximately 1 mm in length along the rostral to caudal axis.

The SCN of the rhesus monkey (Fig. 4.9E) also have a complex three-dimensional shape. The anterior poles are not as broad as the corresponding portion of the squirrel monkey SCN, but the overall mass of the nuclei is greater. The posterior poles converge at the midline and may fuse. These poles also show the dorsal orientation of the squirrel monkey SCN.

HUMAN SUPRACHIASMATIC NUCLEI

Comparisons of the brains of New World and Old World primates and human fetal, child, and adult brains have revealed a cluster of neurons in the human brain (Fig. 4.10) that seems to be homologous to the nonhuman primate SCN (Lydic, Schoene, et al., 1980). Phylogenetic changes in the dimensions of the third ventricle mean that the ap-

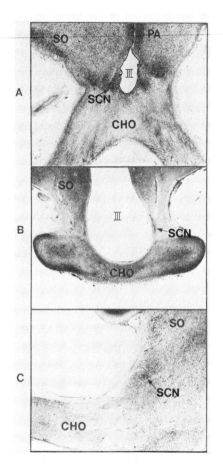

Fig. 4.10 The human SCN in (*A*) horizontal; (*B*) coronal; and (*C*) sagittal sections. The neighboring structures are labeled as in Fig. 4.6. (After Lydic, Schoene, et al., 1980; reprinted by permission of Raven Press.)

parently homologous cluster of neurons is more laterally placed in human brains than in nonhuman primates. Furthermore, the neuronal cluster is more diffusely organized in humans than in lower species, particularly rodents and New World primates.

The very existence of the SCN in man was seriously questioned in a number of authoritative works on the hypothalamus (Clark, 1938; Ingram, 1940; Defendini and Zimmerman, 1978). There appear to be several reasons for this debate. First, this small, diffuse cluster of neurons has been described by at least 12 different names, as listed below, since 1888, leading to some confusion in their description and identification.

Twelve Names of the Suprachiasmatic Nuclei*

Nucleus basalis	Bellonici, 1888
Noyau principal	Ramon y Cajal, 1911
Nucleus preopticus parvicellularis	Rothig, 1911
Ventral nuclei of tuber cinerium	Friedemann, 1912
Nucleus infundibularis medialis	Winkler and Potter, 1914
Suprachiasmatic nucleus	Spiegel and Zweig, 1919
Noyau accessory supraoptique	Foix and Nicolesco, 1925
Nucleus ovoideus	Rioch, 1929
Nucleus preopticus periventricularis pars suprachiasmatica	Loo, 1931
Nucleus infundibulares post. et caudalis	Rose, 1935
Nucleus suprachiasmaticus ventralis	Koikegami, 1938
Nucleus prothalamicus periventricularis ventralis inferior	Brockhaus, 1942

* Compiled with the assistance of Dr. R. Lydic.

Most of the names have never since been utilized, although nucleus ovoideus (Rioch, 1929) has been more generally used. As we have discussed, this name is misleading; the SCN are somewhat ovoid in shape in certain rodent species (Fig. 4.9), but they are far from ovoid in primates (Lydic et al., 1981b).

Second, in the human hypothalamus there are significant phylogenetic differences in gross morphology, such as the expanded third ventricle, which puts the SCN in a more lateral position. Furthermore, in most mammalian species the optic chiasm is contiguous with the ventral surface of the hypothalamus, but in humans the chiasm is largely separate from the ventral floor of the brain, and only its most posterior portion is directly contiguous with the hypothalamus. The definition of this area is therefore more difficult in humans since much of the suprachiasmatic region is not directly adherent to the chiasm.

Third, the function of the suprachiasmatic nuclei has been recognized only since 1972, and their role in nonhuman primates was not clear until even more recently (Fuller et al., 1981). Most of the documentation of the anatomy of the human hypothalamus was undertaken in the early part of this century, when there was no reason to look specifically for this nucleus in man; it was easily missed because of its diffuseness.

Finally, studies of the human anterior hypothalamus by consecutive serial sections are rare. Most early studies examined only every twentieth section or so. This practice of only staining and examining intermittent sections sometimes over a millimeter apart could clearly account for previous failures to identify the SCN, since the entire structure in man extends only about a third of a millimeter in the anterior–posterior plane.

Many of the problems of identifying the human SCN relate to the limitations of studying human subjects. One approach to confirming their presence would be to demonstrate that the same neural inputs exist for the putative SCN of man as for the previously documented SCN of primates. In particular, it would be important to determine if humans have a retinohypothalamic tract (RHT) that impinges on the SCN, as one would expect, since light–dark cycles are effective entraining agents for humans (Czeisler, Richardson, Zimmerman, et al., 1981). The RHT has been consistently observed in all mammalian species up to the great apes (Moore, 1973; Tigges, Bos, and Tigges, 1977), but it has not been demonstrated in humans because documentation by autoradiography requires administering tritiated amino acids to the retina before death, which is obviously not ethically feasible in humans. New techniques must therefore be developed before the presence of the RHT in humans can be confirmed.

Another approach that has been attempted is to define the cytochemistry of the suprachiasmatic cluster of neurons in humans and compare it to similar cytochemical studies in other mammalian species. For example, various peptides such as somatostatin, LHRH (King et al., 1981), vasopressin, and oxytocin (Watkins, 1976) have been localized by immunocytochemistry in the mammalian SCN. Using similar techniques, it has been found that vasopressin and oxytocin are similarly distributed in the human SCN (Dierickx and Vandesande, 1977, 1979). There are problems, however, of interpretation, for peptides such as LHRH are not contained within SCN neurons, but instead are located in axons traversing through the SCN (King et al., 1981).

However, the two approaches mentioned above can only indicate an anatomical homology. Proving a functional homology requires showing that damage to the human SCN has a functional effect similar to that obtained by destroying the SCN in primates. To obtain this evidence, the investigator must rely on experiments of nature. Much work needs to be done in correlating various pathophysiologies of the circadian system (such as disruptions of the sleep–wake cycle) with the anatomical location of damage in the brains obtained from patients at postmortem.

Some pathological evidence dating back to the turn of the century at least suggests such a correlation in man. A number of clinicians, including Fulton and Bailey (1929) and Gillespie (1930) recognized that tumors which damage the anterior tip of the third ventricle and optic chiasm (the location of the SCN) can cause sleep disorders. Furthermore, cerebral tumors involving the third ventricle induced sleep disorders very early in the course of the illness, whereas tumors located elsewhere in the brain sometimes eventually caused excessive sleepiness, but only after intracranial pressure was elevated, suggesting an indirect effect. In patients with third-ventricle tumors who repeatedly fell asleep at any hour of the day (like SCN-lesioned animals), sleep was indistinguishable from normal in that they could be easily aroused at any time, whereas late-onset sleepiness from other tumors was often characterized by more coma-like behavior (that is, arousal was difficult). Fulton and Bailey (1929) in a carefully documented article reported on a number of such patients. Figure 4.11 shows one example, a sarcoma in one of their patients which brought on as an early symptom a profound sleep–wake disorder. The area it destroyed includes the suprachiasmatic region of the hypothalamus.

The problems with these accounts are, of course, that the functional data needed to rigorously demonstrate concomitant circadian arrhythmias were never collected, and the pathological data lack the specificity to determine for certain whether the SCN were destroyed. The investigators were working well before the time when the function of the SCN, or even the concept of the circadian timing system, had been elaborated. In future studies the pathological data must specify the

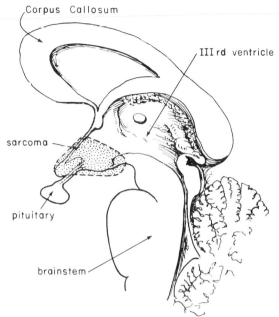

Fig. 4.11 Tumor causing disruption of the sleep–wake cycle in a 28-year-old woman. Sagittal section of the brain at postmortem revealed a sarcoma (shaded area) which involved the anterior tip of the third ventricle and optic chiasm and thus probably destroyed the SCN. (From Fulton and Bailey, 1929.)

exact anatomical area that is damaged, and this evidence must be correlated with functional data precisely documenting the timing of sleep and wakefulness to determine if there is an abnormal arrhythmic pattern.

CYTOARCHITECTURE

Each suprachiasmatic nucleus typically consists of a cluster of small neurons, each neuron 5–15 microns in diameter. Because of their small size, the cells of the SCN are quite distinguishable from the much larger cell bodies in the neighboring supraoptic and paraventricular nuclei. In the rat it has been estimated that each SCN nucleus contains approximately 10,000 neurons packed in a volume of about 0.05 mm^3 (Guldner, 1976; Riley and Moore, 1977). The density of neurons is usually greater on the ventral-medial border than on the dorsal and lateral aspects of the nuclei.

The cells of the rat SCN typically have only a few dendrites branching once or twice (Krieg, 1932; Moore and Lenn, 1972). The extensive ultrastructural studies of Guldner (1976) and Guldner and Wolff (1974, 1978a, b) have demonstrated a wide variety of synaptic connections among the SCN neurons. Commissural connections also exist between the two SCN nuclei with symmetry maintained so that anteriorly placed neurons in one nucleus project mainly to anterior neurons in the other nucleus, and posterior neurons similarly project to their posterior counterparts (Moore, Marchand, and Riley, 1979).

Electron microscopy (Moore, Card, and Riley, 1980; Card, Riley, and Moore, 1980) reveals that the rat SCN are composed of at least two cell types (Fig. 4.12). In the first, an elongated type of cell, concentrated in the dorsal and medial aspects of the SCN, the cell nucleus fills most of the cell body, which has few if any invaginations. The mitochondria and other cell organelles are sparsely distributed in the cytoplasm. In the second and more spherical type of cell, principally located in the lateral and ventral portions of the SCN, the nucleus also fills most of the cell body, but the nucleus has multiple invaginations. In these cells the cytoplasm has a richer collection of organelles and rough endoplasmic reticulum. A striking feature of the second type of cell is that it is often clustered along the walls of the blood capillaries that course through the SCN. The direct apposition of these cells to the circulation suggests that they may release substances into the blood which might rhythmically drive target tissues or that they may act as receptors, sensing hormonal signals from elsewhere.

These ultrastructural studies give hints about the possible functions of the different cells, but many questions remain to be answered. We need to know whether either of these cell types is the principal generator of circadian rhythmicity; whether one cell type might act as a receptor for the other; and whether the rhythmicity is produced by an interaction between the cells or is just a simple summation of their outputs.

INPUT AND OUTPUT PATHWAYS

The most important pathways for environmental inputs to the SCN are the optic tracts, which convey information about the light–dark

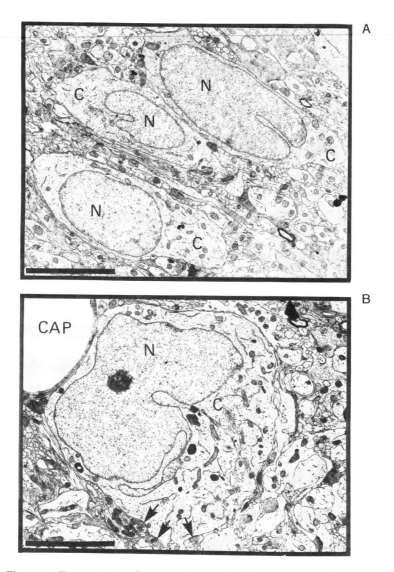

Fig. 4.10 Transmission electron micrographs illustrating typical neurons in the (A) dorsomedial segment and (B) ventral and lateral segments of the rat SCN. N = nucleus; C = cytoplasm; CAP = capillary. Arrows indicate axosomatic synapses. Marker bar — 5μm. For details, see text. (Electron micrographs kindly provided by J. P. Card and R. Y. Moore, Department of Neurology, State University of New York at Stony Brook, copyright 1982 by Card and Moore.)

cycle. The free-running circadian rhythms of blinded mammals cannot entrain to light–dark cycles (Richter, 1968). However, the visual pathways involved in entrainment are separate from those involved in other aspects of vision. Lesions can be placed in the primary visual system which interfere with light-induced behaviors and with vision itself (Dark and Asdourian, 1975; Stephan and Zucker, 1972a; Whishaw, 1974) but still allow entrainment of the circadian system to occur.

Receptors In adult mammals the only functional sites of light reception are the retinae of the eyes (Richter, 1965). One way to determine which of the retinal receptors transduces the light–dark cycle is to examine the effectiveness of different wavelengths of light in entraining the circadian system. McGuire, Rand, and Wurtman (1973) found that rats were most sensitive to light in the green part of the spectrum (approximately 500 nanometer wavelength), and that it took progressively brighter light to entrain the animal to a light–dark cycle when the wavelength was either shorter or longer than 500 nanometers. For example, a hundred-fold increase in light intensity was necessary to entrain animals to light in the blue or yellow part of the spectrum as compared to wavelengths corresponding to green. In Figure 4.13, the action spectrum for entrainment is compared with the action spectrum of rhodopsin, the visual pigment in the rods of the retinae. The correspondence between these spectra might suggest that it is rhodopsin, and therefore the rods of the retinae, that mediate the entrainment of rats to the light–dark cycle. However, the cones are also probably involved; in rats in whom the rods have been destroyed by exposure to bright light (Cicerone, 1976), entrainment can still be achieved by light–dark cycles (Dunn, Dyer, and Bennett, 1972).

Although in adult mammals there are no known photoreceptors outside of the retina that can mediate circadian light–dark entrainment, extraretinal photoreceptors may exist in very young animals (Zweig et al., 1966). In newborn rats environmental lighting can influence the pineal rhythm of serotonin content even after the animals are blinded, but this response can be prevented by placing a hood over the animal's head so light cannot penetrate into the brain. By 27 days of age the rat loses this capability to respond to light via receptors other than those in the eye. The location of these photoreceptors is unknown

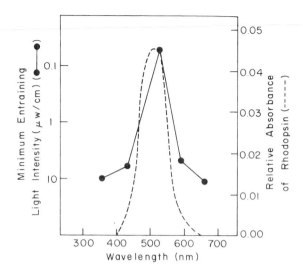

Fig. 4.13 The action spectrum for entrainment of the circadian body temperature rhythm in rats. Light–dark cycles using low-intensity illumination of different colors were provided in separate experiments. The estimated light intensity for entraining 50% of the animals (solid line) is indicated for each wavelength of light. The action spectrum for rat rhodopsin is shown by the dashed line. (From McGuire, Rand, and Wurtman, 1973.)

at this time. The existence of extraretinal photoreceptors has been demonstrated in adult birds; the circadian timekeeping system of blinded birds can be entrained to a light–dark cycle when the light is bright enough to penetrate to extraretinal photoreceptors within the brain (Menaker, Takahashi, and Eskin, 1978).

Visual pathways   The most important visual input to the SCN is via the retinohypothalamic tract (RHT), which consists of small fibers that project exclusively to the SCN. Even if the SCN are destroyed in neonatal rats, the RHT will not innervate any other nuclei in the hypothalamus (Mosko and Moore, 1978a). This specificity of innervation is in contrast to the greater plasticity observed in other components of the visual system. The RHT has been shown to project to the SCN in all major orders of mammals (Moore, 1973; Thorpe, 1975; Kawamura and Ibuka, 1978), except humans, where this has not been demonstrated because ethical considerations bar the application of autoradiographic techniques.

The RHT innervates the ventral lateral portion of the posterior three-fourths of the SCN (Moore, 1978). In rats, approximately two-thirds of this innervation is from the contralateral eye, so most of the fibers must cross over at the optic chiasm. Some estimates place this contralateral component as high as 80% (Wenisch, 1976). There appear to be species differences, however, because the great apes have a pre-dominantly ipsilateral projection of the RHT with only a few fibers crossing over (Tigges, Bos, and Tigges, 1977).

Besides the retinohypothalamic tract, a number of other neural pathways appear to play a role in circadian entrainment to light–dark cycles (Fig. 4.4). The primary optic nerves, which arise from the ganglion cells in each retina, traverse through the optic chiasm, where a portion of each nerve bundle crosses over. All fibers associated with the right field of vision become grouped together on the left side, and those carrying left visual field information are grouped on the right side of the chiasm. The separated fibers carrying right and left visual field information then run in the primary optic tracts, which terminate in clusters of neurons called the right and left lateral geniculate nuclei. Fibers arising from the lateral geniculate nuclei then course caudally through the optic radiations to terminate in the visual portion of the occipital cortex.

Visual information reaches the suprachiasmatic nuclei from the primary optic tract via the lateral geniculate nuclei (Swanson, Cowan, and Jones, 1974; Ribak and Peters, 1975). Autoradiographic techniques have been used to demonstrate that tritiated proline introduced into the ventral portion of the lateral geniculate body traces out projections to the suprachiasmatic nuclei on both sides of the hypothalamus as well as to several sites in the brainstem.

More recently other routes which might mediate light–dark entrainment of the SCN have been demonstrated. Silver and Brand (1979) used Golgi and horseradish peroxidase techniques to show that neurons within the SCN as well as a broad region of the preoptic hypothalamus have dendrites that project into the optic chiasm, as Figure 4.14 shows. Electron microscopy was used to confirm the existence of synapses between these dendrites and the axons coursing through the optic chiasm. Furthermore, Card, Riley, and Moore (1980) have shown

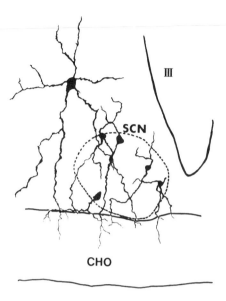

Fig. 4.14 Neuronal processes from the SCN and other regions of the hypothalamus enter the optic chiasm. The Golgi-stained neurons appear to form synaptic contact with axons passing through the optic chiasm. The boundaries of the SCN (dashed line), the third ventricle, III, and the optic chiasm, CHO (solid lines) are shown. (From Silver and Brand, 1979.)

by electron microscopy that there is considerable interdigitation between the processes of SCN nerve cells and the nerve axons coursing through the optic chiasm. Some SCN cell bodies may actually lie in the dorsal portion of the optic chiasm. These studies emphasize the rich interaction between the visual pathways and the nerve cells of the SCN, suggesting that there may be a built-in redundancy to ensure entrainment of the SCN by light–dark cycles. Light–dark information may be conveyed not only to the SCN but to other areas of the hypothalamus as well.

There may prove to be other pathways, such as the accessory visual pathways, that provide inputs to the circadian timing system (Fig. 4.4). A major accessory pathway projects to the superior colliculus, a layered structure on the brainstem that is involved in the orientation of the head and neck. Primary visual afferents from the retinae terminate here and give rise to secondary fibers going to a variety of other structures in the brainstem, including the pretectal area, which mediates visual reflexes. This area also independently receives direct input from the retinae and from the visual cortex. The output of the pretectal cells projects to a number of areas of the brain. Another accessory group of visual pathways, called the accessory optic system, traverses the medial forebrain bundle to end in the terminal nuclei. The physiological significance, if any, of these pathways for the circadian system remains to be determined.

In addition to the direct and indirect inputs from the retinae, there are other neural inputs to the SCN. For example, a group of serotonin-containing terminals within the SCN can be shown to originate from the nucleus centralis superior of the median raphe (Fuxe, 1965; Aghajanian, Bloom, and Sheard, 1969). Additional inputs to the SCN come from descending projections from the anterior hypothalamus and paraventricular nucleus (Stephan and Zucker, 1972a; Conrad and Pfaff, 1976), and ascending fibers reach the SCN from the tuberal hypothalamic area (Saper, Swanson, and Cowan, 1976). Two additional inputs suggested by Moore (1979) include a brainstem projection of norepinephrine-containing neurons to the SCN and a medial hypothalamic projection arising from the hippocampal formation. Each of these can be shown to pass immediately adjacent to the SCN and could provide an input if the dendrites of SCN neurons extend beyond the boundaries of the SCN itself, as the work of Silver and Brand (1979) suggest they do.

Output pathways   As befits their putative role as a pacemaker timing a wide variety of physiological rhythms, the SCN disperse information to a number of other tissues. Connections have been documented between the SCN and a variety of areas of the hypothalamus and the rest of the central nervous system, including the pituitary, pineal, and various brainstem areas. These connections make it conceivable that the SCN are supplying temporal information to most control systems in the body.

Major efferent pathways leave the SCN in primarily dorsal and caudal directions. The terminations of many of these fibers have not yet been rigorously identified. Injections of tritiated material into the vitreous humor of the eye cannot load the optic axons sufficiently so that enough labeled material crosses the synapses in the SCN to label the efferent projections. However, Swanson and Cowan (1975) injected tritiated material directly into the SCN of a single rat and demonstrated axons projecting dorsally and caudally in the paraventricular area and retrochiasmatic region and into the tuberal hypothalamus as far as the caudal border of the ventromedial nucleus. Secondary fibers have been shown to project to a region of the lateral hypothalamus that projects directly to both the brainstem and the upper thoracic spinal cord (Saper, Swanson, and Cowan, 1976). This indicates that the outputs of the SCN reach areas responsible not only for hypothalamic-pituitary function but also for functions influenced by brainstem and spinal cord projections.

Function of the Suprachiasmatic Nuclei

Ascribing a function to any area of the brain is not a simple matter. Even if an animal changes its behavior after a portion of the brain is destroyed, that is no more indicative of the function of the destroyed area than is the result of removing a transistor from a radio indicative of the transistor's function. The announcer's voice may stop in mid-sentence, but that does not mean the announcer or his voice was in the transistor.

Destroying the SCN by placing lesions and finding a loss of circadian rhythmicity does not alone prove that the SCN are a circadian pacemaker. For example, they could just contain fibers passing from a pacemaker located elsewhere. To be sure, we have some confidence in the interpretation that the SCN are a pacemaker because of observations that only lesions which specifically destroy the SCN result in a loss of rhythmicity and not those destroying other areas of the hypothalamus or brain. While some new evidence from other approaches, such as electrophysiological recording, appear to support this interpretation of the SCN's role, most of the evidence pointing to their pacemaking function has come from lesion and knife-cut studies.

LESION AND KNIFE-CUT STUDIES

When the SCN are destroyed by ablation or isolated by knife cuts, circadian rhythmicity is lost in a wide variety of physiological and behavioral variables. However, because of the experimental design and conditions, these studies must be interpreted carefully. For example, those investigators who examined only animals maintained in a light–

dark cycle and never in constant conditions cannot separate the exogenous masking effects of the light–dark cycle from the responses of the internal pacemaking system. Similarly, if data from groups of animals are pooled, then independent but differently phased rhythms in each individual animal may not be detected, since the group mean may be arrhythmic (see Fig. 1.10).

After the first lesioning studies of the SCN were reported in 1972 (Moore and Eichler, 1972; Stephan and Zucker, 1972b), several groups reported loss of rhythmicity in a number of physiological functions after SCN lesions. At that time it was suggested that the SCN were the only circadian pacemaker or even "the circadian clock." Since then it has become clear that the story is not that simple; while the SCN are a major pacemaker driving many rhythms, at least one additional pacemaker located elsewhere within the organism appears to be responsible for the timing of certain other variables. We will therefore examine the effects of lesioning and knife-cut experiments on a number of physiological and behavioral variables in turn.

Activity rhythms   As we discussed in Chapter 2, much of the information on the characteristics of circadian clocks in mammals has been derived from the study of rhythms such as drinking and wheel-running. There is little doubt that the SCN play a key role in timing these rhythms. This has been confirmed by many investigators since Richter (1967) first reported that hypothalamic lesions resulted in arrhythmicity of these behaviors (Fig. 4.2), and Stephan and Zucker (1972b) showed that locomotor activity and drinking rhythms were lost after lesions which specifically destroyed the SCN. (For a review, see Rusak and Zucker, 1979).

Some of the most informative studies of the role of the SCN in controlling behavioral rhythms were undertaken by Rusak (1977, 1979), who used the hamster's wheel-running rhythm, normally a precisely

Fig. 4.15   Double-plotted wheel-running actographs of two hamsters SCN-lesioned on days 10 and 9 (black squares), respectively. Lighting regimens are indicated on the right, and the times of darkness are indicated by the boxes on the top of each figure. In LD only weak, if any, coupling is seen to the light–dark cycle; in constant light, multiple ultradian components are visible in the data, with no persisting circadian rhythm. (From Rusak, 1977.)

Time of Day (hrs)

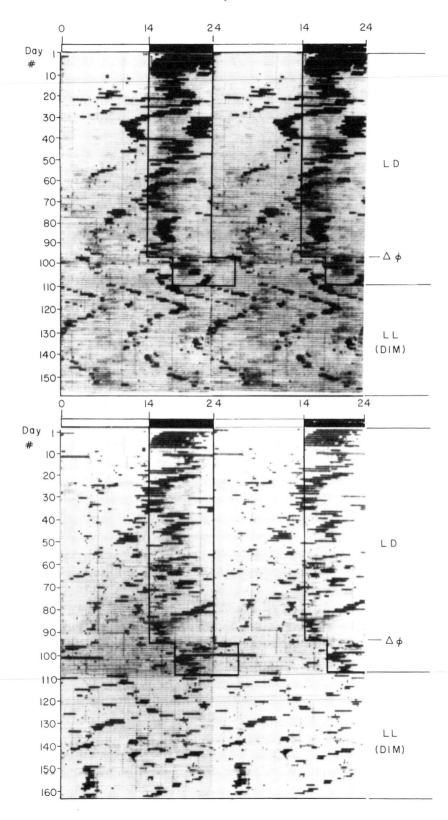

timed behavior within the animal's day. After total destruction of the SCN, the animals were maintained in a light–dark cycle, and some phase control was occasionally seen (Fig. 4.15), although it was quite different from that seen in intact hamsters. When the animals were re- leased into constant dim illumination, the pattern often broke up into a much more even distribution over the course of a day. But wheel-run- ning still tended to be clustered into two to three peaks per day, sug- gesting that the rhythm was in fact driven by a number of secondary oscillators. Without the SCN these oscillators could not maintain their internal coupling to one another and hence could not generate an overt circadian rhythm of wheel-running activity.

The concept of a pacemaker which coordinates a number of second- ary oscillators is also supported by our studies of SCN lesions in pri- mates. Figure 4.16 shows the effects of SCN lesions on the drinking rhythm in the squirrel monkey. Prior to lesions being placed, drinking is mostly confined to the subjective day (Fig. 4.16A). In constant illumi- nation the rhythm free-runs with a period of approximately 25 hours. After SCN lesions are placed (Fig. 4.16B, C), the circadian rhythmicity in drinking is disrupted (Fuller et al., 1981a). Drinking eventually be- comes randomly spread throughout day and night with no temporal organization (Fig. 4.16C). However, for several months before this occurs, totally SCN-lesioned squirrel monkeys can show persisting circadian rhythms in drinking (Fig. 4.16B) which have a gradually de- caying circadian amplitude (Albers et al., 1982). This suggests that the SCN normally play the role of a pacemaker or coordinator of a group of weakly coupled secondary oscillators which drive drinking behav- ior. It is important to note that the total amount of water drunk per 24 hours is unchanged after SCN destruction. Thus, the circadian system acts to organize the timing of drinking, while other centers in the body predominantly determine how much is drunk to maintain homeo- stasis.

One of the problems in interpreting the result of SCN lesions is the masking effect of LD cycles. When squirrel monkeys are studied in constant illumination many months after SCN lesions, drinking is ar- rhythmic, but when the animals are placed in an LD 12:12 cycle, dark- ness depresses drinking so that most is confined to the day, creating a pattern not dissimilar to that of the intact animal. That this is not cir- cadian entrainment can be demonstrated by reducing the photoperiod

Time of Day (hrs)

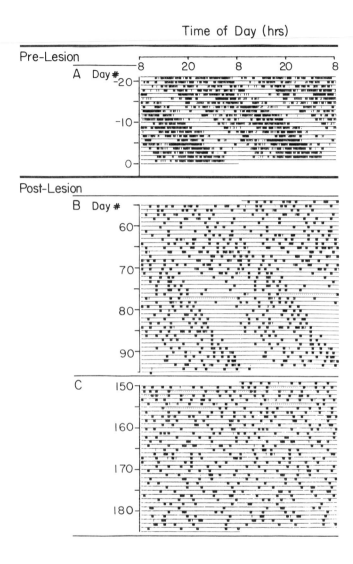

Fig. 4.16 Double-plotted drinking record from a squirrel monkey before (A) and after (B, C) receiving a histologically verified total SCN lesion. The approximately 25-hour drinking rhythm prelesion (A) persisted with a reduced amplitude for over 90 days postlesion (B) before finally decaying into arrhythmia (C). (Copyright 1981 by Albers, Lydic, Gander, and Moore-Ede.)

so less and less light is provided per day. There is a surge of drinking activity whenever the light is on, but the balance is distributed throughout day and night, and no residual rhythm is observed after an acute phase shift in the timing of the light pulse. In contrast, an intact monkey always remains rhythmic in such an experiment. Thus the passive response to light and dark which is not mediated by the SCN can be separated from endogenous rhythmicity, which is disrupted after SCN lesions.

Sleep–wake cycles   Closely related to the animal's activity rhythm is the sleep–wake cycle, because obviously an animal cannot feed, drink, or be active while it is asleep. Polygraphic recordings of the brain's electrical activity can determine the state of consciousness and thus separate lack of movement from actual sleep (see Chapter 5). Ibuka and Kawamura (1975) were the first to report the loss of circadian rhythmicity in the sleep–wake cycle in SCN-lesioned rats, as shown in Figure 4.17. As with the other behavioral functions, the circadian distribution of the sleep–wake cycle was disrupted while the total amount of sleep per 24 hours remained normal. The researchers followed these animals for as many as 80 days after lesioning, but interpretations were complicated because all studies were performed in a light–dark cycle. In a later study Ibuka, Inouye, and Kawamura (1977) confirmed that blinded, SCN-lesioned animals also demonstrated a loss of the circadian sleep–wake cycle. They further observed that animals blinded by primary optic tract lesions caudal to the optic chiasm, so that they lost visual perception, still maintained entrainment to the 24-hour light–dark cycle. This indicated that in the absence of visual perception the intact RHT could provide entrainment to the SCN and the sleep–wake cycle. Since that time other studies on SCN-lesioned rodents have been performed (Coindet, Chouvet, and Mouret, 1975; Stephan and Nunez, 1977; Mouret et al., 1978; Yamaoka, 1978) and have confirmed the loss of circadian rhythms in the timing of sleep.

Sleep, however, is not a single state. It is composed of various stages, the most simple divisions of which are slow-wave sleep and paradoxical or REM (rapid eye movement) sleep—the stage associated with dreaming (see Chapter 5). Some investigators have reported a persistence of the circadian rhythm of REM sleep but a loss of rhythmicity in slow-wave sleep after SCN lesions (Stephan and Nunez, 1977;

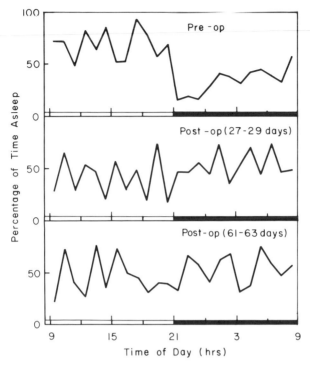

Fig. 4.17 The circadian sleep–wake cycle in an intact rat (*top*); 27–29 days after SCN lesions (*middle*); and 61–63 days after SCN lesions (*bottom*). Percentage of time spent asleep is plotted against clock time. The rat was maintained throughout in an LD 12:12 cycle, with lights on from 0900 to 2100 hours. Before surgery most sleep occurred during the light period. The day–night difference in the percentage of time spent asleep was abolished after the lesions. (From Ibuka and Kawamura, 1975.)

Mouret et al., 1978; Yamaoka, 1978). This raises the possibility that another pacemaker outside the SCN determines the rhythm of REM sleep. However, these experiments were performed in a light–dark environment in which the active and passive effects of light were very pronounced. Thus, the observed dissociation between slow-wave and paradoxical sleep could be a result of the passive influences of the 24-hour environment rather than the output of some other pacemaker.

Food anticipatory rhythms   Ever since the turn of the century, when August Forel first observed that bees would go out and search for food at specific times of day, it has been apparent that one of the major functions of the circadian system is to determine when an animal seeks food. When rats are provided with a bar to press for food pellets that is electrically connected to the pellet dispenser only at certain times of day, the animals will show anticipatory behavior; their bar-pressing activity will start to increase in the hour or so just before food is due (Bolles and Stokes, 1965; Boulos, Rosenwasser, and Terman, 1980). That this is a circadian-timed activity is illustrated in Figure 4.18. If food is presented within the circadian range of entrainment at 23- or 25-hour intervals, the animals readily anticipate the timing of food and show rapid lever pressing, which escalates at the time food is actually delivered (Fig. 4.18A, B). However, if food is presented at 18- or 30-hour intervals, outside the circadian range of entrainment, specific food anticipatory behavior is not seen, presumably because the rats are no longer able to estimate the time intervals. This food antici-patory behavior will continue for at least five cycles, even if it is not reinforced by the actual delivery of food each day (Fig. 4.18E). The rhythmic behavior does gradually extinguish, but a clear peak of antic-ipatory behavior can be seen, free-running with a period shorter than 24 hours when the animals are in constant conditions.

The SCN do not appear to be the pacemaker generating this circa-dian variation. SCN lesions result in arrhythmicity of the activity pat-terns of these animals, but the food anticipatory rhythm is not dimin-ished at all (Phillips and Mikulka, 1979; Stephan, Swann, and Sisk, 1979a, b; Boulos, Rosenwasser, and Terman, 1980). For example, the SCN-lesioned rats of Boulos, Rosenwasser, and Terman (1980) could still anticipate 23- and 25-hour cycles but not 30-hour cycles of food administration (Fig. 4.18C, D). The food anticipatory behavior rhythm persisted for several cycles in constant conditions even when no food was administered to reinforce the behavior (Fig. 4.18F). One apparent change after SCN lesions was that the animals were able to anticipate shorter periods, suggesting that the free-running period of the residual pacemaker may have decreased or its range of entrainment increased. These findings are one of the most striking pieces of evidence pointing to the existence of another pacemaker besides the SCN.

Estrous cyclicity   The circadian timing system plays an important role in reproductive function, particularly in animals with estrous

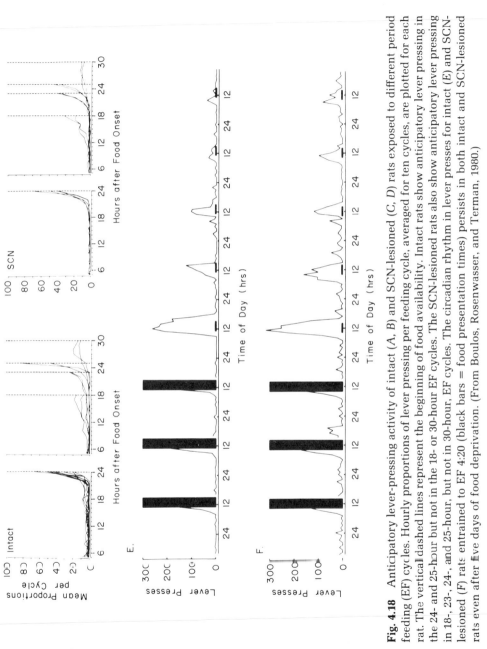

Fig. 4.18 Anticipatory lever-pressing activity of intact (A, B) and SCN-lesioned (C, D) rats exposed to different period feeding (EF) cycles. Hourly proportions of lever pressing per feeding cycle, averaged for ten cycles, are plotted for each rat. The vertical dashed lines represent the beginning of food availability. Intact rats show anticipatory lever pressing in the 24- and 25-hour but not in the 18- or 30-hour EF cycles. The SCN-lesioned rats also show anticipatory lever pressing in 18-, 23-, 24-, and 25-hour, but not in 30-hour, EF cycles. The circadian rhythm in lever presses for intact (E) and SCN-lesioned (F) rats entrained to EF 4:20 (black bars = food presentation times) persists in both intact and SCN-lesioned rats even after five days of food deprivation. (From Boulos, Rosenwasser, and Terman, 1980.)

cycles and those that are seasonal breeders. A number of investigators have studied the effects of SCN lesions on estrous cycles of rats and hamsters. Mosko and Moore (1978b) found that after lesioning the SCN in newborn female rats, as adults they failed to develop estrous cycles. In all cases in which the SCN of adult animals have been lesioned (Stetson and Watson-Whitmyre, 1976; Nunez and Stephan, 1977; Raisman and Brown-Grant, 1977; Shivers, Harlan, and Moss, 1979), estrous cyclicity has been lost. Furthermore, Stetson and Watson-Whitmyre (1976) have shown that the ability of male hamsters to detect changes in the duration of daylight (the basis of seasonal breeding cycles in many species, as Chapter 5 will show) is lost after SCN lesions. Comparable studies in female hamsters have not yet been reported. These findings indicate that at least in rodents a functioning SCN is necessary to regulate reproductive cyclicity. However, no information is available on the role of the SCN in mammals that have menstrual cycles.

Plasma corticosteroid rhythms The study of rhythms in plasma adrenal corticosteroids (primarily cortisol in primates and corticosterone in rodents) is complicated by a number of experimental considerations. First, because corticosteroid secretion is highly responsive to stress, its level may be changed in an SCN-lesioned animal solely as a response to the surgery. The process of blood sampling itself can induce changes in cortisol. Furthermore, corticosteroids are secreted episodically (Weitzman et al., 1971; Czeisler et al., 1976), and if blood samples are taken at infrequent (once every several hours) intervals, the peak and trough of the circadian plasma corticosteroid rhythm may be missed and incorrect conclusions reached about the presence or absence of the rhythm.

It is therefore not surprising that some confusion has arisen as to the effect of SCN lesions on the plasma corticosteroid rhythm. Moore and Eichler (1972) were the first to suggest that the circadian adrenal corticosterone rhythm was lost after SCN lesions. However, they sacrificed several groups of rats for each data point and collected samples only at 6-hour intervals. As we discussed in Chapter 1 (Fig. 1.10), arrhythmia could have been caused by a group of rhythmic rats becoming out of phase with one another. Allen, Kendall, and Greer (1972); Greer, Panton, and Allen (1972); Raisman and Brown-Grant (1977); and Abe and co-workers (1979) have also reported a loss of the diurnal rhythm in plasma corticosterone concentration in rats after isolation of the medial basal hypothalamus, which cuts off communication between the SCN and the pituitary-adrenal axis. However, inspection of Wilson and Critchlow's (1975) data from rats with isolation of the medial basal hypothalamus shows that there were some very major excursions in plasma corticosterone, which could be interpreted either as ultradian periodicities or possibly as free-running circadian rhythms no longer entrained by the environmental light–dark cycle. Circadian corticosteroid rhythms have been reported to persist after SCN lesions (Krieger, Hauser, and Krey, 1977; Szafarczyk et al., 1979); and Phelps and colleagues (1977) have shown that persisting corticosteroid rhythms may be highly variable from animal to animal.

The studies of corticosteroid rhythms are complicated still further by the fact that this function eventually recovers after knife cuts are made. Lengvari and Liposits (1977) investigated the persistence of the diurnal rhythm of plasma corticosterone of rats after frontal isolation of the medial basal hypothalamus. They found a loss of the circadian corticosteroid rhythm up until the thirtieth postoperative week. However, by the thirty-sixth week the rhythms in these deafferented animals had returned. This recovery of function points to inadequacies in many experimental studies in which the animals are studied too soon postoperatively. It is possible that other oscillators in the system may take some time to express themselves after SCN lesions, but before one can reach such conclusions it will be necessary to repeat these experiments in the absence of the light–dark cycle. Since environmental illumination can influence corticosteroid levels, it is hard to interpret experiments unless they have been conducted in conditions of constant illumination.

Other endocrine rhythms   Several other endocrine rhythms appear to be driven principally by the suprachiasmatic nuclei. For example, the circadian variation in pineal N-acetyltransferase is eliminated in rats after SCN lesions (Moore and Klein, 1974; Moore, 1974; Moore and Eichler, 1976; Raisman and Brown-Grant, 1977), and fluctuations in thyroid-stimulating hormone (Phelps et al., 1977) and thyrotropin activity (Abe et al., 1979) are lost following SCN destruction. Growth hormone also has a circadian variation derived principally from the

SCN. Rice, Abe, and Critchlow (1978) reported that medial preoptic lesions, which included the SCN, led to a loss of the circadian variation in the stress response of growth hormone, while lateral preoptic lesions did not have this effect. Willoughby and Martin (1978) confirmed that lesions of the SCN eliminated the circadian variation in growth hormone rhythm and its synchronization with the light–dark cycle. However, the ultradian rhythm in growth hormone secretion persisted, driven by an oscillator probably in the ventromedial hypothalamus (Martin, 1979).

Body temperature rhythm   The circadian rhythm of core body temperature provides the most suggestive evidence for the existence of another pacemaker outside the SCN. Stephan and Nunez (1977), who measured brain temperature in rodents, and Saleh, Hard, and Winget (1977), who recorded body temperature in rodents, both in a light–dark cycle, reported the elimination of the core temperature rhythm after SCN lesions. Others, however (Dunn, Castro, and McNulty, 1977; Powell et al., 1977), have reported the persistence of a rhythm in body temperature in rodents after SCN lesions. Again, these studies were performed in a light–dark cycle, and body temperature is known to be very sensitive to the levels of light intensity (Fuller, Sulzman, and Moore-Ede, 1978a). However, Nakayama, Arai, and Yamamoto (1979) have also observed persistence of a body temperature rhythm in SCN-lesioned rodents that were blinded and thus not influenced by the lighting regimen. It is still too early to determine if the controversy over the temperature rhythm in rodents is a function of subspecies differences or is related to the experimental technique. As Miles (1962) has shown, the body temperature rhythm can be suppressed in rodents as a result of the handling they receive during temperature measurement.

Although the evidence in rodents is controversial, it is much more clear-cut in primates. We have examined this issue in the squirrel monkey, which has a very prominent and stable body temperature rhythm. After total bilateral SCN lesions, feeding and drinking behaviors eventually lost their circadian rhythmicity, as documented in Figure 4.16, but the body temperature rhythm was found to persist when studied over a year post lesion. Figure 4.19 shows the drinking and body temperature rhythms of two monkeys after total bilateral

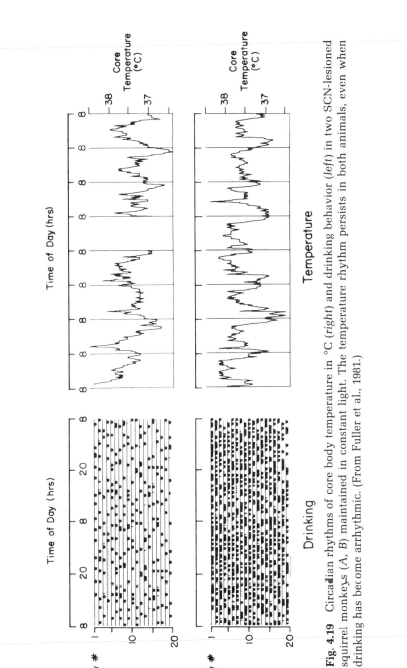

Fig. 4.19 Circadian rhythms of core body temperature in °C (*right*) and drinking behavior (*left*) in two SCN-lesioned squirrel monkeys (*A, B*) maintained in constant light. The temperature rhythm persists in both animals, even when drinking has become arrhythmic. (From Fuller et al., 1981.)

SCN lesions. The persisting body temperature rhythm is readily visible after drinking behavior had become arrhythmic (Fuller et al., 1981a).

These studies suggest that at least in primates there is another oscillator outside the SCN which is responsible for generating the body temperature rhythm. In Chapter 6 we will review in some detail the evidence that this oscillator may serve as a second major pacemaker in the circadian timing system.

ELECTROPHYSIOLOGY

There are some serious limitations to the conclusions that can be reached from lesioning studies alone. Showing that SCN lesions result in a loss of circadian rhythmicity in some physiological function such as wheel-running activity does not prove that the SCN is a pacemaker. The SCN could be, for example, just a way station transmitting temporal information from another structure, the true pacemaker. In that case SCN lesions would be interrupting a pathway between the true pacemaker and the observed rhythm.

To claim the SCN as a circadian pacemaker, two criteria must be satisfied. First, the SCN must be shown to continue its spontaneous rhythmicity when it is isolated from all inputs, and second, it must be demonstrated to drive other rhythmic functions within the organism. As we discussed in Chapter 1, these are the two key features of any clock—the ability to measure time independently and the use of this information to time other events.

The elegant studies of Inouye and Kawamura (1979) have taken a major step toward fulfilling the first of these prerequisites. They demonstrated that continuously recorded electrophysiological activity from the SCN had a marked circadian rhythmicity (Fig. 4.20). The rate of neuronal activity of the SCN showed a twofold variation in the course of the day. They then isolated the SCN with a series of knife cuts so that it had no neural inputs or outputs, and when they recorded from within the hypothalamic island, the circadian rhythmicity was found to persist. Of course, the possibility still exists that the SCN within the island were responding to a rhythmic hormonal signal generated from a pacemaker outside the SCN. Although that seems unlikely, final confirmation awaits the study of circadian rhythms of SCN neural activity *in vitro,* perhaps using the techniques developed by Groos and Hendriks (1979).

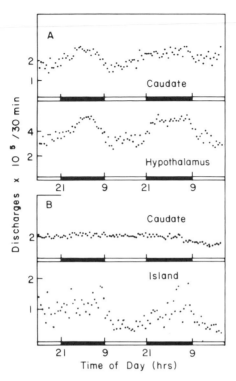

Fig. 4.20 Circadian rhythm of neuronal activity of the brain before and after isolation of the hypothalamus. In an intact rat (*A*), multiple unit activity has a circadian rhythm in both hypothalamic and caudate regions. After the hypothalamus was isolated by knife cuts (*B*), recordings from within the hypothalamic island indicated persisting rhythmicity, while no rhythmicity was evident outside the island, thus suggesting the presence of a hypothalamic circadian pacemaker. (From Inouye and Kawamura, 1979.)

Inouye and Kawamura (1979) have attempted to fulfill the second criterion by studying the role of the SCN's rhythmicity in timing the neural activity of other areas of the brain. Several investigators have reported circadian rhythms in electrophysiological activity of various hypothalamic and extrahypothalamic nuclei (Schmitt, 1973; Koizumi and Nishino, 1976; Nishino and Koizumi, 1976). However, the demonstration of circadian rhythmicity in these various neuronal structures does not necessarily mean that they contain pacemakers. Rather it appears that some of them are passively responding to rhythmic inputs from the SCN. As shown in Figure 4.20, Inouye and Kawamura demonstrated that rhythmic activity in several structures such as the caudate nucleus was eliminated after the SCN were isolated. However, rigorous confirmation of the pacemaking role of the SCN will require the simultaneous study of overt rhythmicity in peripheral tissues of animals with and without hypothalamic islands.

Another area of investigation, still in its infancy, is to determine how the optic pathways convey light–dark information from the retina and entrain the spontaneous rhythmic activity of the SCN. Electrically stimulating the optic nerve (Nishino, Koizumi, and Brooks, 1976) or stimulating the retina by light (Lincoln, Church, and Mason, 1975) excites some of the cells of the SCN while inhibiting others. Groos and his colleagues (Groos and Mason, 1978; Groos and Hendriks, 1979) have confirmed that light has both facilitatory and inhibitory influences on the electrical activity of SCN neurons, and the response was found to be a function of the level of environmental illumination. Unlike visual processing centers, however, the receptive fields of the SCN neurons are large, with no antagonistic centers surrounding them (Groos and Mason, 1980). Thus the SCN appear to be responding more to the overall intensity of environmental illumination than to the spatial patterns of light and dark which are essential to visual perception.

The hypothalamus has always been a very difficult area from which to obtain electrophysiological information, and the SCN are no exception. The nature of the neural networks within the nuclei remains to be determined. Whether each cell is independently circadian or whether the generation of circadian rhythmicity relies on complex interactions of the whole cell population remains an open question. New insights may be provided by techniques such as *in vitro* culture of slices of hypothalamic tissue containing the SCN, developed by Groos and Hendriks (1979).

Electrical stimulation An important tool in electrophysiological research which has been underutilized in the study of the circadian timing system is electrical stimulation. It is particularly useful for examining the influence of neural inputs on physiological function and for determining the influence of the input frequency and its patterning. Additionally, when coupled with electrophysiological recording techniques, stimulation can be a powerful tool for tracing neural pathways between two loci in the brain.

This latter approach has been used to a limited extent to trace some of the output pathways from the SCN to other structures. For example, Oomura and colleagues (1979) have demonstrated that there are connections between the SCN and the ventral medial nucleus of the hypothalamus, and Nishino, Koizumi, and Brooks (1976) have shown that stimulation of the SCN strongly inhibits the electrical activity of the central sympathetic nervous system in a way that is similar to direct inhibition by light or optic nerve stimulation. This suggests that the effects of light on the pineal gland may involve the SCN as a way station between the retina and the sympathetic innervation to the pineal.

To our knowledge, only a few preliminary studies on circadian entrainment have been performed by electrical stimulation, and the results have not yet appeared in print. However, a wide variety of questions could be addressed using such techniques, and this approach deserves further attention.

PHARMACOLOGY

Relatively little is known about the chemical substances in the nerve cells of the SCN, whether the chemicals are synthesized there or derived from another location. Neuropeptides, which act as hormones or releasing factors, have received the most attention. For example, vasopressin and its carrier, neurophysin, have been localized in the rat and mouse SCN (Vandesande, Dierickx, and Demey, 1975; Swanson, 1977; Buijs et al., 1978). The SCN also appear to be rich in luteinizing hormone releasing hormone (LHRH) (Barry and Dubois, 1976; Setalo et al., 1976), and a key neurotransmitter, serotonin (Meyer and Quay, 1976a, b).

Some preliminary information has been obtained on the neuropharmacological properties of SCN neurons. Nishino and Koizumi (1977) have examined the responses to iontophoretic injections of several putative neurotransmitters, including glutamine, dopamine, catecholamine, noradrenaline, and serotonin. In general, catecholamine was excitatory for SCN neurons, as was glutamine. Serotonin, noradrenaline, and dopamine, however, were predominantly inhibitory. Perkins and Whitehead (1978) have also shown that dopamine and noradrenaline were inhibitory in the preoptic region of the hypothalamus.

It is not at all clear as yet what functional roles are played by the various types of neurons and the various neurotransmitters which have been shown to influence SCN function. For example, when a depletor of brain serotonin, parachlorophenylalanine (PCPA), was administered to rats, it resulted in the loss of circadian rhythms in both locomotor activity and plasma corticosterone (Honma, Watanabe, and Hiroshige, 1979). After the rhythms reappeared several days later, the free-running period was unchanged, but the rhythm was phase shifted by about five hours. However, another depletor of serotonin, 5, 6-dihydroxytryptamine, had no effect on the circadian rhythms of the animal. In another study by Honma and Hiroshige (1979), catecholamine levels were depleted by the administration of 6-hydroxydopamine (6-OHDA), but this chemical had no effect on the circadian rhythm of locomotor activity, which free-ran in constant light with normal period and amplitude. However, the 6-OHDA-treated rats displayed a loss of the circadian corticosterone rhythm in constant light and reverted to an ultradian periodicity of 6 to 8 hours in length. Zatz and Brownstein (1979) demonstrated that carbachol, a cholinergic agonist, administered to the SCN can mimic the effects of light on the circadian rhythm of pineal serotonin N-acetyltransferase activity. However, each of these studies is difficult to interpret until more is known about the functional role of the input and output pathways of the SCN.

METABOLISM

The cells of the SCN show a prominent circadian variation in their metabolic activity. A circadian rhythm in glucose uptake has been demonstrated in rats with an analogue of glucose, 2-deoxy-D-glucose, which can be used to measure the rate of glucose utilization by cells (Schwartz and Gainer, 1977; Schwartz, Davidsen, and Smith, 1980). These investigators showed (Fig. 4.21) that glucose utilization reached a maximum during subjective day when the animal was at rest, then fell to a minimum during subjective night when the animal was active. This relationship to the rest–activity cycle was maintained when animals were entrained to a light–dark cycle or were free-running after being blinded.

The technique used by Schwartz and his colleagues has some particularly useful features for the probing of pacemaker function. In histological sections of the brain the SCN, compared to neighboring struc-

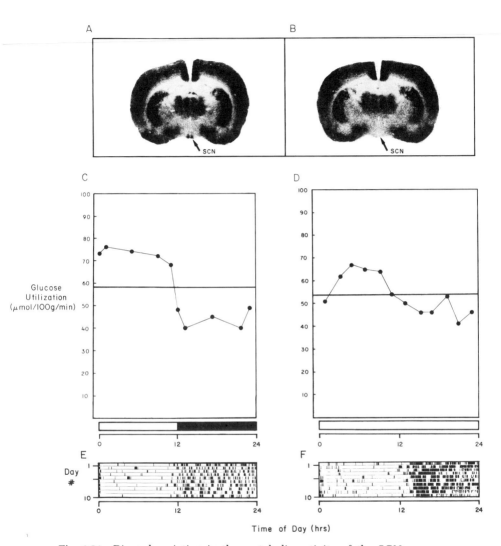

Fig. 4.21 Diurnal variation in the metabolic activity of the SCN, measured using a glucose analogue, 2-deoxy-D-glucose. (A) Coronal section of the rat brain shows the darkly staining accumulation of 2-deoxy-D-glucose in the SCN during the light phase. However, during the dark phase (B), the SCN does not stain darkly, indicating low metabolic activity. Circadian rhythms of glucose utilization in both LD (C) and after blinding (D) show a circadian maximum during the rest phase of the animal's activity cycle, indicated at the bottom of the figure (E, F). (From Schwartz, Smith, and Davidsen, 1979.)

tures, stands out as an area that is labeled more or less densely, depending on the time of day. This independent measure of SCN phase from histological sections can be correlated with other events in the brain.

THE ONTOGENY OF SCN AND RHT FUNCTION

Studies of newborn animals have demonstrated that the SCN start their oscillatory behavior before input or output connections are made. Using the 2-deoxy-D-glucose technique (Schwartz and Gainer, 1977), Fuchs and Moore (1980) demonstrated that although there was no circadian difference in SCN glucose utilization in rats before birth, a significant circadian difference was visible on the first day after birth, even in rats maintained in constant darkness (Fig. 4.22). The circadian rhythmicity in rat SCN function thus commences well before the retinohypothalamic tract becomes fully developed at three or four weeks of age (Campbell and Ramally, 1974).

A number of investigators have shown that in rodents overt rhythms are not observed in many physiological functions until about three weeks of age. For example, the onset of the diurnal variation in serum corticosterone (Campbell and Ramally, 1974), as well as in corticotrophin releasing factor (CRF) activity (Hiroshige and Sato, 1970) occurs at three weeks of age. These findings suggest that the output pathways from the SCN may take some time to become established and that the first developmental step is the formation of the pacemaker function.

Less attention has been given to input pathways other than the retinohypothalamic tract. So, Schneider, and Frost (1971) have used both anteriograde and retrograde techniques to describe the formation of the retinal projection to the lateral geniculate body of hamsters. These fibers are present but sparse by day 3 and completely formed by day 8 of the hamster's life. To our knowledge, however, nobody has investigated the time-course development for the fibers that traverse from the lateral geniculate body to the SCN itself.

THE STUDY of the circadian pacemakers is therefore at an exciting stage. One putative pacemaker has been identified and another shown to exist, although its location is not certain. In the next chapter we will see how these pacemakers keep circadian time in various physiological systems, but to get there we must make a conceptual jump. We

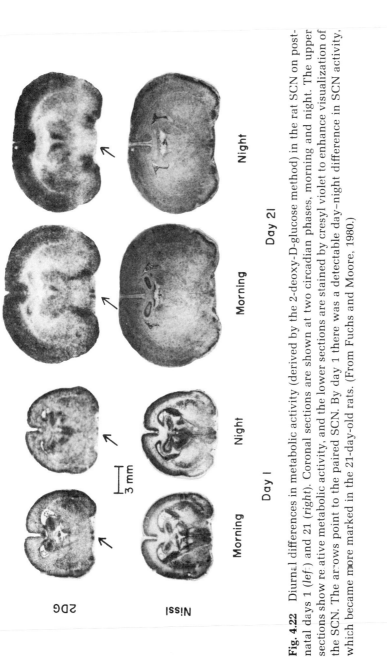

Fig. 4.22 Diurnal differences in metabolic activity (derived by the 2-deoxy-D-glucose method) in the rat SCN on postnatal days 1 (*left*) and 21 (*right*). Coronal sections are shown at two circadian phases, morning and night. The upper sections show relative metabolic activity, and the lower sections are stained by cresyl violet to enhance visualization of the SCN. By day 1 there was a detectable day–night difference in SCN activity, which became more marked in the 21-day-old rats. (From Fuchs and Moore, 1980.)

know very little about the neural and endocrine pathways that convey temporal information from the pacemakers to the overt rhythms in physiology and behavior. The hands of the clock are well characterized, and we are beginning to have some appreciation of the clock mechanism. But the transmission process is largely unknown.

Circadian Timing
of Physiological Systems

The precise scheduling of behavioral and physiological events is one of the most critical services performed by the circadian timing system. An animal's survival in a highly periodic day–night environment depends on the appropriate timing of its responses. We will now turn our attention to the question of how the central circadian pacemakers govern the timing of other body functions.

A considerable number of circadian rhythms have been documented in a wide range of behavioral and physiological variables. It is not the purpose of this chapter to catalog every physiological system with reported rhythmicity, as this would take a book in itself. Rather, believing that the "essence of physiology is regulation" (Grodins, 1963), we will focus on those physiological systems of which most is known about their circadian control. The timing of sleep and wakefulness; feeding and drinking behaviors; thermoregulation; and endocrine, renal, and reproductive function, taken together, more than adequately illustrate what is known about the general mechanisms of circadian regulation. We will not discuss in any detail the respiratory, cardiovascular, or reticuloendothelial systems; although each displays circadian rhythms, less is known about the responsible mechanisms.

Some common themes will become apparent. Not only does the circadian timing system control the basal level of physiological systems, it also influences the responsiveness of each system to challenges at different times of day. While circadian rhythms have been documented in the basal levels of many physiological variables, there has as yet been relatively little study of circadian variation in dynamic responses to environmental challenges. In some senses these may be more important, because an animal requires such dynamic responses to cope with its environment. If the daily timing of the challenge is highly predictable, an internal circadian clock can enable the animal to place the appropriate homeostatic mechanisms "in a state of alert." Such anticipatory actions may be particularly valuable when synthesis of enzymes or hormones is required; without advance notice there would be a delay of several hours in achieving the corrective response.

Another theme is the interaction, in physiological systems, of the active responses of circadian oscillators to stimuli and the passive responses of nonoscillatory elements. Part of the job of the experimenter is to tease out and separate the time-of-day-dependent effects from those that are independent of the time of day. To date, most physiologic studies have been concerned with events which are presumed to be independent of the time of day, or at least the investigators have tried to negate the effect of circadian timekeeping. However, our increasing knowledge of the dynamics of oscillating systems should make possible a new level of understanding of physiology in the time domain.

Sleep–Wake Cycles

The circadian rhythm of which we are most aware is our daily cycle of sleep and wakefulness. The dramatic change from consciousness to unconsciousness dominates our lives like no other rhythm in body function. We are much more aware of a change in the timing of sleep and wakefulness because of nocturnal insomnia or daytime sleepiness than of a comparable change in the rhythms of endocrine or renal function. Similarly in the animal world, the nocturnal or diurnal nature of an animal dominates its behavior. In many ways the adoption of a specific temporal niche for sleeping and waking is as ecologically important as its spatial niche.

It is important to distinguish the circadian sleep–wake cycle from the rest–activity cycles which have been discussed throughout this book. Much of our knowledge of circadian timekeeping in mammals has been gained by recording rest–activity patterns. Yet rest is not synonymous with sleep (one can lie awake at night), and activity is not synonymous with wakefulness (movements regularly occur during sleep and may even include sleepwalking). The importance of not equating rest with sleep has been emphasized in studies in which sleep has been polygraphically recorded at the same time as the rest–activity cycle. The two rhythms are found to not always correlate with each other (Mitler et al., 1977). To reliably distinguish sleep from wakefulness in the laboratory, the electrical activity of the brain must be measured as an electroencephalogram (EEG), muscle activity recorded as an electromyogram (EMG), and eye movement detected by electrooculogram (EOG). Relatively few circadian studies have done this, however.

Not all animals show the consolidated circadian sleep–wake pattern of humans, with sleep essentially uninterrupted for 8 or so hours each day and wakefulness similarly uninterrupted for 16 hours. At the opposite end of the scale, the domestic cat, as we discussed in Chapter 2 (Fig. 2.3) has short bouts of sleep and waking throughout day and night, with very weak expression of the circadian sleep–wake cycle (Sterman et al., 1965; Moore-Ede et al., 1981). Many animals, such as the rat, fall between the two extremes, with a preponderance of sleep at certain phases of the day–night cycle and lesser amounts at other times of day.

The percentage of each 24 hours occupied by sleep in various mammals also shows a very wide range, from the donkey, which spends only about 13% of the time asleep, to the opossum, which sleeps 80% of the time (Snyder, 1969). Similarly, the phase of the light–dark cycle at which most sleep is taken is highly variable in different species. Some animals are diurnal, others nocturnal, and some crepuscular—awake only around dawn and dusk.

Within an individual the distribution and percentage of time asleep is a function of age. Figure 5.1A shows that the newborn human infant sleeps approximately two-thirds of the time and that the proportion of wakefulness then slowly increases throughout life. Also it takes the first two years or so of life for sleep to become fully consolidated in the nocturnal hours (Fig. 5.1B), although adult humans may nap during the day in certain cultures (for example, the siesta) or in advanced age. In Chapter 2 we discussed Kleitman and Engelmann's (1953) elegant demonstration of how the short bouts of sleep throughout day and night in

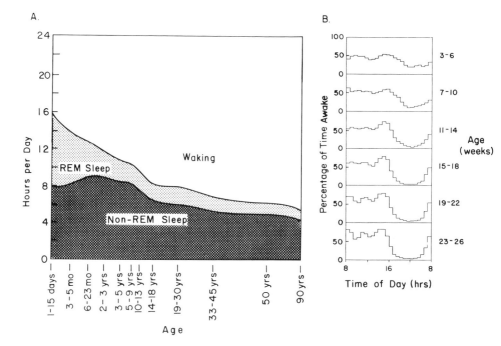

Fig. 5.1 (*A*) Change with age in total amount of daily sleep and proportion of REM sleep in humans. The amount of REM sleep drops sharply
after the first year, from 8 hr at birth to less than 1 hr per night in old age.
The change in nonREM sleep is much less marked, falling from 8 hr to
about 5 hr over the lifespan. (*B*) The mean percentage of time spent
awake over each hour of the day in six successive 4-week observation periods of a group of human infants. (*A* after Roffwarg, Muzio, and Dement;
copyright 1966 by the American Association for the Advancement of Science. *B* after Kleitman, reprinted by permission of the University of Chicago Press, copyright 1963, The University of Chicago.)

the newborn infant gradually coalesce into a more consolidated
sleep–wake rhythm (Fig. 2.10).

ULTRADIAN RHYTHMS IN SLEEP STAGES

Sleep itself is not a single state. Shortly after methods to record the
electrical potentials of the human brain cortex (EEG) were developed
by Hans Berger in 1930, it was recognized that variations in the electrical potentials were associated with changes in the depth of sleep

(Loomis, Harvey, and Hobart, 1935). The researchers recognized specific changes in the frequency and amplitude of the EEG waveform during sleep. However, the most marked changes in EEG, EMG, and EOG occur during dreaming. Aserinsky and Kleitman (1953) reported that bursts of rapid eye movements (REM) periodically occur during sleep, and Dement and Kleitman (1957) linked these episodes of REM sleep to dreaming. Subjects awakened during REM sleep were usually in the midst of a vivid dream, whereas when awakened in nonREM sleep, they rarely reported vivid dreams. REM sleep is accompanied by a loss of muscle tone, because the activity of spinal motor neurons is inhibited (Morrison and Pompeiano, 1965). This atonia may be what prevents us from acting out our dreams. In animals, when the inhibitory mechanisms are blocked by specific lesions in the brainstem, animals will undertake a variety of active behaviors during episodes of REM sleep (Hendricks, Bowker, and Morrison, 1976).

By recording the EEG, eye movements, and muscle tone, it is possible to distinguish clearly between REM and nonREM sleep. As Figure 5.2 shows, REM sleep can be recognized from the rapid eye movements and loss of muscle tone. Because the EEG patterns during REM sleep do not appear markedly different from the patterns of the waking state, REM sleep is often referred to as "paradoxical" or "desynchronized" sleep. NonREM sleep in humans has been divided into four stages, according to the classification originally proposed by Dement and Kleitman (1957). Because it is characterized by EEG waves of increasing magnitude and slower frequency as sleep becomes deeper, the name "slow-wave" or "synchronized" sleep is often used. Muscle tone, however, is maintained, and thus most of the movements during sleep occur in the nonREM stages. In young adult humans approximately 20–25% of sleep is spent in REM, while in other mammals this percentage ranges from as little as 7% in goats and sheep to 33% in the opossum (Snyder, 1969).

The alternation between REM and nonREM sleep is cyclic, with an approximately 90 100-minute period in man (Dement and Kleitman, 1957; Williams, Agnew, and Webb, 1964; Webb and Agnew, 1969). Figure 5.3 shows the typical sleeping pattern in an adult human, a cat, and a rat. Humans normally start with nonREM sleep; the first REM episode occurs 70 to 90 minutes later and recurs thereafter at approximately 90–100 minute intervals. However, in the first part of the night

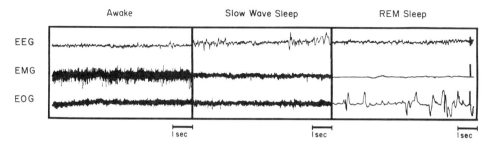

Fig. 5.2 Polygraphic recording of typical changes in bioelectrical potentials during wakefulness and sleep, showing EEG, EMG, and EOG. The patterns are distinctly different during wakefulness (*left*), slow-wave sleep (*center*), and REM sleep (*right*). During REM sleep, for example, there is a loss of tone in the neck muscles, eye movements at irregular intervals, and a low-amplitude, rapid EEG record. Vertical bars on right = $50\mu v$ calibration. (Replotted from Jouvet, 1962.)

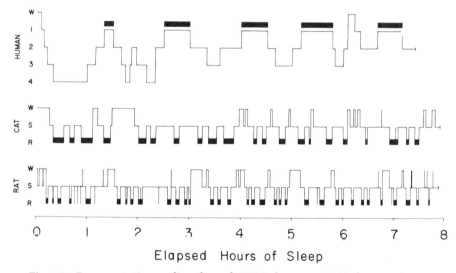

Fig. 5.3 Representative cycling through REM sleep, nonREM sleep, and wakefulness in a human, a cat, and a rat. Eight hours of continuous observations are plotted for each species, with REM periods indicated by dark bars. The plots for the cat and rat show the timing of shifts between wakefulness (W), nonREM sleep (S), and REM sleep (R). For the human, 4 stages of sleep are distinguished by the EEG. The period of the REM–nonREM cycle increases with the size of the animal. (From Dement et al., 1969.)

nonREM sleep predominates, and the bouts of REM are short, whereas later the REM episodes lengthen so that they encompass a larger fraction of sleep. The period of the REM–nonREM cycle varies among species, as Figure 5.3 shows, with shorter periods in smaller animals (Dement et al., 1969). At the extremes of the range, the elephant has a 120-minute REM–nonREM cycle, while the mouse has a cycle of 20–30 minutes. A corresponding change in period is seen with age. In human infants the REM–nonREM cycle has a 50–60 minute period, which gradually lengthens by adulthood. Hobson (1975) has shown that both species- and age-related differences in the REM–nonREM cycle length are related to differences in brain size.

ENDOGENOUS NATURE OF SLEEP–WAKE CYCLES

Both the circadian sleep–wake cycle and the ultradian REM–nonREM cycle are endogenous rhythms which persist in the absence of environmental variation (Fig. 5.4). Studies of polygraphically recorded sleep have been performed with monkeys in constant illumination (McNew et al., 1972) and with humans isolated in caves (Chouvet et al., 1974) or in isolation chambers (Webb and Agnew, 1974; Czeisler, 1978; Weitzman, Czeisler, and Moore-Ede, 1979a; Zulley, 1979). The circadian sleep–wake cycle has a typical free-running period of about 25 hours, although in most human subjects studied over a long enough time, the period eventually lengthens to 30–50 hours (Czeisler, Weitzman, et al., 1980). The ultradian REM–nonREM cycle persists but its period does not lengthen proportionately in free-running conditions. It thus does not show any obvious subharmonic synchronization to the circadian cycle (Czeisler, Zimmerman, et al., 1980), and the ultradian and circadian cycles are thus essentially autonomous rhythms.

Although the sleep–wake cycle is timed by the circadian timekeeping system, other influences may modify the timing and organization of sleep. For example, increased daytime physical activity will lead to relatively more slow-wave sleep. Similarly, it is a matter of common observation that animals (and humans) tend to become sleepy after eating. The relationship between food and sleep is more than coincidental. Kukorelli and Juhasz (1977) found that satiated cats slept more than fasting cats, and that electrical stimulation of the intestine led to longer sleep episodes in the starved animals. The neural pathways that were stimulated apparently projected to the solitary tract in the brain

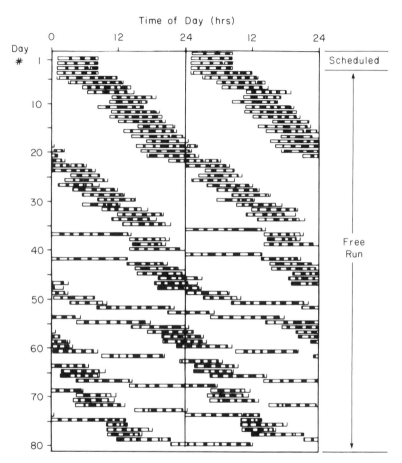

Fig. 5.4 Double plot of the sleep–wake rhythm of a human living in temporal isolation. The horizontal bars show the times of polygraphically recorded sleep on individual days, with the solid bars = REM sleep, and the open bars = slow-wave sleep. Times in bed spent awake are shown by a thin horizontal line. On days 1–4 the subject followed a 24-hour schedule. On day 5 he retired at the usual time but was allowed to arise and go to sleep whenever he wanted for the remainder of the experiment. Both the circadian sleep–wake cycle and the REM–nonREM cycle persisted during the free-run. (After Czeisler, Weitzman, et al., 1980, reprinted by permission of Raven Press.)

medulla, an area which when stimulated directly will lead to synchronized slow-wave EEG patterns characteristic of nonREM sleep (Magnes, Moruzzi, and Pompeiano, 1961). Furthermore, destruction of this area resulted in arousal (Berlucchi et al., 1964).

Another mechanism that may account for the relationship between feeding and sleep was studied by Rubinstein and Sommenschein (1971). They showed that cholecystokinin, a hormone released when nutrients enter the small intestine, promotes sleep when injected into animals, even if the gut has been denervated. However, all these effects are only modifiers of the sleep–wake cycle and not the primary determinants; when fed continuously throughout day and night both animals and humans continue to show a circadian sleep–wake cycle.

Ambient temperature can also influence the amount and type of sleep. Maximum sleep durations in cats occurred when they were kept within an ambient temperature range of 10–25° C (Parmeggiani, Rabini, and Cattalani, 1969; Parmeggiani and Rabini, 1970). When the ambient temperature was below or above this range, sleep duration decreased. Both slow-wave and REM sleep were reduced, although REM sleep appeared to be more affected. Similar findings in rats have been reported (Schmidek et al., 1972). Intermediate environmental temperatures, however, have little influence on the sleep–wake rhythm.

Similarly, posture may modify sleep. When a person is tired, it is harder to stay awake lying down, and standing or walking can keep one from falling asleep. Although a certain amount of nonREM sleep can be achieved while standing, REM sleep cannot, because it is accompanied by muscle atonia. In fact, an animal can be selectively deprived of REM sleep by standing it on an inverted flower pot surrounded by water.

It is possible for humans to exert conscious control and override the desire to sleep for a night or two. Animals may occasionally forego some hours of sleep when hunting for food or in times of extreme danger, but humans often deprive themselves of sleep for more abstract reasons, such as studying for an exam or trying to beat a record in the *Guinness Book of World Records*. It is extremely difficult to stay continuously awake without outside help for more than two or three days. The 264 hours (11 days) without sleep by a 17-year-old boy in San Diego in 1966 were accomplished only with a lot of help from other people in keeping him awake (Gulerich, Dement, and Johnson, 1966).

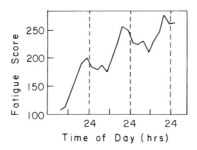

Fig. 5.5 The mean circadian rhythm of fatigue, subjectively assessed, in 63 humans kept awake and active for three days. Even though the subjects were active both day and night, they felt less fatigue during the day than during the night. The vertical dashed lines are at 24-hour intervals. (After Conroy and Mills, 1970; data from Froberg and Levi.)

Thus, although we have some capability to postpone sleep, we can never eliminate it. Indeed, there is a continuing circadian rhythm in the desire for sleep, which persists even if we stay awake. Anyone who has stayed up all night will recognize the alertness that comes with morning. Figure 5.5 shows the circadian rhythm in the subjective rating of fatigue by a group of subjects who stayed awake for 72 hours. Each night fatigue increased at the time they would have been in bed, and each morning they became relatively more alert.

Such studies of sleep deprivation have indicated that the duration of uninterrupted sleep does not depend primarily on the length of prior wakefulness, as might be intuitively supposed. When the record-seeker who spent 11 days without sleep finally went to bed, he slept for only 14 hours and thereafter returned to his normal 8 hours a night (Gulerich, Dement, and Johnson, 1966). Other subjects have had similarly short lengths of recovery sleep after sleep deprivation (Patrick and Gilbert, 1896).

Human subjects who are not sleep-deprived, but living in an isolation facility where their circadian rhythms are free-running, may show an alternation between long (15–20 hours) and short (6–10 hours) sleep durations with no consistent relationship to the duration of prior wakefulness. Aschoff and his colleagues (Aschoff, 1965; Wever, 1979) have long recognized that when internal desynchronization occurs in humans, the rest–activity cycle may free-run with a period that is much longer than that of the body temperature rhythm (see Chapter 3). Under these conditions the length of the rest–activity cycle is modulated as it passes through the various phase relationships with the body temperature rhythm.

Czeisler and colleagues (Czeisler, 1978; Czeisler, Weitzman et al., 1980) polygraphically recorded sleep in such internally desynchronized patients and demonstrated that sleep duration depends not on how long a person has previously been awake, but instead on the circadian phase of the body temperature rhythm when he went to sleep. This is illustrated in Figure 5.4, in which the sleep episodes of a subject in a self-scheduled environment were double-plotted in a standard circadian actogram. There were wide variations in sleep duration, especially after day 36, when the sleep–wake cycle became internally desynchronized, so that the sleep–wake cycle and the body temperature rhythm free-ran with independent periods. Short sleep episodes usually began just at or after the mid-trough of the temperature cycle, whereas long sleep episodes occurred when sleep began with the body temperature rhythm above its mean value.

The mean sleep durations of a group of five subjects studied for a total of 192 days are shown in Figure 5.6. The dependence of the average sleep length on the circadian phase of the core body temperature rhythm is clearly evident. When subjects went to bed at the trough of the averaged temperature cycle, sleep episodes were short and wake times occurred in the rising phase of the temperature cycle. When subjects went to bed at or after the peak of the temperature rhythm, sleep was extended, and wake times again occurred on the next upslope of the temperature curve. In fact, 86% of all wake times in these subjects occurred in the rising phase of the temperature rhythm (18–6 hr CT). Thus it appears that sleep duration is the product of an interaction between the separate pacemakers driving the sleep–wake cycle and the body temperature rhythms (see Chapters 4 and 6).

No significant correlation was observed in these experiments between sleep duration and the length of prior wakefulness, provided that the subjects had been awake for at least 14 hours. The occasional subject with a sleep–wake cycle much shorter than 24 hours continued to show sleep episodes of 6–10 hours duration with the usual phase relationship to the temperature rhythm. However, sleep episodes begun at the maximum of the temperature rhythm were very short (less than 4 hours) instead of very long (more than 12 hours). Thus, length of prior wakefulness appears to be a factor in determining whether sleep episodes of abnormal length are very short or very long (Czeisler, Weitzman, et al., 1980). This is in agreement with common experience,

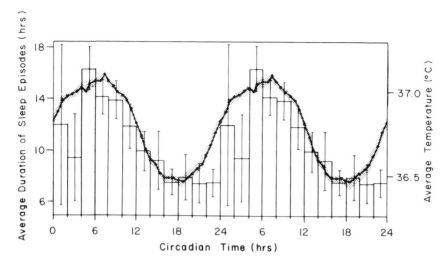

Fig. 5.6 Relationship of the average duration of sleep episodes to the phase of the body temperature rhythm when subjects went to bed (192 days of recordings from five human subjects). The solid line represents the mean core temperature (stippled area = ± SEM); the vertical bars indicate the average sleep length for sleep episodes begun within each phase interval. Only those sections of data in which the average period of the rest–activity cycle differed from that of the temperature cycle were used in this analysis. The rhythms for all subjects were normalized to circadian time with CT 0 = time of upward crossing through the mean temperature. (From Czeisler, Weitzman, et al., 1980, copyright 1980 by the American Association for the Advancement of Science.)

in that an afternoon nap some 6 hours after awakening in the morning, taken at the maximum of the body temperature rhythm, usually does not last more than an hour or so. However, if the person had stayed awake all the previous night, sleep begun at that time of day might continue until the next morning.

Not only is the length of sleep primarily dependent on the phase of the averaged temperature cycle, but Czeisler, Weitzman, and colleagues (1980) also found that all free-running subjects chose to go to bed at certain phases of the temperature cycle much more frequently than at others. The bedtime chosen most frequently was just after the temperature cycle minimum (18–24 hr in Fig. 5.6), a phase that also corresponded to the nadir of the subjects' alertness assessment rhythm. The internal structure of sleep also varied with the phase of the temperature cycle. When sleep began just after the trough of the temperature cycle, the first 50 minutes of REM sleep were accumulated an average of 2 hours earlier, compared to sleep beginning just after the temperature cycle maximum (Czeisler, Zimmerman, et al., 1980). The researchers concluded that the long-recognized variation in the amount of REM sleep with the "time of day" (Czeisler and Guilleminault, 1980) is based on a close relationship between REM sleep and the body temperature rhythm.

In internally desynchronized subjects, both the propensity for REM sleep (which cannot, of course, be expressed as overt REM sleep when the subject is awake) and the body temperature rhythm oscillate together with a different period from that of the sleep–wake cycle itself. Furthermore, when desynchronized subjects are released into free-running conditions, there is an internal phase delay of slow-wave sleep with respect to both body temperature and REM sleep (Czeisler, 1978; Weitzman, Czeisler, and Moore-Ede, 1979). Recognition of this REM sleep propensity rhythm also explains, in part, the occurrence of REM episodes at the onset of sleep in normal free-running subjects (Chouvet et al., 1974). This is otherwise a very rare phenomenon which is diagnostic of patients with narcolepsy (Rechtschaffen et al., 1963). Yet sleep-onset REM episodes often occurred in normal subjects when they chose to go to bed just after the temperature cycle trough (Czeisler, 1978; Czeisler, Weitzman, et al., 1980). The increase in the amount of REM sleep at certain phases of the temperature cycle is not a consequence of any changes in the period of the ultradian REM-nonREM cycle. Instead, as shown in Figure 5.4, the duration of each REM episode is increased (Czeisler, Zimmerman, et al., 1980).

These relationships are also relevant to the normal 24-hour, scheduled sleep–wake cycle, as Akerstedt and Gillberg (1980) have shown. They studied unrestricted sleep in normal subjects kept awake for a variable number of hours, such that their scheduled bedtimes occurred at systematically varied clock hours throughout day and night. Despite the potential confounding influence of sleep deprivation inherent in such a design, they found a similar relationship among sleep length, REM tendency, sleepiness, and core body temperature at different times of day as Czeisler, Weitzman, and colleagues (1980) described in free-running subjects.

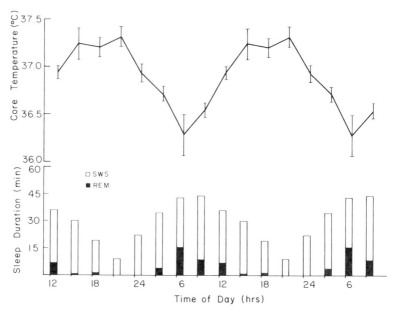

Fig. 5.7 Sleep patterns of human subjects maintained on a cycle of 2 hrs of activity and 1 hr of rest throughout day and night. (*Top*) The educed body temperature rhythm and (*bottom*) the average amounts in minutes of total sleep (vertical bars) and REM sleep (dark bars) in each 1-hr bed rest episode are double-plotted against the time of day. A clear circadian variation is evident. (After Czeisler, 1978, data from Weitzman et al., 1974.)

Even in subjects maintained on a 3-hour cycle of 2 hours of activity and lights on and 1 hour of bed rest and lights off, these circadian rhythms in sleep duration and organization were maintained (Fig. 5.7). Although 60 minutes was available for sleep, and the subjects after several days of this routine felt fatigued, there was a marked circadian variation in both the fraction of the hour in bed spent asleep and in the amount of REM sleep (Weitzman et al., 1974). Reanalysis of these data has revealed that the percentage of time spent in REM sleep bears the same relationship to the body temperature cycle under these experimental conditions as in the studies conducted in constant environments (Czeisler, Zimmerman, et al., 1980).

These studies, taken together, demonstrate unequivocally that the phase of the circadian rhythm rather than the duration of prior wakefulness is the major determinant of the length of sleep in normal humans. These findings also have major consequences for the design of shift work schedules and for understanding the effects of jet lag, as we will discuss in Chapter 7.

PHYSIOLOGICAL BASIS OF SLEEP

Before we consider the mechanisms responsible for the circadian timing of the sleep–wake cycle, we need to review briefly what is known about the genesis of sleep. Kleitman, in his classic treatise *Sleep and Wakefulness* (1963), pointed out the difficulties in defining the exact anatomical and physiological basis of sleep. The available evidence indicates that the cerebral cortices are not necessary for sleep or wakefulness and that therefore the regulation of sleep and arousal is a subcortical function. Several different loci in the hypothalamus and brainstem have been implicated, but no clear "sleep center" has been defined, and it may not even exist. Instead, sleep appears to be the result of interaction among clusters of nerve cells at several sites within the brain, including the reticular formation, the raphe, and the locus coeruleus (see Fig. 5.8).

The reticular formation, a loosely organized collection of neurons and their interconnections, running longitudinally in the upper brainstem, plays a major role in arousal from sleep. Bremer (1935) reported that transection of the cat's brainstem above the reticular formation at the level of the midbrain resulted in continuous sleep. That there were arousal mechanisms located below this brain level in the reticular formation was eventually proved by studies in which stimulation of the reticular formation led to arousal from sleep (Moruzzi and Magoun, 1949; Lindsley, Bowden, and Magoun, 1949; Lindsley et al., 1950). Lesions destroying the reticular formation led to loss of wakefulness.

The midbrain raphe nuclei, which run through the core of the brainstem from the medulla to the midbrain appear to contribute to the induction of sleep by inhibiting the reticular system. Bremer (1935) showed that transection of the brainstem below the raphe did not interrupt the normal sleep–wake cycle. However, lesions destroying large sections of the raphe inhibited the occurrence of sleep in cats (Jouvet and Renault, 1966). Since that time, Jouvet (1974) has implicated serotonin, a major neurotransmitter of the raphe, in the control of sleep. Depletion of this neurotransmitter leads to an inhibition of

A

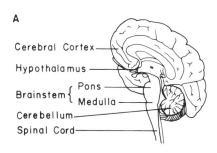

Cerebral Cortex
Hypothalamus
Brainstem { Pons
Medulla
Cerebellum
Spinal Cord

B

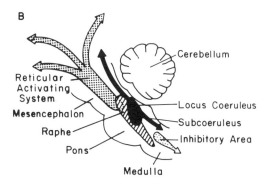

Cerebellum
Reticular Activating System
Mesencephalon
Raphe
Pons
Locus Coeruleus
Subcoeruleus
Inhibitory Area
Medulla

C

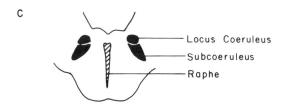

Locus Coeruleus
Subcoeruleus
Raphe

Fig. 5.8 Anatomical sites in the brain involved in the regulation of sleep. (A) Sagittal section through the brain showing the location of the pons, medulla, and hypothalamus. (B) An expanded sagittal schematic view of the brainstem. The cells that seem to be important in initiating and maintaining wakefulness are located mainly in the mesencephalon. Cells that probably play a role in nonREM sleep are located in the raphe nuclei in the pons and the posterior mesencephalon. The nucleus locus coeruleus and the nucleus subcoeruleus have been implicated in the coordination of REM sleep. The medullary inhibitory area is thought to be the common pathway for all motor inhibition. (C) A cross section of the brainstem at the level of the pons, showing the raphe in the midline of the brainstem and the locus coeruleus placed more laterally. (After Dement, 1972.)

sleep in the animal, similar to that resulting from a lesion of the raphe, although sleep recovery eventually occurs after prolonged serotonin depletion (Dement, Mitler, and Henriksen, 1972).

Although serotonergic pathways may induce sleep, adrenergic neural pathways appear to be involved in arousal, especially the pathways from the locus coeruleus, a brainstem nucleus that utilizes norepinephrine as its primary neurotransmitter. Lesions of this nucleus lead to an increased amount of sleep (Jones, Bobillier, and Jouvet, 1969), and stimulation of this area produces arousal (Moruzzi and Magoun, 1949). Inhibition of catecholamine synthesis with alpha-methyl-p-tyrosine also suppresses arousal. There is an important interaction between the locus coeruleus and the raphe; Pujol and colleagues (1973) showed that transections of pathways from the locus coeruleus to the raphe led to increased activity of the raphe and an increase in both REM and non-REM sleep.

Certain areas of the hypothalamus have also been implicated in the regulation of sleep and wakefulness. Von Economo (1929), from studies of the brains of human patients with sleep disorders, and Nauta (1949), from lesion studies in rats, both suggested that a sleep center existed within the hypothalamus. Hess (1957) found that stimulation of the dorsal anterior hypothalamus resulted in sleep, whereas stimulation of the ventral-posterior hypothalamus led to awakening. Stimulation of the preoptic area of the hypothalamus, in particular, has been shown to produce synchronization of the EEG similar to that occurring in slow-wave sleep (Sterman and Clemente, 1962), and lesions of the posterior portions of the hypothalamus have been shown to produce sleep (Naquet, Denavit, and Albe-Fessard, 1966). The mechanisms accounting for the sleep produced by hypothalamic manipulations are, however, poorly understood.

How do changes in the activity of certain brainstem and hypothalamic nuclei produce changes in overall activity of the whole central nervous system? This question remains unanswered. However, some possible mechanisms have been identified. For example, there is evidence that certain small-molecular-weight, sleep-promoting substances accumulate in the brain and cerebrospinal fluid during wakefulness. If these substances are infused into another animal, they will induce sleep (Monnier et al., 1972; Pappenheimer, 1976). It appears that these substances become more concentrated during sleep deprivation.

Their chemical structure and how they act on the central nervous system, however, are not yet understood.

NEURAL CONTROL OF THE SLEEP–WAKE CYCLE

It is possible to separate the mechanisms responsible for the genesis of sleep from those responsible for the timing of sleep within the 24-hour day. As we discussed in Chapter 4, lesions that destroy the suprachiasmatic nuclei (SCN) of the hypothalamus result in a disruption of the circadian sleep–wake cycle. The same total amount of sleep occurs per 24 hours, but the consolidation to any particular phase of the cycle is lost. Ibuka and Kawamura (1977), for example, studied rats before and after SCN lesions were made. Before lesions were placed, the rats' sleep was confined mostly to the daytime hours, but after lesions the distribution of sleep became essentially random throughout day and night, although the total accumulated duration of sleep per 24 hours was virtually unchanged.

The SCN thus cannot be said to produce either sleep or wakefulness; they do, however, influence the distribution of sleep within each 24 hours. But whereas slow-wave sleep appears to be timed by the SCN, the timing of REM sleep may not be. As we have discussed, REM sleep in humans is closely correlated with the body temperature rhythm (Czeisler, Zimmerman, et al., 1980), which, from studies in monkeys, appears to be generated by a pacemaker outside the SCN (Fuller et al., 1981). Furthermore, Stephan and Nunez (1977), Mouret and co-workers (1978), and Yamaoka (1978) have reported the persistence of a circadian rhythm in REM sleep in SCN-lesioned rodents in which the circadian variation of slow-wave sleep was lost. The most plausible model at the current time would thus consist of two separate circadian pacemakers: the SCN, modulating the circadian distribution of nonREM sleep, and the other, as yet undefined, modulating the timing of REM sleep.

GENESIS OF THE REM–NONREM CYCLE

Elegant work by Hobson, McCarley, and Wyzinski (1975) and McCarley and Hobson (1975), using cats, has shown that the REM–nonREM cycle is apparently the result of a periodic interaction between the locus coeruleus and a specific part of the reticular formation known as the gigantocellular tegmental field, or FTG. The cells of the FTG show a buildup in neural activity during REM sleep (McCarley, Hobson, and Pivik, 1972), whereas the nucleus subcoeruleus, a group of cells in the posterior part of the locus coeruleus (Hobson et al., 1973), as well as the raphe (McGinty and Harper, 1976) show an inversely correlated decrease in activity. During slow-wave sleep the distribution of activity is reversed, with increasing activity in the locus coeruleus and companion areas and reduced FTG cell activity.

McCarley and Hobson (1975) proposed that there are excitatory connections within the FTG and excitatory connections from the FTG to the locus coeruleus, but inhibitory connections within the locus coeruleus and from the locus coeruleus to the FTG (Fig. 5.9). Thus, at the start of each REM sleep episode the firing of each giant cell within the FTG initiates activity in the other giant cells, and the feedback increases the firing rate of the original cells. The buildup in FTG activity subsequently starts to increase the activity within the locus coeruleus. However, the locus coeruleus activity eventually inhibits the FTG activity, and slow-wave sleep then occurs. Using the Volterra-Lotka equations, originally developed to describe the periodic buildup of predator and prey populations in a closed system, McCarley and Hobson (1975) have developed a model of the behavior of these two cell populations. This scheme suggests that the ultradian REM–nonREM cycle is not the product of a discrete "pacemaker" nucleus such as the SCN. Instead, it appears to be the product of interreactions between neurons distributed in various areas of the brainstem, none of which by itself acts as the pacemaker.

Feeding

The gathering and ingestion of food requires that an animal be awake and active. Hence, it is not surprising that in most animals feeding is rhythmic and is normally coupled to the circadian sleep–wake and rest–activity cycles. The close correlation between the feeding and activity rhythms was first precisely documented by Richter (1927, 1965). He showed the close temporal correspondence between the locomotor rhythm and the rhythms of ingestive behavior in rats both in entrained and free-running conditions (Fig. 5.10).

However, feeding does not passively follow the rest–activity cycle; indeed, the reverse may sometimes be true. In nature, food may be

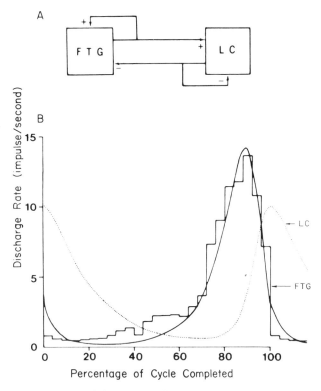

Fig. 5.9 Interactions of the giant cells of the reticular formation (the gigantocellular tegmental field, or FTG) and the locus coeruleus (LC). (A) Model of the proposed coupling and feedback interactions of the FTG and the LC. (B) Plot of the neural activity of the FTG (solid histogram) and the predicted discharge of the LC (dotted line), showing that when one is active the other tends to be inactive. (After McCarley and Hobson, 1975, copyright 1975 by the American Association for the Advancement of Science.)

available only periodically or may be safe to obtain only at a certain time of day when predators are asleep. We have already discussed in Chapter 1 the relationship between the flower clock of Linnaeus (with each flower species opening at a particular time of day) and the *zeitgedachtnis*, or time sense, of bees, which enables them to conserve energy by restricting their forages for pollen to the time of day when a given flower is open. Similarly, mammals have evolved in a world with

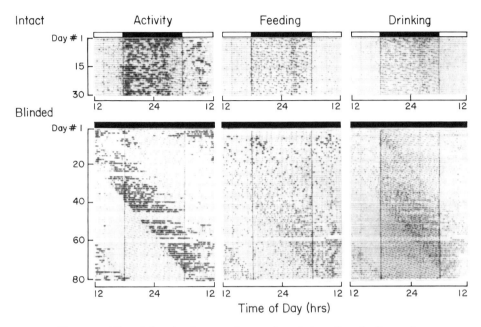

Fig. 5.10 Plot of the simultaneously monitored circadian rhythms of activity, feeding, and drinking in a rat. (*Top*) When entrained to an LD 12:12 cycle, all three rhythms showed similar phase relationships. (*Bottom*) The three rhythms are shown free-running after a rat was blinded, with stable phase relationships similar to those in entrainment. (From Richter, 1965, reprinted by permission of Charles C. Thomas, publisher.)

periodic food availability and periodic variations in the danger of obtaining it. Indeed, the food source and the difficulties in obtaining it are major determinants of whether a species is nocturnal, diurnal, crepuscular, or has no marked circadian rhythmicity in its behavior. For example, although field mice could gather corn at any time of day or night, it is safe to obtain it only when it is too dark for predatory hawks to see them. However, the division of feeding into day and night may vary in extent between species. In the laboratory, rats eat 75% of their food in the dark phase (Zucker, 1971) while squirrel monkeys eat over 90% of their food during the day (Sulzman, Fuller, and Moore-Ede, 1978c).

Anticipation of the time when food will be available may be one of the most important features of circadian feeding rhythms. Anyone who wants to get maximum value from a ticket to the zoo should visit just before feeding time; although they cannot read the clock on the wall, the animals are active and pacing around their cages. In nature, such anticipation is important because it may take some time to travel from the home burrow to the place where food is available. The anticipatory mechanisms serve to fully alert the animal for its foray into the outside world.

The importance of the circadian timing of feeding behaviors is underlined by the fact that food availability is an effective zeitgeber to the circadian system in certain species, as Curt Richter (1922) first demonstrated. As we discussed in Chapter 2, food made available only at a particular time of day will entrain the circadian rhythms of most mammalian species. The synchronization may be only partial if the light–dark cycle provides conflicting time cues (Sulzman et al., 1978) but is complete in the absence of other zeitgebers (Sulzman, Fuller, and Moore-Ede, 1977a, b).

EXOGENOUS INFLUENCES

Although circadian rhythms of feeding persist in constant environments, in nature they are modified by environmental conditions as well as being entrained by zeitgebers. Light, for example, has an inhibitory effect on feeding in nocturnal species, while darkness is inhibitory in diurnal species. Borbely and Huston (1974) followed the feeding rhythm of rats in LD 1:1 cycles and showed that the circadian rhythm persisted, but when the lights were on during the subjective night, there was much more feeding than when the lights were off. We observed analogous results with diurnal primates, as discussed in Chapter 3 (Fig. 3.20). Monkeys maintained in an LD 1:1 cycle showed a free-running feeding rhythm, because this high-frequency LD cycle is far outside the range of entrainment of the circadian system. Superimposed on this endogenous rhythm, however, was the passive response to the LD 1:1 cycle. When the 1-hour periods of darkness fell in the monkeys' subjective day, there was a marked inhibition of spontaneous feeding. Such passive responses may also modify the waveform of the feeding rhythm when the animal is in a regular LD 12:12 cycle, leading to a more square-wave feeding pattern, as compared to the near-sinusoidal pattern seen in constant conditions (Fig. 5.11).

Depriving an animal of food will, of course, modify the rhythm, be-

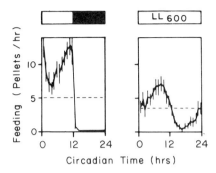

Fig. 5.11 Average waveforms of the feeding rhythm of squirrel monkeys maintained in LD 12:12 (*left*) and LL (*right*) with L = 600 lux. The mean number of pellets/ hr (± SEM) is plotted at hourly intervals based on 30–40 days of data from four monkeys. The horizontal dashed line indicates the average daily food intake in each condition. (After Sulzman, Fuller, and Moore-Ede, 1979a.)

cause the hungry animal will eat whenever food is supplied. However, the endogenous circadian rhythm will resume whenever food becomes continuously available. This feeding rhythm is not determined by the rate of absorption of the nutrient from the gut nor by the rate of its metabolism. Until the beginning of this century it was commonly accepted that the drive for feeding originated in the stomach with spontaneous contractions (hunger pangs). However, individuals with denervated stomachs, who cannot sense these contractions, still eat normally. Furthermore, Sherrington (1900) noted that hunger persists even in a person who does not have a stomach, and Tsang (1938) demonstrated that gastrectomized rats were still readily motivated by food. Similarly, the amount of energy used up between meals is not related to the time of the subsequent meal (Bernstein, 1975). The relationships between meal size and postmeal interval also show no correlation (Le Magnon and Tallon, 1966), suggesting that the rate of utilization of food is not directly related to the timing of eating.

PHYSIOLOGICAL REGULATION

Before we examine how the circadian system might time feeding, we will briefly consider what is known about the physiological mechanisms that regulate feeding behavior. Three mechanisms appear to be involved—neural, nutritional, and alimentary.

With the advent of stereotaxic surgery, it was quickly established (Hetherington and Ranson, 1939) that lesions of the ventromedial hypothalamus produced overeating and obesity. Later, Amand and Brobeck (1951) showed that lesions in the lateral hypothalamus of rats rendered them uninterested in eating or drinking. Additionally, Miller (1957) demonstrated that electrical stimulation of the lateral hypothalamus increased eating and that ventromedial hypothalamus stimulation suppressed eating. It was therefore at first commonly accepted that the lateral hypothalamus contained a feeding or hunger center and that the ventromedial hypothalamus contained a satiety center, which inhibited the feeding center.

However, this neural model is undoubtedly too simple, as the Dahlem Conference on Appetite and Food Intake noted (Silverstone, 1975). The lateral hypothalamus may be involved in general arousal rather than specific food-directed activity. The ventromedial hypothalamus may be somehow connected with lipogenesis, so that rather than acting directly on the lateral hypothalamus, it may act quite indirectly via areas outside the central nervous system.

Such arguments have led to the proposition that the neural feeding centers respond to the body's nutritional needs. It has long been known that a decrease in blood glucose level induces hunger, and this led to the glucostatic theory of feeding (Mayer, 1953). This theory says that when the blood glucose concentration becomes too low, an animal will increase feeding until the glucose concentration returns to normal. Two types of evidence support the theory. First, Drachman and Tapperman (1954) showed that the ventromedial nuclei have the ability to concentrate glucose. They demonstrated that a heavy metal analogue of glucose, goldthioglucose, would concentrate in the ventromedial hypothalamus in rats and thereby damage the tissue and lead to obesity. Second, Amand and colleagues (1964) showed that when blood glucose levels increased, there was an increased rate of firing of the ventromedial hypothalamus and a decreased rate of neural activity in the lateral hypothalamus. These data provide support for the glucostatic model, but there is not a simple relationship between blood glucose and feeding. For example, Van Itallie and Hashim (1960) have reported that when people are fed a low-carbohydrate diet, the swings of blood glucose levels are minimal or absent, but feeding is normal.

There are additional theories to account for hunger. The lipostatic theory was proposed by Kennedy (1953), who argued that feeding is inversely proportional to fat content. In this scheme, some product of fat metabolism can feed back on the lateral hypothalamus to inhibit feeding. Malinkroft and associates (1956) observed an inverse relationship between amino acid levels and food intake and suggested that these nutrients may also be involved in regulating feeding.

The alimentary tract is also involved in the regulation of food intake, but unlike the mechanisms described above, it appears to regulate only the cessation of feeding, not the initiation. Oral receptors monitor the amount of food that passes through the mouth. Thus, because of oral metering an animal with a fistula in the esophagus is relieved of his hunger after eating a normal-size meal, even though all of the food passes out of the body through the fistula and none enters the stomach (Janowitz and Grossman, 1949). Distension of the gastrointestinal tract also inhibits feeding activity. The vagus nerve and the sympathetic nerves of the upper gastrointestinal tract mediate this response. Additionally, simple stretching of the abdominal cavity inhibits food intake.

CIRCADIAN CONTROL

We can now consider how the circadian rhythm of spontaneous feeding is generated. The evidence is incomplete as yet, but it suggests that two separate pacemakers are involved, the suprachiasmatic nuclei (SCN) and another pacemaker located elsewhere in the organism, perhaps more integrally involved in the mechanisms of feeding behavior.

Spontaneous feeding in constant environments (LL, EE) is timed largely by the SCN, a circadian pacemaker that is not an integral part of the food regulatory mechanisms. Destruction of the SCN results in the loss of feeding rhythms in animals provided with food *ad libitum* (Nagai et al., 1978; Rusak and Zucker, 1979). In contrast, lesions in either the ventromedial or lateral hypothalamus do not appear to eliminate the spontaneous feeding rhythm. However, these studies have not been conducted under identical circumstances. For example, Balagura and Davenport (1970), who showed that the feeding rhythm persisted with a reduced amplitude in rats with lesions of the ventromedial hypothalamus, conducted their studies in an LD cycle, so the contribution of passive responses to light and dark could not be assessed.

Rats with lesions in the lateral hypothalamus cease to eat immediately after surgery. If these animals are maintained by tube feeding, they will pass through well-defined stages of recovery and eventually eat spontaneously (Teitelbaum, 1971). Kissileff (1970) has shown that in LD cycles these rats still have a rhythm of feeding. In fact, the amplitude of their feeding rhythm is greater than that of intact controls, because lesioned animals feed in the light portion of the LD cycle less than intact rats do. Rowland (1976a) has examined the feeding rhythm of recovered lateral-hypothalamic-lesioned rats in constant conditions and has shown that these animals also have persisting rhythms in DD, with a much sharper delineation between feeding and nonfeeding times. In constant bright light the rhythm damps out within a few days but is easily reinitiated by LD cycles. These data on lateral hypothalamus lesions indicate that this area is not essential for the generation of feeding rhythms.

However, in environments where food is available only at circadian intervals, a circadian periodicity persists in both intact (Richter, 1922; Boles and Stokes, 1965) and SCN-lesioned animals (Stephan, Swann, and Sisk, 1979a, b; Boulos, Rosenwasser, and Terman, 1980). As we discussed in Chapter 4, the animal is able to anticipate the daily time of food availability even if it has been starved for several days. Thus, anticipatory feeding behaviors appear to be driven by an oscillator outside the SCN that responds to periodic feeding cues. In Chapter 3 (Fig. 3.11) we discussed how food cues are coupled into the circadian system independently of light–dark cycle cues. When an EF cycle is applied with a different period or phase from that of the LD cycle, some of the animal's circadian rhythms entrain to the EF cycle while others entrain to the LD cycle. Thus, this second pacemaker has its own food-mediated coupling to the environment and apparently drives a set of metabolism-oriented rhythms.

As yet, the anatomical locus of this second pacemaker has not been determined. However, there is some evidence implicating the ventromedial nuclei of the hypothalamus (VMH). Recent experiments by Krieger (1980) show that rats with VMH lesions cease to be entrained by EF cycles. Unfortunately, these studies were conducted in LD cycles only, so it is not possible to determine if the VMH is a pacemaker or is only part of the mechanism that transduces the temporal information from LD and EF cycles. There are, however, extensive neural connections between the SCN and the VMH (Oomura et al.,

1979), a prerequisite for pacemakers in a tightly coupled circadian system, and circadian rhythmicity in neural activity has been documented in these nuclei (Koizumi and Nishino, 1976).

Secondary oscillators in peripheral tissues may also be directly coupled to the timing of food intake. Both liver and intestinal cells are involved in the physiological processing of nutrients, and it has been shown that both can generate self-sustained circadian rhythms in tissue culture (Bünning, 1973; Hardeland, 1973). Each of the various physiological processes that process nutrients and regulate food intake, then, may have some endogenous rhythmicity which can be synchronized by periodic food administration, provided the intervals between meals are within the circadian range of entrainment.

From an evolutionary standpoint it does seem advantageous for animals to have evolved a separate oscillator or pacemaker to time feeding behavior. While the SCN keeps other behaviors appropriately synchronized with a precise phase relationship to the light–dark cycle, the timing of food availability in nature may change considerably within the day–night cycle. It would seem advantageous, therefore, to be able to adjust food-seeking behavior independently. This would allow activation of the behaviors specific to food gathering and processing in anticipation of food availability without resetting any unrelated physiological rhythms. A prerequisite for such a pacemaker is that it be entrainable by cycles of food ingestion so that it can be reset whenever food becomes predictably available at a specific time of day.

Drinking

The timing of drinking is closely related to the circadian activity and feeding rhythms. Each have comparable patterns in both entrained and free-running conditions, as was shown in Figure. 5.10. Animals drink only when they are awake, often in association with eating. However, the circadian mechanisms that time drinking behavior can at times act independently from those that time rest–activity and feeding. Hence, it is appropriate to consider the circadian control of drinking separately.

One important difference between the temporal organization of feeding and of drinking is that while food availability can act as an effective time cue for the circadian system, water availability cannot.

Restricting water to a few hours each day will not synchronize the free-running circadian rhythms of squirrel monkeys (Sulzman, Fuller, and Moore-Ede, 1977b) or rats (Moore, 1980). This means that animals are not able to anticipate when water will become available, and their behavior is not organized to enable them to look for water at any given time of day.

EXOGENOUS INFLUENCES

Circadian rhythms in drinking behavior will persist with a free-running period in constant environments whether or not the animal is restrained or is fasting. However, there is evidence that drinking behavior in normal day–night environments is more sensitive to exogenous influences than either activity or feeding rhythms.

Drinking rhythms are more influenced by light intensity than are feeding rhythms. The phasing of the drinking rhythm in an LD 12:12 cycle is thus even more tightly controlled than feeding. Nocturnal rats typically drink 85% of their water during the night, while diurnal squirrel monkeys drink 95% of their water during the day. Zucker (1971) has reported that when nocturnal rodents are transferred from an LD 12:12 cycle to constant illumination (150 lux), there is an immediate and sustained drop in water consumption of more than 20%; eating is only slightly suppressed (6% less food consumed per day) in constant illumination. This photoinhibition of the drinking rhythm is achieved via the retinohypothalamic tract. Stephan and Zucker (1972a) showed that bilateral ablation of the primary optic tracts or the inferior accessory optic tracts in rats did not interrupt the suppressive effects of light on drinking.

The rest–activity cycle influences the timing of drinking because animals cannot drink during sleep. However, activity rhythms do not drive drinking rhythms; they can be dissociated from each other. Using hamsters, Zucker and Stephan (1973) showed that following an inversion of the LD cycle, the activity rhythm resynchronized to the new phase of the LD cycle within six to nine days. The drinking rhythm, on the other hand, took more than two weeks for complete resynchronization.

In rats there is a very close temporal correlation between the rhythms of feeding and drinking, with about 70% of drinking associated with meals. Fitzsimons and LeMagnen (1969) have also shown

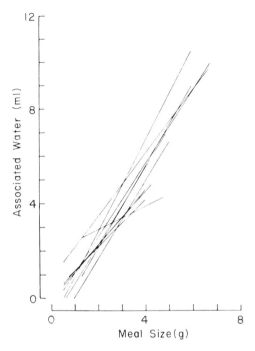

Fig. 5.12 Relationship between the amount of water drunk with a meal and the size of the meal. Each line shows the regression for an individual rat. The left- and right-hand limits of each regression line represent the limits of the size of meals for that rat. (From Fitzsimons and LeMagnen, 1969, reprinted by permission of the authors and publisher, copyright 1969 by the American Psychological Association.)

that there is a highly significant correlation between the amount of water drunk with a meal and the size of the meal (Fig. 5.12). However, drinking rhythms can be separated from those of feeding. In the complete absence of food, rats continue to be substantially nocturnal drinkers (Morrison, 1968). Furthermore, Oatley (1971) has been able to dissociate the circadian drinking and eating pattern by feeding rats identical meals at regular intervals (either 1.0 or 2.4 hours) while they are maintained in an LD 12:12 cycle. Under these conditions the rats ate equal amounts in the light and in the dark period, but most of the drinking still occurred in the dark. The Oatley experiment, however, did not consider the passive effect of light on the drinking behavior, since the animals were maintained in an LD cycle. We have conducted similar experiments with monkeys who were fed identical meals at hourly intervals but in both LD and LL (Sulzman et al., unpublished observations). Even in constant light the feeding and drinking patterns became dissociated, with most of the drinking occurring during the subjective day, again indicating that the drinking rhythm is not generated by the feeding rhythm.

The feeding and drinking rhythms can also be dissociated when rats are transferred from free-running conditions to LD 12:12 cycles. Borbély and Huston (1974) followed the rate of resynchronization of these rhythms to the LD 12:12 cycle and showed that the amplitude of the drinking rhythm was stabilized within 1 to 2 days, while it took more than a week for the amplitude of the feeding rhythm to become stable. However, at least some of this apparent difference in the rate of resynchronization may be caused by the greater sensitivity of the drinking rhythm to the masking effects of light.

PHYSIOLOGICAL REGULATION

Before considering the mechanisms of the circadian rhythm of drinking, we should briefly review what is known about the physiological regulation of water balance. Terrestrial mammals are faced with the constant problem of maintaining body fluids in a relatively dry, gaseous environment. To maintain water balance, they must drink to replenish the water lost by excretion and evaporation.

The fluid in the body is in two compartments: intracellular (that is, the water within the cells) and extracellular (the lymph, the plasma of the blood, and the interstitial fluid that bathes the cells). According to the double depletion hypothesis of Epstein (1973), depletion of either the extracellular or the intracellular compartment will induce drinking. These responses appear to be separately controlled by distinctly different mechanisms.

It has been known for many years that cellular dehydration stimulates thirst (Smith, 1951). If intravenous injections of hypertonic solutions are given which cause the plasma and interstitial fluid to become hypertonic, water moves out of the cells into the extracellular fluid down the osmotic gradient. The drinking response to injection of hypertonic sucrose, for example, is greater than the response to equally hypertonic urea, because sucrose, which does not enter cells, creates a greater osmotic gradient and thus more water movement than does urea, which can enter cells. Wolf (1950) has calculated that cells need to be shrunk by only 1–2% of their original volume for thirst to be triggered.

Cellular dehydration which induces drinking specifically involves the lateral preoptic area of the anterior hypothalamus. Cells in this area appear to be sensitive to their own dehydration. Simple rehydration of these cells, either by water ingestion or infusion of hypotonic fluids locally into the lateral preoptic area, results promptly in cessation of drinking (Fitzsimons, 1961). There is a linear relationship between the degree of osmotic stress stimulating thirst and the rehydration sufficient to suppress thirst. Additionally, electrophysiological studies have characterized the importance of the lateral preoptic area in drinking activity. Bilateral lesions to this region eliminate drinking in response to cellular dehydration (Peck and Novin, 1971; Blass and Epstein, 1971).

Extracellular dehydration is probably more important than intracellular dehydration. First, for proper cardiac function it is necessary to maintain circulatory volume; second, the extracellular fluid compartment is more vulnerable because its volume is less than half the volume of the cellular compartment; and third, the extracellular compartment is the principal area for exchange with the external environment.

From antiquity it has been known that thirst may ensue from extracellular fluid loss caused by hemorrhage, vomiting, or diarrhea. This must be related to extracellular dehydration since the fluid loss is essentially isotonic. The reduction of extracellular fluid volume and especially the reduction in blood volume is detected by receptors located mainly in the atria of the heart (Kappagoda, Linden, and Snow, 1972). These initiate reflex changes in drinking via vagal afferents as well as appropriate renal responses to conserve fluid balance (Epstein, 1978a). Alterations in the circulation that do not affect the total body fluid balance can cause drinking because the volume receptors are located mostly in the atria of the heart. Thus, obstructing the blood flowing toward the atria in the inferior vena cava causes a reduction in venous return and therefore a fall in atrial pressure. Within an hour after such a procedure, rats begin to drink substantial amounts of water. This drinking occurs well before any overall changes in fluid or electrolyte balance have occurred (Fitzsimons, 1964). Conversely, expanding the atrial blood volume by causing a fluid shift from the lower extremities to the thorax results in a reduction of drinking (Kass et al., 1980a).

The conservation of water balance is not the only drive for drinking behavior. The 24-hour water intake of rats with unlimited access to water is about 12 ml per 100 grams of body weight. If, however, the water intake is restricted to 7.5 ml per 100 grams over 24 hours, food intake remains the same, and there is no loss of body weight (Dicker and Nunn, 1957). The main response to this reduction in water intake is reduction in urinary water excretion. Furthermore, intragastric infusion of water suppresses drinking less than would be expected if the need for water was the only cause for drinking. Fitzsimons (1971) showed that uninfused rats drank 3 ml in the light and 16 ml in the dark, but if they were infused with a total of 12 ml of water per 24 hours, drinking declined only to 1.5 ml in the light and 12 ml in the dark. Thus, the rats drank about 50% more than they required.

The stimulus may be partly related to dryness of the mouth. In intact rats most drinking is postprandial, that is, after meals. If rats are surgically desalivated, the drinking pattern becomes prandial; a small amount of food is eaten, followed by a draft of water, then more eating, then drinking (Epstein et al., 1964; Vance, 1965). After recovery from lateral hypothalamus lesions, rats are behaviorally desalivated, in that saliva production is impaired and prandial drinking is evident when the animals eat dry food (Epstein, 1971).

It is interesting that unlike feeding, drinking is apparently not orally metered. Maddison and co-workers (1980) have reported that if the esophagus is cannulated, so that water entering the mouth cannot reach the stomach, monkeys drink excessive amounts of water. Thus, it appears that there are no receptors which directly monitor the amount of fluid drunk. Instead, the regulation of fluid intake begins after the duodenum.

The kidney plays a major role in regulating not only water excretion but also water intake. Linazosoro and colleagues (1954) first suggested that the kidney might play a role in thirst. He observed that nephrectomized rats drank more after kidney extracts were administered. Fitzsimons (1964, 1969) later showed that the kidney secretes renin, an enzyme which catalyzes the formation of a small polypeptide hormone, angiotensin. Several lines of evidence indicate that angiotensin mediates thirst as well as having its earlier discovered pressor effect of raising blood pressure. First, extracts having pressor and thirst actions are found only in the renal cortex; second, it is impossible to separate the drinking and pressor activities of kidney extracts; third, inhibiting the conversion of renin to angiotensin reduces the drinking-inducing and pressor activities of kidney extracts; and fourth, physiological doses of angiotensin injected into rats cause them to drink water.

In rats, injection of only a few picomoles of angiotensin into the hypothalamus causes immediate vigorous drinking that is quite specific and behaviorally normal compared to the behavior of untreated rats. The anatomic site of action of angiotensin-induced drinking is the subfornical organ, which is on the roof of the third ventricle under the fornical commissure. The subfornical organ has no blood–brain barrier and so is well situated to respond to circulating angiotensin levels. Evidence for its involvement includes: (1) it is uniquely sensitive to angiotensin; (2) electrical stimulation produces drinking behavior; (3) angiotensin is unable to elicit drinking when the subfornical organ is ablated; and (4) pharmacological inhibition of the subfornical organ gives specific and reversible inhibition of angiotensin-induced drinking (Epstein, 1973).

CIRCADIAN TIMING

The pacemaker responsible for timing the circadian rhythm of drinking appears to be the SCN, since both in rats (Stephan and Zucker, 1972b) and in monkeys (Fuller et al., 1981a) the drinking rhythm is disrupted by SCN lesions. However, relatively little is known about the coupling of the SCN to the centers which control drinking.

Circadian rhythms have been demonstrated in plasma renin activity and angiotensin concentration (Breuer et al., 1974; von Mayersbach, 1974; Aschoff, 1979). However, they do not appear to be the primary mechanisms mediating the drinking rhythm. Plasma angiotensin peaks during nocturnal sleep in humans (Breuer et al., 1974; Aschoff, 1979a), and this is not the phase relationship we would expect if plasma angiotensin levels were involved in the drinking rhythm. In rodents, although the peak of plasma renin activity does coincide with the maximum drinking behavior at the beginning of the dark phase (von

Mayersbach, 1974), there is evidence that drinking rhythms persist after surgical removal of the kidneys even though the animals can produce no renin (Fitzsimons, 1960).

Whether the lateral hypothalamus mediates the drinking rhythm is also unclear. Damage to the lateral hypothalamus has a more permanent effect on the drinking rhythm than on the feeding rhythm. However, recovered animals do show drinking rhythms that are entrained by LD cycles and also persist in DD (Rowland, 1976b). It is quite possible, though, that the drinking rhythm of an animal with a lateral hypothalamic lesion is being passively driven by the feeding rhythm. Salivary production in these rats is impaired, and drinking provides lubrication for the swallowing of dry food. The recovered animals show a characteristic pattern of interrupting their feeding after every few mouthfuls for a small draft of water. They drink only when they eat and apparently meet their water needs only in this way. It would therefore be most interesting to see whether such animals would show a circadian variation in drinking if fed meals at regular intervals throughout their subjective day and night. If the drinking rhythm did not persist, this would suggest that the lateral hypothalamus is involved in the generation of the autonomous circadian drinking rhythm.

Thermoregulation

Mammals are homeotherms; that is, they can maintain their body temperature at relatively constant levels in varying environmental temperatures. However, body temperature is not entirely constant, and in most mammals it demonstrates a highly regular circadian rhythm. The diurnal variation in body temperature was first recognized over 200 years ago (Hunter, 1778). In the early observations, using the newly invented thermometer, it was noted that body temperature was approximately 1.5° F lower at night when a person was sleeping than when he was awake. It was not until the middle of the next century (Gierse, 1842; Ogle, 1866) that the complete 24-hour body temperature in humans was documented.

The first extensive circadian temperature studies in animals were undertaken by Simpson and Galbraith (1906), who documented the circadian variation in body temperature in monkeys under a variety of

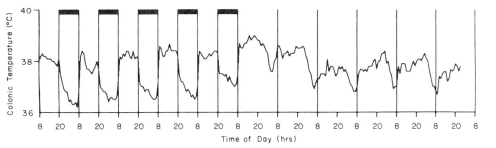

Fig. 5.13 The circadian rhythm of colonic temperature (° C) of a squirrel monkey kept in an LD 12:12 cycle for 5 days and then in constant light (600 lux) for 6 days. The time of darkness is shown by the dark bars at the top. Time of day is plotted in hours EST. (After Fuller, Sulzman, and Moore-Ede, 1979a.)

conditions. As was described in Chapter 1, they demonstrated not only that there was a circadian variation in body temperature but also that the temperature rhythm was entrained by light–dark cycles and would resynchronize after a phase shift. They also demonstrated the persistence of the rhythm in both constant light and constant darkness.

An example of the body temperature rhythm of a squirrel monkey living first in a light–dark cycle (LD 12:12) and then in constant conditions (LL) is shown in Figure 5.13. In these day-active primates, body temperature reaches a daytime plateau of approximately 38.3° C shortly after lights-on and begins to fall at the time of lights-out. Body temperature reaches a minimum of 36.4° C late in the dark phase, then begins to rise again, anticipating the time of light onset. The rhythm persists in LL but shows a free-running period and a somewhat reduced circadian range.

Most mammals have prominent circadian variations in body temperature, including bats (Menaker, 1959), rats (Abrams and Hammel, 1965), camels (Schmidt-Nielson, 1975), primates (Simpson and Galbraith, 1906; Smith and Wekstein, 1969; Crowley et al., 1971, 1972; Fuller, Sulzman, and Moore-Ede, 1979a), and man (Aschoff, 1955, 1970; Colin et al., 1968; Czeisler, 1978). The body temperature rhythms of dogs and cats, on the other hand, are rather obscure and of small amplitude, as are their other rhythmic behaviors (Hawking et al., 1971). The maximum body temperatures normally occur during an animal's daily active phase, and the minimum temperatures during the rest phase. Thus, in nocturnal rodents body temperature peaks at night (Abrams and Hammel, 1965) whereas in the diurnal squirrel monkey it peaks during the day (Fuller, Sulzman, and Moore-Ede, 1979a).

EXOGENOUS INFLUENCES

The endogenous nature of the body temperature rhythm is demonstrated by its persistence in constant conditions with a free-running period and its inability to entrain to day–night cycles with period lengths outside of a circadian range (Kleitman, 1963; Mills et al., 1977; Wever, 1979). This endogenous rhythm is, however, normally modulated by a number of exogenous factors.

Light, for example, has a direct effect on body temperature, an effect that is quite separate from the normal phase control by the light–dark cycle. Figure 5.14A illustrates the average body temperature rhythm in monkeys maintained in either an LD 12:12 cycle or in a higher-frequency cycle (LD 2:2). It can be seen that whenever the lights (600 lux) are switched on, body temperature becomes elevated, and when the lights are switched off, temperature falls. Further, the effect of light is much more marked during the animals' subjective night. During subjective night, 600 lux increases body temperature by more than 1° C, but during subjective day the change may be as small as 0.2° C.

These results help explain the reduced amplitude of the body temperature rhythm in constant bright light (LL, 600 lux; Fuller, Sulzman, and Moore-Ede, 1979a). Figure 5.14B shows that the rhythm in LL 600 lux followed the upper boundary of the range of body temperatures seen during the LD 2:2 cycle (Fig. 5.14A). In other words, when light of 600 lux intensity was applied during the subjective night in the LD 2:2 regimen, body temperature rose to the same level as shown at that phase while free-running in constant illumination of the same intensity. This indicates that in an LD 12:12 (600:<1 lux) cycle half the amplitude of the circadian body temperature rhythm is due to the endogenous rhythm and half is due to the exogenous masking effects of the light–dark cycle. Because of the masking effect of light, body temperature has a lower mean level in LL 60 lux than in LL 600 lux, with the greatest differences occurring during subjective night (Fig. 5.14B).

Ambient temperature has only a small effect on diurnal variation in the core body temperature. Changes in ambient temperature do pose

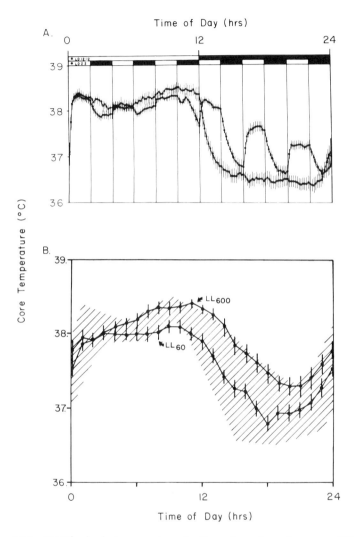

Fig. 5.14 (A) The body temperature rhythm of monkeys (mean ± SEM) in an LD 12:12 cycle (squares) and an LD 2:2 cycle (circles). Light during the subjective day has little effect on temperature, but during the subjective night light produces a large increase in body temperature. (B) Temperature patterns of monkeys maintained in constant light at 600 lux and 60 lux. The stippled area shows the range of temperatures from (A). Part of the range of the body temperature rhythm in LD cycles can be accounted for by these passive responses to environmental illumination. (From Fuller, Sulzman, and Moore-Ede, 1981.)

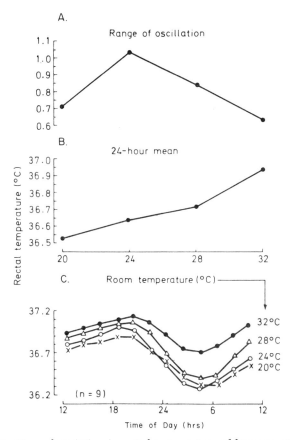

Fig. 5.15 Diurnal variation in rectal temperature of humans at four different ambient temperatures. The curves are the averages of data from nine male subjects. (A) The range of the oscillation, and (B) the 24-hr mean values as a function of ambient temperature. These data are calculated from (C), the mean 24-hr colonic temperature rhythms at each ambient temperature. (From Aschoff et al., 1974.)

significant challenges to the thermoregulatory system and cause major changes in the strategies for maintaining body temperature, by changing levels of heat production and/or heat loss. However, the circadian rhythm in body temperature appears to be generated by periodic variations in body heat loss and heat production, whether the ambient temperature is cold, neutral, or warm (Aschoff and Heise, 1972; Aschoff et al., 1974; Fuller, Sulzman, and Moore-Ede, 1980). Figure 5.15

shows the response of the body temperature rhythm in humans when ambient temperatures were raised from 20° C to 32° C (Aschoff et al., 1974). Although the mean level of body temperature was elevated, the rhythm showed only a small change in amplitude, mostly caused by alteration in the nighttime level (as also occurred with the exogenous effects of light). Thus, the effects of ambient temperature on the body temperature rhythm are minimal, especially when compared to the much more pronounced effects of light intensity.

Food intake has long been recognized as having a specific dynamic action, leading to an increase in metabolism and heat production, which in turn can lead to an increase in body temperature (Kleiber, 1975). Although body temperature typically is elevated at the phase when feeding occurs, the thermodynamic action of food does not play a major role in the normal diurnal temperature variation. Fasting subjects, for example, will continue to show circadian rhythms in body temperature (Fig. 5.16A; Wever, 1979). However, the passive response to feeding shows some variability among species. In the rat, for example, the effects may be relatively large, although the increases in body temperature caused by feeding are smaller during subjective night than during subjective day (Heusner, 1956; Honma and Hiroshige, 1978). In contrast, squirrel monkeys, when daily food intake is confined to a three-hour interval (either early or late in the subjective day), show only a very small temperature increase (0.1–0.2° C) in response to the once-a-day meal (Sulzman et al., 1978).

Although vigorous muscular activity can elevate body temperature, the normal rest–activity cycle has relatively little influence on the temperature rhythm. This was convincingly demonstrated more than 100 years ago by Juergensen (1873), who studied the human body temperature rhythm during 48 hours of continuous bed rest with the subject either eating normal meals or completely fasting (Fig. 5.16A). The response to exercise varies markedly with time of day (Kanecko et al., 1968; Wenger et al., 1976). Exercise typically raises body temperature looo during the day than at night. Moreover, the level at which thermoregulatory responses occur (that is, the temperature at which sweating begins) also is lower at night than during the day, even if the subjects are always fully aroused and dressed and have showered before exercising.

The body temperature rhythm can also persist independently of the sleep–wake cycle, as Patrick and Gilbert first reported in 1896. Figure

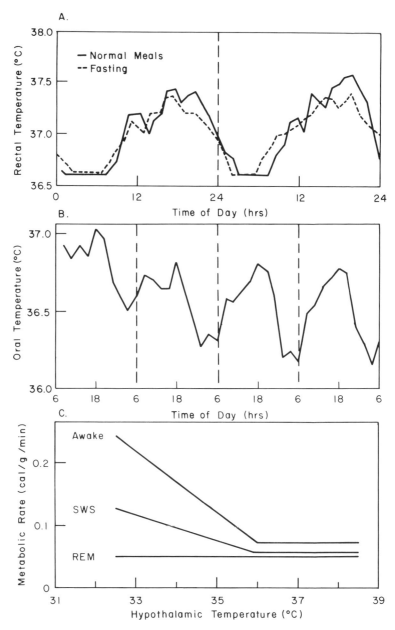

Fig. 5.16 Persistence of the body temperature rhythm during (A) continuous bed rest and normal meals (solid line) or fasting for 48 hours (dashed line), and (B) 98 hours of sleep deprivation (mean data from 15 subjects). (C) Response of metabolic heat production to reductions in hypothalamic temperature in a kangaroo rat during wakefulness, slow-wave sleep (SWS), and REM sleep (REM). The thermoregulatory response is impaired especially during REM sleep, but this is not a major contributor to the observed circadian rhythm of body temperature (A from Wever, 1979, data from Juergensen, 1873; B from Kleitman, 1963, reprinted by permission of the University of Chicago Press, copyright 1963, The University of Chicago; C from Glotzbach and Heller, 1976, copyright 1976 by the American Association for the Advancement of Science.)

5.16B shows the mean body temperature rhythm of 15 subjects who were kept awake for 96 hours (Murray, Williams, and Lubin, 1958). However, thermoregulation is influenced by sleep and different sleep states (Fig. 5.16C). Although a drop in brain temperature will induce an increase in heat production in awake animals, the response is reduced in animals during slow-wave sleep and markedly compromised in REM sleep (Glotzbach and Heller, 1976).

To summarize, then, the circadian rhythm of core body temperature is generated by an endogenous internal timekeeping system and is not simply a passive response to the environment or the behavior of the animal. However, these exogenous factors can modify the expression of the body temperature rhythm, so it is essential for the researcher to define both the environmental conditions and the animal's activities at the time of temperature measurement.

PHYSIOLOGICAL REGULATION

Before considering temporal control over body temperature rhythms, we should briefly review the mechanisms of mammalian temperature regulation. The temperature at any site in the body is a function of the heat content of that specific site. Changes in temperature over time reflect modifications in heat content because of heat either entering or leaving the site. In general, the thermoregulatory system maintains a relatively stable temperature of the body core, but to do this, mammals have evolved a large repertoire of effector mechanisms for altering the heat content of the body. Some mechanisms increase the level of heat production by causing shivering in skeletal muscle or by increasing the metabolic rate in other body tissues (non-shivering thermogenesis). Other mechanisms increase evaporative heat loss via sweating and/or panting or increase convective and radiative heat loss via modifications of skin temperature. Changes in behaviors, such as postural adjustments or wetting the fur during grooming, can also modify rates of heat loss.

These effectors are coordinated by the thermoregulatory system so as to maintain the body's optimal temperature level or heat content. This is accomplished by a control system involving negative feedback; the body temperatures sensed by thermoreceptors are, in effect, compared with reference levels and are modified to bring the sensed temperatures to the appropriate level. Changes in body temperature are detected at many diverse sites, including the skin, spinal cord, hypothalamus, and gut. This information is transmitted to the posterior hypothalamus for processing (Hammel et al., 1963; Hardy, 1973; Adair, 1974; Horowitz, Fuller, and Horwitz, 1976). Then the appropriate heat production and/or heat loss effectors produce any necessary temperature changes. For example, if core temperature is sensed as being too low, heat loss will be reduced. Blood will be shunted away from the skin, sweating will cease, and the animal might make postural adjustments, such as curling up into a ball, to reduce surface area. At the same time, heat production mechanisms will be activated; the animal will start to shiver, and nonshivering thermogenesis will also increase.

The effectors that the animal uses for thermoregulation are a function of the ambient temperature (Kleiber, 1975). When an animal is in a cold environment, heat loss mechanisms operate at a minimal level, and body temperature is regulated solely by changing the level of heat production. Conversely, if the animal is in a warm or thermally neutral environment, the metabolic level is maintained at a minimum, and heat loss mechanisms (first dry, then evaporative) maintain body temperature.

CIRCADIAN TIMING

The timing of both heat loss and heat production contributes to the circadian rhythm of core body temperature. The fall in temperature in each cycle is caused in part by increased skin blood flow and therefore increased loss of heat by the body (Bloch, 1964; Smith, 1969; Aschoff and Heise, 1972; Gauthrie, 1973; Fuller, Sulzman, and Moore-Ede, 1979a). Similarly, a rhythm in metabolism or heat production has been demonstrated in several species of mammals (Kreider, Buskirk, and Bass, 1958; Kreider and Iampietro, 1959; Aschoff and Pohl, 1970; Fuller, Sulzman, and Moore-Ede, 1979a). In addition, behavioral changes at certain phases of the cycle may contribute to control of the body temperature rhythm (Ingram, Walters, and Legge, 1975; Cabanac et al., 1976; De Castro, 1978).

The interaction of heat loss and heat production in generating the circadian rhythm of body temperature is illustrated by the data from squirrel monkeys in Figure 5.17. The fall in body temperature at the end of the subjective day is a combination of a fall in heat production (as estimated by oxygen consumption), starting before the onset of

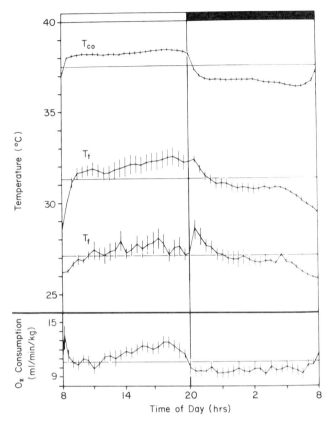

Fig. 5.17 Temporal relationships of some thermoregulatory rhythms which contribute to the circadian variation in core body temperature (T_{co}) in squirrel monkeys. Tail temperature (T_t); foot temperature (T_f), and oxygen consumption ($\dot{V}_{O_2}$) (mean ± SEM) are shown from monkeys maintained in an LD 12:12 cycle at approximately 26° C. The morning rise in T_{co} is achieved by an increase in $\dot{V}_{O_2}$ and reductions in T_t and T_f, whereas the evening fall is mediated by decreases in V_{O_2} and increases in skin temperatures. (Copyright 1982 by Moore-Ede, Sulzman, and Fuller.)

darkness, and a sharp rise in heat loss (indicated by the increase in foot skin temperature) immediately after the onset of dark. Similarly, the increase in core body temperature in anticipation of lights-on is accompanied by a fall in skin temperature and heat loss at the same time as a sharp rise in oxygen consumption and heat production occurs.

Furthermore, the skin temperature stays low until the daytime plateau in body temperature has been reached.

There is evidence that heat production and vasomotor heat loss may be timed by separate oscillators within the circadian system. In squirrel monkeys studied in the absence of environmental time cues, the rhythms of both core body temperature and skin temperature persist with free-running periods. However, as opposed to the very precise temporal relationship between heat loss and core body temperature during light–dark entrainment, these two rhythms showed highly varied phase relationships in constant conditions (Fuller, Sulzman, and Moore-Ede, 1979a). Figure 5.18, a phase plot of the colonic and tail temperature rhythms of four squirrel monkeys studied in LL, shows the wide variations in phase relationships. Furthermore, monkeys free-running in constant conditions may occasionally develop an ultradian ("split") rhythm in the core body temperature while continuing to have a circadian rhythm in skin temperature. These rhythms can also on occasion become internally desynchronized in humans studied in temporal isolation. The skin temperature rhythm follows the sleep–wake cycle, whereas the core body temperature rhythm shows an independent periodicity (Czeisler et al., 1977; Czeisler, 1978).

The circadian rhythm of core body temperature appears to be driven predominantly by a pacemaker located outside the SCN, as we have discussed in Chapter 4. Presumably, the pacemaker mainly influences the timing of heat production, since the circadian variation in skin temperature (and therefore heat loss) may at times oscillate independently of the core body temperature rhythm. Because the skin temperature rhythm maintains a consistent relationship with the sleep–wake cycle, this suggests that heat loss rhythmicity is predominantly controlled by the SCN.

Temperature regulation thus appears to be influenced by both of the major pacemakers of the circadian timing system. It is possible that the two pacemakers act at different sites, because the components of the thermoregulatory system are separately located. For example, in rats nonshivering thermogenesis is controlled by neurons in the spinal cord (Fuller, Horowitz, and Horwitz, 1977), while behavioral thermoregulatory responses are controlled by the lateral hypothalamus (Satinoff, 1978). It may be that the secondary circadian oscillators are actually located in the separate thermoregulatory subsystems, as we

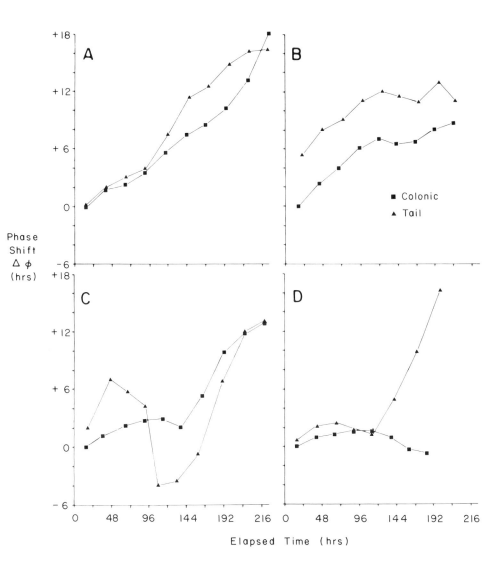

Fig. 5.18 Phase plots of colonic (squares) and tail (triangles) temperatures of four squirrel monkeys (A, B, C, D) free-running in LL. The phase of the colonic temperature maximum on day 1 was normalized to 0 hr. The change in phase over the course of the experiment is plotted versus elapsed experimental time. It is evident that the phase relationships between the two temperatures can vary considerably. (From Fuller, Sulzman, and Moore-Ede, 1979a.)

discussed in Chapter 3. Much needs to be done, however, both in determining the anatomical location of the pacemaker(s) which drive the core body temperature rhythm and in determining the mechanisms by which these pacemakers regulate thermoregulatory responses.

Endocrine Rhythms

Hormones transmit information from one body site to another. Transported in the bloodstream, the hormone secreted by an endocrine gland will activate biochemical processes in the various "target" cells in the body which have specific receptors for that hormone. Not surprisingly, the circadian timing system uses circadian rhythms in plasma hormone levels to transmit temporal information and ensure the internal synchronization of the multitude of rhythmic functions in the various body tissues. Most hormones have circadian rhythms in plasma concentration with a peak at one characteristic time of day and a trough at another.

Circadian rhythms have now been documented in a wide variety of hormones. However, for technical reasons, up till the last 10 years many of these rhythms had not been precisely characterized. Before the development of radioimmunoassay by Berson and Yalow in the mid-1950s (see Yalow, 1978, for review), hormones had to be measured either by bioassay, which assessed the response of biological tissue or whole animal to a plasma sample, or by chemical analyses (Pincus, 1943). Bioassays required large samples and were notoriously hard to standardize; chemical assays were more standardized but were still less sensitive than radioimmunoassays. Modern radioimmunoassay techniques now require only a few microliters or even picoliters of plasma for the quantitative determination of hormone concentration.

The development of highly accurate and specific radioimmunoassays using very small samples of blood plasma has enabled investigators to overcome another hurdle in determining hormone circadian rhythms. As Weitzman and co-workers (1966, 1971; Hellman et al., 1970) demonstrated, many hormones are secreted with episodic patterns. Rather than releasing the hormone into the bloodstream continuously, most endocrine glands appear to secrete only intermittently; the plasma concentration of the hormone typically consists of a series of peaks, during secretory episodes, and troughs as the hormone is metabolized with no further secretion. Because the half-life of hormones varies from a few minutes to an hour or so, low plasma levels may be reached between secretory episodes, which accounts for the highly varied values observed when samples were taken infrequently (say every 3 to 4 hours). Nowadays it is recognized that the frequent sampling technique introduced by Weitzman and colleagues (1966) is required to determine the true periodic temporal pattern of a hormone. Episodes of secretion typically occur every 60–120 minutes, and thus the pulsatile pattern is often considered to have an ultradian (less than circadian) rhythm.

The circadian rhythmicity of many hormones is now well characterized (Aschoff, 1979). Figure 5.19 shows the rhythms of cortisol, growth hormone, aldosterone, prolactin, testosterone, thyrotropin, luteinizing hormone (LH), and follicle-stimulating hormone (FSH), averaged from studies by many different investigators in which human subjects were living on a regular 24-hour day–night cycle and the interval between measurements did not exceed one hour. Despite somewhat varied experimental conditions, the hormone rhythms documented by these various laboratories are sufficiently similar for a standardized "normal" rhythm for each hormone to be recognized. Rhythms have been categorized for a wide variety of other hormones, including renin (Bartter, Chan, and Simpson, 1979), insulin (Jarrett, 1979), melatonin (Klein, 1979) and a number of gonadal steroids such as pregnenolone and progesterone (Rebar and Yen, 1979).

Figure 5.19 shows clearly the differences among hormone rhythms in phase and amplitude. Some, such as cortisol and growth hormone, have high-amplitude rhythms with maximum deviations of nearly ± 100% from the 24-hour mean. Others, such as aldosterone and prolactin, show approximately ± 50% deviation from the mean, while testosterone and thyrotropin have rhythms with maximum deviations of no more than 20%. In contrast, LH and FSH, the reproductive hormones released by the pituitary, show little or no circadian variation in the male. The predominant periodicity of LH and FSH in female mammals is instead estral or menstrual, but even the timing of the preovulatory LH surge is under the influence of the circadian system.

Similarly, there are marked differences in the phase relationships of these rhythms to the day–night cycle. Growth hormone typically reaches a maximum during the first two hours of sleep, a time when

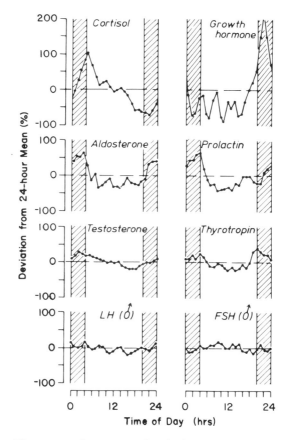

Fig. 5.19 The temporal patterns of eight hormones in human plasma, showing mean values from studies in which at least hourly samples were obtained. The shaded areas represent approximate times of sleep. (From Aschoff, 1979a, reprinted by permission of Raven Press.)

plasma cortisol concentration is at its minimum. Cortisol rises to a maximum at or just before the time of waking, as do aldosterone and prolactin.

The episodic pulsatile pattern of secretion of the hormones cannot be fully appreciated from an examination of Figure 5.19. Because the episodic patterns are not precisely synchronized to the 24-hour day–night cycle, a given episode does not occur at a precise time of day, so the patterns are obscured in the averaging of the 24-hour

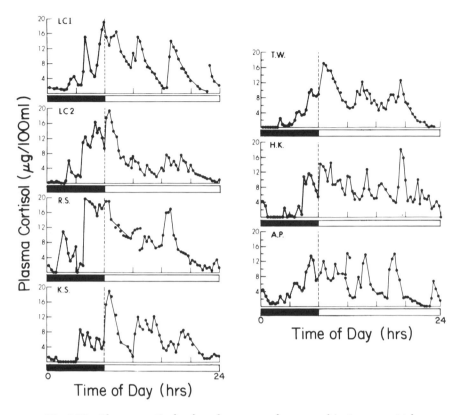

Fig. 5.20 Plasma cortisol values from seven human subjects over a 24-hr period, with samples obtained every 20 min. The dark bar indicates time of bed rest, which was provided at the subject's usual nocturnal sleep time. (From Weitzman and Hellman, 1974, reprinted by permission of John Wiley and Sons.)

rhythm (Weitzman and Hellman, 1974; Kronauer, Moore-Ede, and Menser, 1978); however, the circadian component, with its tight synchronization to the day–night cycle, is readily visible. With hormones such as LH, the episodic pattern is more prominent than the circadian. To determine the episodic nature of most hormones, however, it is necessary to examine the patterns of individual subjects, as shown in Figure 5.20.

The circadian hormonal rhythms of other mammalian species differ considerably in amplitude and in the strength of the episodic ultradian component. Furthermore, there are major differences between nocturnal and diurnal species in the phasing of circadian rhythms with respect to the light–dark cycle. Plasma corticosteroids, for example, peak just before the onset of dark in the rat (Krieger, 1979b) and just before the onset of light in man. However, the phase relationship to the daily onset of activity is much the same in both species (Gibbs and Van Brunt, 1975; Czeisler et al., 1976).

Because of Krieger's (1979a) comprehensive volume on endocrine rhythms, it is unnecessary in this chapter to discuss systematically the temporal control of every hormone. Instead, we will take one hormonal rhythm, that of plasma corticosteroid concentration, and consider in some detail how it is generated and controlled. Probably more is known about the corticosteroid rhythms than any other hormonal circadian rhythm, so it should serve to illustrate the temporal order in the body's endocrine systems.

CIRCADIAN RHYTHMS IN PLASMA CORTICOSTEROIDS

The first indication of circadian rhythmicity in adrenal function was provided by Pincus (1943), who showed a daily pattern of 17-keto-steroid excretion in the urine of healthy adults, with significantly higher amounts present in the morning and lower amounts in the evening. Once 17-hydroxycorticosteroids could be directly measured, a circadian rhythm of plasma corticosteroids was soon recognized (Migeon et al., 1956). With the advent of radioimmunoassays, the specific corticosteroids such as cortisol and corticosterone could be assayed, and species differences in their preponderance noted. In humans there are prominent circadian rhythms of plasma cortisol concentration. The levels of corticosterone are much lower, but the rhythm is similarly timed, so that an approximately 13:1 ratio of cortisol to corticosterone is maintained throughout each 24 hours (Peterson, 1957). In rats, corticosterone is the predominant adrenal steroid. However, because these animals are nocturnal, the phase relationship with the light–dark cycle is 180° different, so that it peaks at the end of the daily photoperiod. In all species the maximum of the corticosteroid rhythm occurs just before or at the onset of activity.

Figure 5.20 shows the individual patterns of plasma cortisol concentration in seven normal human subjects. Both circadian and episodic components are readily visible because the plasma samples were taken at 20-minute intervals from an indwelling catheter during both sleep and wakefulness. Plasma cortisol concentration starts to climb from baseline levels about four to five hours prior to the time of waking, reaching a peak of 15–20 μg/100 ml at or near the time of waking. Over the course of the day it falls, reaching low or undetectable levels an hour or two before bedtime. Six to nine discrete episodes of secretion occur per day, each inducing a rapid rise in cortisol concentration.

That the episodic pattern was a result of secretory episodes was confirmed by Hellman and colleagues (1970), who showed that radioactively labeled cortisol was diluted when it was injected intravenously just before the cortisol pulse. Between secretory episodes, no further dilution of labeled cortisol was observed, and the concentration fell at the rate predicted from the half-life of cortisol in plasma. Calculations by Weitzman and co-workers (1971) have shown that active secretion occurs during only 25% of each 24 hours and that it is the clustering of secretory episodes in the late night and the early half of the day which accounts for the circadian rhythmicity in plasma cortisol concentration.

Not only is there circadian rhythmicity in baseline corticosteroid levels, there is also a rhythmicity in the system's responsiveness to external challenges. A circadian rhythm in the response of plasma corticosterone in rats briefly exposed to ether vapor, immobilization, or novelty stress has been documented, showing that responses are smallest when the stress is administered at the peak of the plasma corticosteroid rhythm and greatest when plasma corticosteroid concentration is low (Zimmerman and Critchlow, 1967; Dunn, Scheving, and Millet, 1972). Similar rhythms of responsiveness have been demonstrated in humans; the greatest effects of pyrogen injection are during the evening when plasma cortisol concentrations are low (Takebe, Setaishi, and Hirami, 1966).

The plasma cortisol rhythm is not present in the newborn, but develops with age. In the rat a significant circadian variation in plasma corticosterone levels is seen by three weeks of age (Hiroshige and Sato, 1970); in humans, the rhythm is not fully established until two to three years (Franks, 1967).

Exogenous influences   The plasma corticosteroid rhythm is normally tightly synchronized to the day–night and sleep–wake cycles, but it is surprisingly little influenced by day-to-day behavioral and environmental changes. If a human subject is maintained on constant bed rest (Katz, 1964; Vernikos-Danellis et al., 1972) or is deprived of sleep for one or two nights (Halberg et al., 1961) or fed identical amounts of a liquid diet at hourly intervals (Erlich, Weitzman, and McGregor, 1974), the circadian as well as the ultradian rhythmicity of plasma cortisol concentration persist.

The independence of the plasma cortisol rhythm is most dramatically illustrated by experiments in which the bed rest–activity cycle is completely reversed. While sleep–wake cycles adjust within approximately a week, it takes one to three weeks for the circadian rhythm of plasma cortisol to invert (Martel et al., 1962; Weitzman et al., 1968). Internal desynchronization of the sleep–wake and cortisol rhythms can also occur spontaneously in subjects isolated from time cues, with the plasma cortisol rhythm always displaying a period close to 25 hours (Czeisler et al., 1980a). Even if a subject maintains a 3-hour sleep–wake cycle, with 2 hours of activity and 1 hour for sleep in a repeating cycle, an essentially normal circadian rhythm of plasma cortisol is observed (Weitzman et al., 1974). In these various experiments, posture, activity, sleep, and wakefulness can all be separated from the plasma cortisol rhythm; therefore, it cannot be directly dependent on any of them.

Although the light–dark cycle is an important time cue, which normally synchronizes the plasma cortisol rhythm to the 24-hour day–night cycle, the rhythm is not passively driven by light and dark. The circadian and episodic cortisol secretory patterns of humans who have been totally blind from birth are similar to those of normal subjects (Weitzman et al., 1972), although some blind people have free-running rhythms, and hence their cortisol patterns are not synchronized to the 24-hour day–night cycle.

Meals are an important time cue but not a direct determinant of the plasma corticosteroid rhythm in many species. The peak of the corticosteroid rhythm typically anticipates the daily initiation of feeding. Food availability may sometimes be a stronger zeitgeber than the light–dark cycle, as Krieger (1974) has shown in rodents fed at different phase relationships to an LD cycle (Fig. 5.21). Not all investigators find such entrainment by feeding cycles (Haus, 1976), but this has been partly explained by Takahashi, Hanada, and Takahashi (1979), who showed that it takes more than a week for the corticosteroid rhythm in

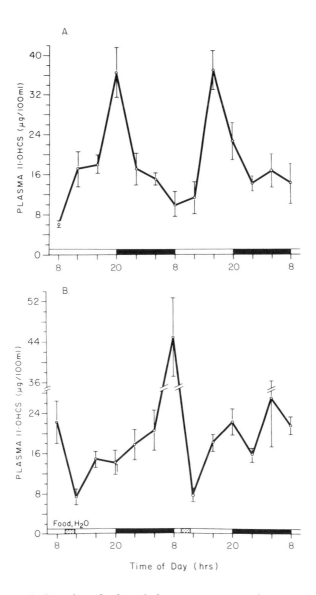

Time of Day (hrs)

Fig. 5.21 (*A*) Circadian rhythm of plasma corticosteroids (mean ± SEM) in female rats with food continuously available. Blood was obtained by tail vein sampling over 48 hr of rats in an LD 12:12 cycle (dark bars indicate dark period). The peak of plasma corticosteroid concentration occurred at the beginning of darkness, just before the time of maximum food consumption. (*B*) Corticosteroid periodicity of the same rats after 15 days during which food and water consumption was restricted to 2 hours each day between 0930 and 1130 hrs (stippled bar). The corticosteroid rhythm shifted with respect to the LD cycle so that it again anticipated the time of feeding. (After Krieger, 1974.)

rats to phase shift after introduction of a schedule of restricted food intake. This is to be expected, especially if the LD cycle provides conflicting temporal information and retards the rate of entrainment of the adrenal rhythms. It remains to be determined how much of the entrainment by light–dark cycles is mediated by the concurrent rhythmicity in *ad lib* feeding.

Sleep per se does have some direct influence on the cortisol pattern, since the first two or three hours of sleep are associated with a suppression of cortisol secretion (Weitzman, Czeisler, and Moore-Ede, 1979). This suppression occurs at the onset of sleep, no matter what the phase of the cortisol rhythm. Normally this is the time of day when the cortisol rhythm reaches its lowest point. However, when the cortisol rhythm and sleep–wake cycle become desynchronized, the sleep-onset suppression of cortisol becomes readily visible, since sleep can occur at any phase of the cortisol rhythm. This passive response or masking effect may account for the claims that the plasma cortisol rhythm follows the rest–activity cycle when subjects are exposed to day lengths outside the normal circadian range of entrainment (Orth, Island, and Liddle, 1967). Presumably in these cases the sleep-onset suppression contributed a secondary periodicity to the plasma cortisol rhythm. However, careful analysis has shown that even after three weeks of exposure, the major circadian component of corticosteroid rhythms did not become synchronized to a 21-hour day–night cycle to which the sleep–wake cycle became entrained (Simpson, Lobban, and Halberg, 1971).

Stress is a well-known stimulus for corticosteroid secretion. At any time of day a sufficiently stressful stimulus will induce an increase in corticosteroid secretion, but, as we have discussed, the magnitude of the response depends on the time of day (Zimmerman and Critchlow, 1967; Dunn, Scheving, and Millet, 1972). However, the level of stress required to induce cortisol secretion may be quite high. Patients on the day before open cardiac surgery may show considerable anxiety, yet most of the time their episodic pattern and circadian rhythm of cortisol secretion appear indistinguishable from those of normal subjects (Czeisler et al., 1976). However, specific stressors, such as preoperative shaving, can induce an episode of cortisol secretion that is superimposed on the basic circadian rhythm (Fig. 5.22). Our daily exposure to stresses cannot account for either the normal circadian rhythm of

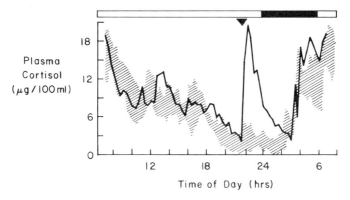

Plasma
Cortisol
(μg / 100 ml)

Time of Day (hrs)

Fig. 5.22 24-hr plasma cortisol concentration pattern of a patient on the day prior to elective coronary artery bypass surgery, superimposed over the mean pattern ($\pm$ SEM) of control subjects not anticipating surgery (shaded area). The time of mean sleep onset in the prior week was used as the common time reference; mean sleep time for the patient is indicated by the black bar. The time of body shaving on the evening before surgery is indicated by the arrowhead. A major cortisol secretory episode occurred at the time of the presurgical shave, but the circadian rhythm was otherwise unchanged. (From Czeisler et al., 1976.)

plasma cortisol concentration nor the pattern of episodic secretory episodes. Most cortisol is in fact secreted in the latter half of the night, when we are normally fast asleep.

Physiological basis The rhythm in plasma concentration of any hormone is determined by a number of factors. The rate of synthesis and secretion, the volume of the body fluid compartments in which it is diluted, the binding (if any) to plasma proteins, the rate of metabolism into conjugated forms, and the rate of excretion may each be rhythmic and contribute to the observed circadian rhythm in plasma concentration. These factors are summarized in Figure 5.23.

There is circadian variation in the rates of both synthesis and secretion. Cortisol and corticosterone, the principal corticosteroids of mammals, are steroid hormones synthesized in the cortex of the adrenal gland in a series of steps starting with cholesterol. The most direct evidence for a rhythm in corticosteroid secretion is provided by experiments in which adrenal glands cultured *in vitro* for several consecutive days have shown circadian rhythmicity in corticosteroid secretion

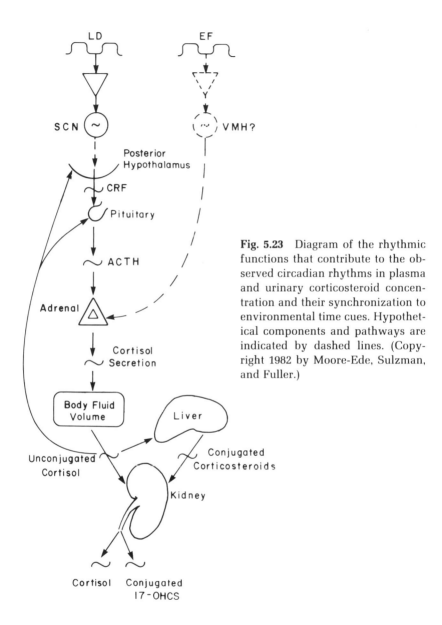

Fig. 5.23 Diagram of the rhythmic functions that contribute to the observed circadian rhythms in plasma and urinary corticosteroid concentration and their synchronization to environmental time cues. Hypothetical components and pathways are indicated by dashed lines. (Copyright 1982 by Moore-Ede, Sulzman, and Fuller.)

(Andrews and Folk, 1964; Andrews, 1968; Shiotsuka, Jovonovich, and Jovonovich, 1974). Since the content of corticosteroids in the adrenal gland has a prominent circadian variation (Ungar, 1967), which reaches a peak at approximately the same time as plasma corticosteroid concentration, this suggests that the synthesis of corticosteroids is probably also rhythmic. However, this has not yet been verified *in vivo*.

Both blood volume (Cranston and Brown, 1963) and extracellular volume (Moore-Ede, Brennan, and Ball, 1975) have a circadian variation, with a maximum during daytime hours, thus resulting in maximum dilution of plasma corticosteroid concentration at this time. The range of the circadian variation in blood volume is only ± 10%, so this cannot be a major determinant of the corticosteroid concentration rhythm. Circadian variation in binding of cortisol to plasma proteins, such as transcortin, does not appear to contribute to the observed circadian variation in free cortisol concentration (De Moor et al., 1962).

The metabolism and conjugation of corticosteroids also influence the circadian rhythm in plasma concentration. Studies examining the fate of intravenously injected radioactive cortisol ^{14}C in humans have shown a circadian variation in the metabolic clearance rate of cortisol from the plasma (De Lacerda, Kowarski, and Migeon, 1973). However, since clearance is greatest at the peak of the plasma rhythm, the clearance rhythm, if anything, damps the plasma rhythm. A circadian variation in the activity of the liver enzymes which degrade cortisol has been demonstrated in rats (Marotta et al., 1975; Graef and Golf, 1979). The liver degrades and conjugates cortisol into its metabolites tetrahydrocortisone and tetrahydrocortisol, each of which has circadian variation in plasma (Gordon et al., 1968). This rhythm of conjugated plasma corticosteroids phase-lags the free plasma corticosteroid rhythm by two to four hours (Brown et al., 1957; Martin, Mintz, and Tamagaki, 1963).

The rate of cortisol excretion by the kidney also shows a circadian variation. Small amounts of unconjugated cortisol are excreted in the urine with a detectable rhythm (Kobberling and von zur Muhlen, 1974). However, this excretion of free cortisol represents only a small fraction of the cortisol secreted by the adrenal, and it cannot contribute significantly to the plasma cortisol rhythm. In any case the urinary cortisol and plasma cortisol rhythms are approximately in phase with each other, and thus the urinary cortisol rhythm can only damp slightly the amplitude of the plasma rhythm. The excretion of conjugated cortisol metabolites also shows a circadian rhythm. However, the urinary 17-hydroxycorticosteroid rhythm is approximately three hours phase-delayed from the plasma cortisol (and urinary) cortisol rhythm but is in phase with the plasma rhythm of conjugated cortico-steroids (Doe, 1965).

Circadian timing   The central nervous system regulates adrenal func-tion indirectly via the hypothalamo-pituitary-adrenal axis (Yates and Maran, 1974). A corticotrophin-releasing factor (CRF), not yet chemi-cally identified, is released from the posterior hypothalamus into a portal blood circulation that perfuses the anterior part of the pituitary gland. This stimulates the release of ACTH, a polypeptide hormone, which in turn stimulates corticosteroid synthesis and secretion from the adrenal gland. Circadian rhythmicity has been demonstrated at each step of this hypothalamo-pituitary-adrenal axis (Fig. 5.23). In mice (Hiroshige, Sakakura, and Itoh, 1969; Ixart et al., 1977) and in rats (David-Nelson and Brodish, 1969; Hiroshige and Sato, 1970; Takebe, Sakakura, and Mashimo, 1972) the circadian rhythm of hypothalamic CRF activity peaks before that of plasma corticosterone, as would be predicted. Rhythms of pituitary ACTH content (Halberg et al., 1965) and plasma ACTH have also been demonstrated in rodents (Cheifetz, Gaffud, and Dingan, 1968; Szafarczyk et al., 1979) and in humans (Gal-lagher et al., 1973). Since the circadian rhythms of CRF, ACTH, and corticosteroids are appropriately phase related, it would be tempting to conclude that plasma corticosteroid rhythms are primarily driven through the pituitary-adrenal axis.

However, the story is not so simple. Temporal information flows up the axis as well as down it. Animals that have been adrenalectomized (Hiroshige and Wada, 1974) or hypophysectomized (Takebe, Saka-kura, and Mashimo, 1972) show a persisting CRF rhythm, but it is phase advanced as compared to controls. Similarly, a phase advance in the ACTH rhythm is seen in adrenalectomized rats (Cheifetz, Gaffud, and Dingan, 1968). Thus, the timing of the CRF and ACTH rhythms are modified through feedback from the lower elements in the axis.

More important, the hypothalamo-pituitary-adrenal axis may not be the only pathway by which circadian information reaches the adrenal gland. Meier (1977) has shown that hypophysectomized rats, im-planted with pellets containing ACTH and thyroxine so that contin-uous ACTH levels are maintained, continue to show a corticosterone rhythm synchronized to the light–dark cycle. Similarly, Szafarczyk and colleagues (1979) report that the circadian rhythm in plasma ACTH is totally eliminated after lesions of the suprachiasmatic nuclei in rats, whereas plasma corticosterone rhythms persist. Dallman and her associates (Dallman et al., 1977, 1978) have provided evidence that there is an alternate, probably neural, input to the adrenal cortex which can regulate corticosteroid secretion separately from plasma ACTH.

How then do the central neural pacemakers of the circadian timing system exert their influence on the adrenal cortical rhythms? One pacemaker, the SCN, apparently exerts phase control via the hypothalamo-pituitary-adrenal axis since plasma ACTH rhythms are eliminated after SCN lesions. This pathway is probably the predominant route by which temporal information about the light–dark cycle is conveyed. However, as we have discussed, food timing exerts a stronger phase control over the adrenal rhythms than does the LD cycle, whether or not the SCN are destroyed. This indicates a separate pathway for temporal information about feeding cycles, possibly involving the ventromedial nuclei of the hypothalamus (Krieger, 1980) and the neural pathway to the adrenal (Dallman et al., 1978).

Adrenal function is not passively driven by either of these pathways. Episodes of ACTH secretion, for example, do not necessarily induce corticosteroid secretion (Krieger, 1979b). Andrews (1968) has shown, furthermore, that pulses of ACTH exogenously administered to rhythmically functioning adrenal glands maintained *in vitro* will produce phase shifts in the corticosteroid rhythm but only when given at certain points in the cycle (see Chapter 3). Thus the pituitary-adrenal axis and the other putative neural pathway probably exert phase control on adrenal corticosteroid rhythms but do not generate them.

The pituitary-adrenal axis thus represents an excellent model for studying the transmission of circadian information in the body. It has the advantage that the components are anatomically quite separate, and at least three separate hormones in series are involved in passing on temporal information. Hence the function and behavior of each part of the system are accessible for study.

Renal Function

Since ancient times people must have recognized their propensity for producing more urine during the day than at night. The observation that there is a circadian variation not only in urine volume but also in its constituents first attracted the attention of investigators more than a hundred years ago (Schweig, 1843; Roberts, 1860; Vogel, 1863; Weigelin, 1868). In studies of normal human subjects the rates of potassium, sodium, and water excretion were shown to fall to a minimum during nocturnal sleep and then rise to a maximum during the early afternoon. Since that time circadian rhythms have been documented in most of the urinary variables that have been appropriately examined, including the renal excretion of water and electrolytes (Bartter, Chan, and Simpson, 1979), metabolites (Conroy and Mills, 1970), and pharmacological compounds (Reinberg et al., 1967). Our discussion here can focus on only a few of these rhythmic variables, so we will choose those that best illustrate the mechanisms of circadian rhythmicity.

Most studies of renal circadian rhythms have been conducted in humans or in nonhuman primates. Figure 5.24 shows some circadian rhythms in renal excretion in the squirrel monkey. Each variable reaches its maximum at a characteristic time of day, and there is usually a two- or threefold range between maximum and minimum values. The few studies that have been conducted with nocturnal rodents show, as would be expected, an inverted phase relationship of these renal rhythms to the light–dark cycle (Cohn, Webb, and Joseph, 1970; Ratte et al., 1974; Hilfenhaus and Hertig, 1979). These circadian rhythms in urinary variables and other aspects of renal function, such as glomerular filtration rate and renal plasma flow (Wesson, 1964), are normally the major contributor to the total daily variance in excretion rate.

The responsiveness of renal mechanisms to homeostatic challenges also varies with the time of day. Stanbury and Thomson (1951) showed that in human subjects the diuretic response to the ingestion of 1 liter of water depended on when it was given. If it was administered at 10:00 A.M., peak urine flow rates of 16 ml/min were produced, whereas if the same challenge was given at 3:00 P.M., urine flow never exceeded 8 ml/min. If these authors had compared night and day responses, one would expect a much greater variation. The diuretic and natriuretic response to volume expansion differs markedly between the middle of the day and the middle of the night.

Vagnucci and co-workers (1969) studied subjects who were kept standing erect for 12 hours and then allowed to lie supine. If the transition from erect to supine posture occurred during the day, there was

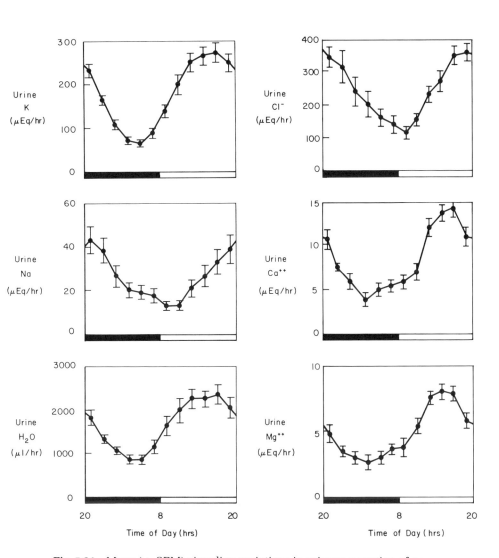

Fig. 5.24 Mean (± SEM) circadian variations in urinary excretion of potassium, sodium, chloride, calcium, and magnesium, in μEq/ hr, and water in μl/ hr of five monkeys, each studied for four days in an LD 12:12 cycle with food continuously available. (*Left*, from Moore-Ede and Herd, 1977; *right*, copyright 1982 by Moore-Ede.)

an immediate marked diuresis and natriuresis. This is because when we stand erect, blood and extracellular fluid tend to pool in the legs. When we lie down, the fluid shift from the legs and lower abdomen results in expansion of the blood vessels in the upper half of the body (Sjostrand, 1952). This central vascular expansion stimulates volume receptors in the atria of the heart and results in reflex increases in sodium and water excretion (Henry-Gauer reflex) (Epstein, 1978). However, when Vagnucci and colleagues (1969) studied the transition to recumbency at 4:00 A.M., no immediate changes in urine water and sodium excretion occurred. Instead, the volume load was excreted later in the day. Figure 5.25 shows a similar response in squirrel monkeys, where fluid shifts to the upper half of the body can be reliably induced by applying positive air pressure on the lower half of the body. There was a prominent circadian variation in the renal response. During the middle of the day, central vascular expansion induced a dramatic fifteenfold increase in sodium excretion, but when the same stimulus was given at night, very little change was observed (Kass et al., 1980a).

Day–night differences in the reflex responses to postural changes are not confined to volume homeostasis. Calcium is lost from the skeleton whenever weight-bearing forces are removed from the bones of the body (Issekutz et al., 1966; Mack and Lachance, 1967). However, the urinary excretion of this calcium is suppressed at night (Moore-Ede and Burr, 1973). Bed rest induced an increase in calcium excretion only when it occurred during the day; recumbency during the normal nocturnal hours of bed rest resulted in no net change in calcium excretion. This nocturnal suppression of calcium excretion was not caused by day–night differences in dietary intake. It occurred whether the subjects were upright or supine, or consuming identical small meals at three-hour intervals throughout day and night (Moore-Ede, Faulkner, and Tredre, 1972).

These findings are in agreement with our everyday experience. Humans are typically supine for 8 hours every night, yet there is no diuresis or increase in sodium or calcium excretion. If anything, there is a reduction in renal water and electrolyte excretion that is independent of day–night differences in dietary intake. Thus there appears to be a switching-off of certain renal reflexes at night, which has obvious

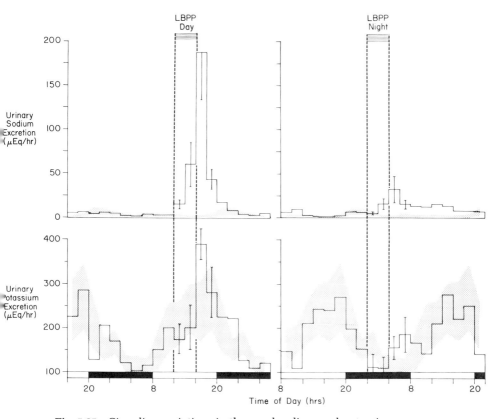

Fig. 5.25 Circadian variations in the renal sodium and potassium excretion responses of squirrel monkeys exposed to 4 hrs of positive pressure (20 mm Hg) applied on the lower body (LBPP). The stippled portions represent control data (mean ± SEM), and the solid lines show the excretory rates on the days and nights when lower body positive pressure was applied. (From Kass et al., 1980a.)

evolutionary advantages. In upright terrestrial animals such as man, fluid and mineral conservation is critical, yet we run the risk of depleting blood volume and bone mass every time we lie down to sleep. The adaptation which occurs at night (and it is not dependent on whether the subject is asleep) thus acts as an important conservation mechanism and prevents the nightly depletion of the hard-won fluid and minerals gained during the previous day.

ACTIVE AND PASSIVE COMPONENTS

There are three possible sources for circadian variation in the excretion of a urinary constituent: exogenous, extrarenal, and intrarenal. We will consider examples of rhythms that are predominantly derived in each of these ways, recognizing that most represent some combination of these various sources.

Very few renal rhythms are directly dependent on exogenous sources such as ingested food or water. Although it is easy to envision an overt urinary rhythm resulting from ingestion of a substance at the same time each day, this simple case is rarely observed. Even the rate of excretion of a foreign substance such as salicylate after the ingestion of an aspirin tablet is dictated not only by the time of ingestion but also by the circadian rhythm of urinary acidity. Salicylate, because of its acid dissociation content (pKa) of 2.97, is excreted far more rapidly in the evening, when urine is relatively alkaline, than during the morning, when urine is acid (Reinberg et al., 1967). Hence, the resultant rhythm of salicylate excretion is a combination of exogenous and endogenous influences.

Rhythms in the excretion of some urinary constituents, such as unconjugated cortisol (Kobberling and von zur Muhlen, 1974), are passive responses to endogenous circadian oscillators located elsewhere in the body. The major determinant of the urinary cortisol rhythm is the circadian variation in plasma cortisol concentration, which is driven principally by endogenous circadian oscillators in the adrenal cortex but is also influenced by other circadian oscillators, as we discussed in the previous section. However, circadian oscillators intrinsic to the kidney may also modulate the rate of cortisol excretion, although in this case they appear to play a relatively minor role compared to the rhythmic secretion from the adrenal gland.

The third possible source of rhythms in urinary excretion are circadian oscillators within the kidney. Because the kidney is bathed in an extracellular milieu that is highly rhythmic in many of its properties and constituents, it is often difficult to define precisely whether the source of the rhythmicity is intrarenal or extrarenal. Only one renal rhythm—the excretion of urinary potassium—has been studied sufficiently to demonstrate that the source of the rhythmicity is intrarenal. The rest of our discussion of renal rhythms will focus on this example, because it illustrates some useful approaches for future studies of other rhythmic functions that may be endogenous to the kidney.

THE URINARY POTASSIUM RHYTHM

Exogenous influences   The circadian rhythm in potassium excretion is little influenced by day–night variations in posture, activity, sleep, or dietary intake. Simpson (1924, 1926) and Norn (1929) were the first to recognize that urinary flow rate and electrolyte excretion would continue to oscillate between a daytime maximum and a nocturnal minimum in the absence of day–night differences in activity and food or fluid intake. Simpson (1924, 1926), for example, fed normal subjects 100 cc of water every hour while keeping them recumbent in bed day and night and found that the circadian rhythms in urine flow rate and electrolyte excretion persisted.

Experiments in which day–night variations in food and fluid intake have been eliminated for up to 40 days (Manchester, 1933; Borst and DeVries, 1950) and in which strict supine posture and minimal muscular activity were maintained at all times (Mills and Stanbury, 1954, 1955; Reinberg et al., 1971; Moore-Ede, Brennan, and Ball, 1975) have firmly established that behavioral regimens play very little role in direct causation of the circadian variations in urinary electrolyte excretion. Figure 5.26 shows the urinary potassium excretion rhythms of two human subjects who were maintained continuously supine in bed and fed an identical liquid meal every three hours throughout night and day for six days (Moore-Ede, Brennan, and Ball, 1975). The five-fold circadian variation is stable and highly reproducible from cycle to cycle despite the lack of change in posture, activity, or diet. This circadian rhythm of potassium excretion has also been shown to persist during fluid loading and during a 48-hour fast (Manchester, 1933; Stanbury and Thompson, 1951) and where subjects followed a regular exercise routine every three hours for 48 hours (Moore-Ede, Faulkner, and Tredre, 1972). The urinary potassium rhythm also persisted unaltered in squirrel monkeys deprived of food or water for 36 hours or fed equal amounts at 2-hour intervals throughout day and night for a week (Moore-Ede and Herd, 1977).

Sleep also has little influence on the circadian variation of electrolyte excretion. Moore-Ede, Faulkner, and Tredre (1972) showed that

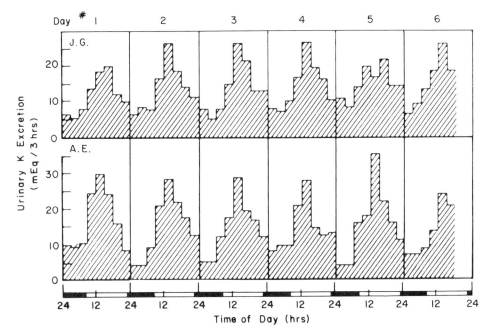

Fig. 5.26 Circadian variation in urinary potassium excretion in two humans who were fed identical meals at 3-hr intervals during constant bed rest for seven days and nights. A prominent rhythm in potassium excretion persisted. Dark bars represent sleep. (From Moore-Ede, Brennan, and Ball, 1975.)

the nocturnal fall of potassium excretion still occurred when subjects were kept awake and active all night. Other studies have shown that the urinary potassium rhythm persists with an approximately 24-hour period in subjects living on 12-hour (Mills and Stanbury, 1952) or 21-hour (Simpson, Lobban, and Halberg, 1970) sleep–wake cycles. During such regimens the maximum of the urinary potassium excretion rhythm from time to time coincided with the periods of sleep, thus ruling out the possibility that the rhythm was passively dependent on the sleep–wakefulness pattern.

Urinary electrolyte circadian rhythms also have been shown not to be passively dependent on the environmental light–dark cycle. Simpson (1924) demonstrated that a subject on an hourly feeding regimen who was kept in a dark room for 24 hours had normal urinary rhythms. Studies in polar regions, where natural illumination in certain seasons varies relatively little over 24 hours (Volker, 1927; Mills, 1951), have also demonstrated unaltered urinary potassium rhythms. Similarly, the potassium rhythm in rhesus monkeys (Hawking and Lobban, 1970) and in squirrel monkeys (Sulzman, Fuller, and Moore-Ede, 1979a) persists in constant illumination.

Not all mammalian species show such persistent renal rhythms. Dogs when isolated in constant conditions either lack or have very damped urinary excretory rhythms of water and electrolytes (Hawking et al., 1971). Although circadian urinary electrolyte rhythms have been occasionally reported in dogs kept on a regular daily feeding and activity regimen (Hartenblower and Coburn, 1973), the circadian rhythm can easily be eliminated by feeding the dog at regular intervals throughout day and night (Moore-Ede, Drake, and Hotez, unpublished observations). It also appears easy to damp out urinary electrolyte rhythms in rats. Feeding rats hourly (Cohn, Webb, and Joseph, 1970) or placing them in constant light (Kellogg, 1953) will, after several weeks, result in a major reduction in the amplitude of the urinary rhythms. In contrast, renal electrolyte rhythms in primates persist under such conditions.

Extrarenal or intrarenal source   Although it is clear (especially in humans and in nonhuman primates) that the source of the urinary potassium circadian rhythm is within the organism, it is less obvious whether the source is extrarenal or intrarenal. The kidney is subject to multiple rhythmic influences, any of which could potentially drive the renal potassium excretion rhythm (Fig. 5.27). For example, circadian variations in plasma potassium concentration could be responsible for the renal rhythm. Plasma potassium concentration usually has a circadian range of about 10% relative to the mean (Wesson, 1964; Stamm, 1967; Morimoto and Shiraki, 1970; Moore-Ede, Brennan, and Ball, 1975). This reaches a maximum at or just before the time of the maximum in urinary potassium excretion even in recumbent subjects under constant conditions consuming meals every 3 hours (Moore-Ede, Brennan, and Ball, 1975). However, Moore-Ede and co-workers (1978) have shown that the kidney's responsiveness to changes in plasma po-

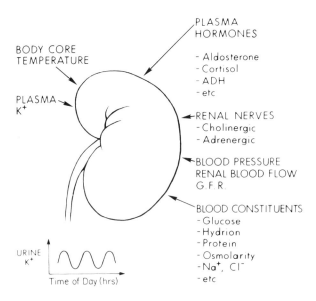

Fig. 5.27 Possible routes of information controlling or synchronizing the urinary potassium excretion rhythm. ADH = antidiuretic hormone; G.F.R. = glomerular filtration rate. (Copyright 1982 by Moore-Ede, Sulzman, and Fuller.)

tassium concentration varies markedly with the time of day, so the urinary potassium rhythm cannot be a passive consequence of variations in plasma potassium.

Potassium excretion and hydrogen ion excretion are often inversely correlated and compete with one another to exchange for sodium in the distal tubule (Berliner, Kennedy, and Orloff, 1951). The circadian rhythms of potassium and hydrion excretion are usually 180° out of phase with one another, so that the rate of titratable acid excretion reaches a maximum during sleep when potassium excretion is at its lowest ebb (Stanbury and Thomson, 1951). However, Longson and Mills (1953) have disrupted the daily rhythm of hydrion excretion by inducing a respiratory acidosis in normal subjects by ventilating them with 5–6% CO_2, and this did not prevent the normal morning rise in potassium excretion. Furthermore, Simpson, Lobban, and Halberg (1970) have shown that the circadian rhythm of urinary pH will follow an imposed 21-hour sleep–wake cycle while the rhythm of urinary potassium excretion continues with a near-24-hour period. Thus acid–base changes in the urine do not appear to be the source of the urinary potassium rhythm.

It is, however, theoretically impossible to rule out all the possible sources of periodic information to the kidney shown in Figure 5.27, because many of them are still unknown. As we discussed in Chapter 3, an ultimate approach would be to study circadian rhythms in renal function in organ or tissue culture *in vitro*. To our knowledge no one has yet succeeded in maintaining a kidney *in vitro* for a sufficient length of time (that is, several days) at a temperature close to body temperature. This is not to say it is impossible; indeed, it might be a most worthwhile investigative task. For the present discussion, however, we are limited to a consideration of *in vivo* approaches.

Physiological mechanisms Before we discuss two *in vivo* approaches which demonstrate the existence of intrarenal oscillators, it is appropriate to examine the mechanisms in the kidney which must be rhythmic to produce a circadian rhythm of renal potassium excretion. Figure 5.28 is a schematic drawing of one of the million or so nephrons in the mammalian kidney. Approximately 90% of the potassium filtered from the blood plasma in the glomerulus is reabsorbed before the tubular fluid reaches the early distal tubule (Brenner and Berliner, 1973). The fraction of filtered potassium arriving at the early distal tubule remains virtually constant under a wide variety of conditions, ranging from very low urinary potassium excretion, as when dietary potassium depletion and metabolic acidosis are induced by administration of ammonium chloride, to maximal urinary potassium excretion, as when high dietary potassium intake and metabolic alkalosis are induced by sodium bicarbonate infusion (Giebisch, 1971). Then, over the length of the distal tubule, secretion or reabsorption by the distal tubular cells (Fig. 5.28) causes the potassium content of the tubular fluid to rise or fall to the levels found in the final urine. The collecting ducts do not contribute importantly to potassium secretion (Bennett, Brenner, and Berliner, 1968), although there may be further potassium reabsorption at this site if the final urinary potassium excretory rate is very low (Malnic, Klose, and Giebisch, 1964). The rate of potassium excretion is thus determined principally by the potassium flux between the cells lining the distal tubule and the lumen of the tubule.

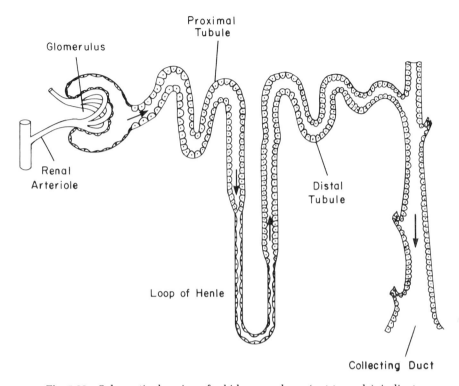

Fig. 5.28 Schematic drawing of a kidney nephron (not to scale), indicating the proximal and distal tubules and the loop of Henle. The cells lining the lumen of the distal tubule are responsible for potassium movement in and out of the tubular fluid. These cells essentially control the potassium concentration of the urine. (Copyright 1982 by Moore-Ede, Sulzman, and Fuller.)

The magnitude and direction of this flux are determined by two factors: (1) the rate of passive potassium flux from the distal tubular cell into the lumen, and (2) the activity of a potassium pump in the luminal membrane which counterbalances the passive potassium flux (Giebisch, 1971). The rate of the passive potassium flux is itself determined by the permeability of the luminal membrane and the electrochemical gradient for potassium between the distal tubular cell and the lumen.

A circadian variation in any one of these factors could cause the circadian oscillation in urinary potassium excretion. In humans this circadian variation is four- or fivefold (Conroy and Mills, 1970), and there must be comparable changes in potassium flux. Although these changes are large enough to detect, to our knowledge this question has not yet been examined, perhaps because the species most commonly used for studies of renal tubule function is the rat, an animal with a much less stable urinary potassium rhythm than primates.

Indirect evidence suggests that circadian variations in the intracellular potassium concentration of the renal distal tubular cells may contribute to the urinary potassium rhythm through variations in the electrochemical gradient. Although the distal tubular cells have not been directly examined, other tissues, such as muscle and red blood cells, show such changes in intracellular potassium concentration (Moore-Ede, Brennan, and Ball, 1975). In humans, potassium moves out of body cells into the extracellular fluid during the morning and then back into the cells at night. Because the circadian rhythm of potassium flux in and out of the body cell mass is synchronized with the circadian rhythm of urinary potassium excretion, it is likely that the renal distal tubular cells function similarly to the other body cells. During the morning, potassium apparently moves from the distal tubular cells into a special extracellular fluid compartment—the distal tubular fluid—resulting in an increase in urinary potassium excretion. In the late evening, when there is a net potassium flux into the body cell mass, urinary potassium excretion is at a minimum, suggesting that the distal tubular cells either have greatly reduced secretion or have reabsorbed potassium from the distal tubular fluid (Moore-Ede, Brennan, and Ball, 1975).

There is evidence to suggest that potassium movement across cell membranes may be an essential component of the basic circadian oscillator mechanism. By perturbing potassium concentration gradients across the cell membrane, it is possible to manipulate circadian oscillations in many diverse biological functions. Valinomycin, which acts as a specific transmembrane potassium carrier, can induce phase shifts in the rhythm of leaf movements in beans (*Phaseolus*; Bünning and Moser, 1972) and in the rhythm of bioluminescence in the alga *Gonyaulax* (Sweeney, 1974). Elevation of the potassium concentration of the culture medium can induce phase shifts in the circadian rhythm of activity of an isolated eye of the sea hare, *Aplysia*, similar to those caused by exposure to light (Eskin, 1972). Application of potassium externally can antagonize the daily closure of excised leaflets of the silk tree *Albizzia;* (Satter, Marinoff, and Galston, 1970). Finally, it has been postulated that potassium fluxes are the basis of the phytochrome response to light in plants as well as being an essential component of endogenous circadian leaf movements (Satter and Galston, 1971).

These lines of evidence, coupled with the findings that circadian rhythms persist in enucleated cells (Sweeney and Haxo, 1961; Schweiger, Wallraff, and Schweiger, 1964) and are resistant to nucleic acid and some protein synthesis inhibitors (Hastings, 1971), have led Njus, Sulzman, and Hastings (1974) to postulate that ionic fluxes across cell membranes act as a basic component of the cellular circadian oscillator. They suggest that cyclical changes in membrane ion permeability and transmembrane ion gradients may act as a feedback system to generate self-sustained circadian oscillations in cells. If this model proves to be correct, then the circadian oscillations in distal tubular potassium flux may be more than just an overt rhythm; they may be a fundamental part of the oscillator itself.

Genesis of the rhythm   Obtaining direct evidence of a renal oscillator driving the urinary potassium rhythm would require long-term (several-day) *in vitro* studies of renal function, which have not yet proved feasible. However, as we discussed in Chapter 3 (see especially Fig. 3.19), there is an indirect way to examine whether a tissue contains a circadian oscillator. Oscillators and passive elements behave in distinctly different ways when the rhythmic inputs that act as mediators to them are manipulated. Using this approach, we have demonstrated that the circadian rhythm of plasma cortisol concentration acts as an internal mediator timing the renal potassium rhythm. Furthermore, the behavior of the potassium rhythm when cortisol rhythms are manipulated suggests that there is an intrarenal circadian oscillator which generates the excretory rhythm.

The adrenal steroids, aldosterone and cortisol, have long been recognized as important determinants of urinary potassium excretion (Barger, Berlin, and Tulenko, 1958; Mills, Thomas, and Williamson, 1960; Fimognari, Fanestil, and Edelman, 1967). Since each of the steroids has circadian variations in plasma concentration that are appropriately phase-related to the urinary potassium rhythm (Conroy and Mills, 1970), it has been suggested that the renal potassium (and sodium and water) rhythms may be passively dependent on adrenal steroid rhythmicity (Borst and DeVries, 1950). This view seemed to be supported by the observation that the circadian rhythms in urinary water and electrolyte excretion are often absent in patients with adrenocortical insufficiency, or Addison's disease (Levy, Power, and Kepler, 1946; Finkenstaedt et al., 1954). However, several groups have shown that urinary electrolyte and water rhythms persisted in Addisonian, adrenalectomized, or panhypopituitary subjects who were given their replacement steroid or ACTH therapy in equal doses at regular intervals over the 24 hours (Garrod and Burston, 1952; Lewis, 1953; Rosenbaum, Papper, and Ashley, 1955; Nabarro, 1956; Doe, Vennes, and Flink, 1960; Liddle, 1966). This has caused many authors to rule out any role for the adrenal steroids in the generation of urinary electrolyte rhythms other than the possible "permissive" one (Ingle, 1952) that is suggested by the damping of electrolyte rhythms in untreated Addisonian patients (Finkenstaedt et al., 1954).

However, several lines of evidence indicate that there is normally a temporal relationship between plasma cortisol concentration and urinary potassium excretion. Thus, after a phase shift of the light–dark cycle, the excretory rhythms of sodium and water adjust their phase within three to four days, whereas the urinary potassium and adrenal steroid rhythms typically take one to two weeks to adjust together (Flink and Doe, 1959; Martel et al., 1962; Elliot et al., 1972). Similarly, studies of subjects living in the Arctic with a 21-hour day (Simpson, Lobban, and Halberg, 1970), showed that the urinary potassium and corticosteroid rhythms maintained a dominant near-24-hour period, whereas other monitored rhythms, such as urinary pH excretion, followed the 21-hour sleep–wake cycle.

Further evidence of a temporal relationship was provided by Doe, Vennes, and Flink (1960), who produced an artificial rhythm of plasma cortisol concentration in Addisonian patients by timing the administration of oral cortisol hemisuccinate therapy. They found that by altering the timing of the artificial cortisol rhythm, they could phase-shift the urinary potassium rhythm. Two other similar reports have appeared. Reinberg and colleagues (1971) demonstrated that if the maximum of the rhythm of urinary 17-hydroxycorticosteroids (17-

OHCS) excretion was delayed by 5 hours by manipulating the timing of cortisol replacement dosage in Addisonian or hypophysectomized patients, then the rhythms of urinary potassium and sodium excretion would be similarly delayed. The circadian rhythms of body temperature and heart rate, however, were unaffected by the delay in the adrenal steroid rhythm. A similar alteration in the timing of prednisone treatment in asthmatic children also can shift the time of the maximum of urinary potassium excretion (Reindl et al., 1969). Prednisone was given once daily at either 1 A.M., 7 A.M., 1 P.M., or 7 P.M. for periods of at least three months. Urinary potassium excretion usually reached a maximum six to seven hours after the time of daily administration.

There have been suggestions that cortisol rather than aldosterone might be the temporally active adrenal steroid. The circadian pattern of urinary aldosterone excretion is poorly correlated with that of potassium excretion (Vagnucci, Shapiro, and McDonald, 1969), and plasma aldosterone concentration is largely dependent on postural changes (Muller, Manning, and Riondel, 1958; Wolfe et al., 1966; Mills, 1973), whereas urinary potassium excretion is not (Conroy and Mills, 1970). In comparison, urinary 17-OHCS and potassium excretion are usually well correlated (Vagnucci, Shapiro, and McDonald, 1969), although the two rhythms can on occasion become uncoupled (Bayliss, 1955; Mills, 1973).

Thus, the circadian rhythm of plasma cortisol could be acting as a mediator conveying temporal information from the light–dark cycle and other internal oscillators to the kidney. However, the kidney might contain either a passive element or an active (secondary) oscillator. As we discussed in Chapter 3 (Fig. 3.19), if cortisol passively induced the urinary potassium rhythm, then (1) phase-shifting the cortisol rhythm must produce an equal and immediate phase shift in the renal potassium rhythm; (2) potassium excretion must cease to oscillate if cortisol is maintained at a constant level; and (3) a change in the plasma level of cortisol must induce an equivalent change in potassium excretion at any time in the 24-hour day.

However, a series of experiments using adrenalectomized squirrel monkeys suggest that the kidney contains a spontaneous circadian oscillator. Each animal was prepared with a chronically implanted catheter so that plasma adrenal steroid rhythms could be artificially generated by infusion (Moore-Ede, Schmelzer, et al., 1977). A single daily pulse of cortisol, administered at the time when the circadian maximum of plasma cortisol concentration would normally occur, maintained the urinary potassium rhythm with the same phase and amplitude as was seen in intact controls (Fig. 5.29A). When the artificial cortisol rhythm was manipulated it was found, first, that phase shifts in the timing of the plasma cortisol rhythm caused a comparable (but not equal) phase shift in urinary potassium excretion, but only after a transient response lasting several days (Fig. 5.29B). Second, elimination of circadian rhythmicity in adrenal steroid administration in adrenalectomized animals resulted in free-running persistent oscillations in urinary potassium excretion (Fig. 5.29C). Third, a pulse of cortisol administered in the evening hours to intact animals induced another cortisol peak approximately 12 hours phase-delayed from the endogenous peak of plasma cortisol concentration and of similar magnitude, but the induced elevation in urinary potassium excretion was not equivalent to that induced by the endogenous cortisol peaks (Fig. 5.29D). These results suggest that the action of cortisol as an internal synchronizer of the urinary potassium rhythm may involve a phase-control mechanism comparable to the well-documented action of light on circadian systems (see Chapter 2). The results also suggest that the circadian rhythm of urinary potassium excretion is the product of one or more autonomous intrarenal oscillators which can persist and free-run in the absence of temporal information from the plasma cortisol rhythm.

It is significant that cortisol rather than aldosterone, a more potent mineralocorticoid, appears to act as the hormonal mediator synchronizing the circadian rhythm of renal potassium excretion with the light–dark cycle. The urinary potassium rhythm was indistinguishable whether cortisol and aldosterone together or cortisol alone was administered to the adrenalectomized monkeys. Thus, cortisol appears to be involved in the circadian synchronization of renal potassium excretion (Moore-Ede, Schmelzer, et al., 1977) while aldosterone is involved in its moment-to-moment regulation. As the components of the circadian timing system and their internal zeitgebers are identified, it may turn out to be a general finding that circadian synchronization is subserved by its own specialized set of hormones and neural pathways.

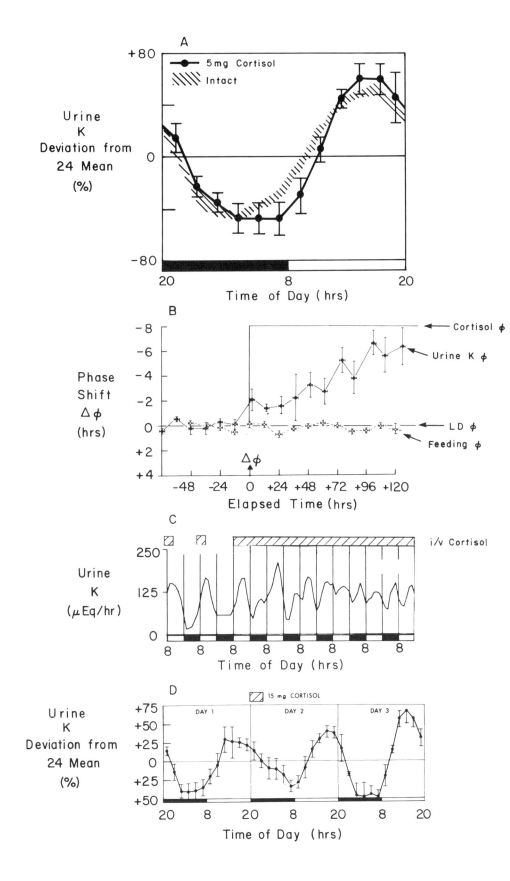

Fig. 5.29 Synchronization of the urinary potassium excretion rhythm by artificially controlled rhythms of plasma cortisol concentration in adrenalectomized squirrel monkeys in an LD 12:12 cycle. (*A*) Potassium excretion of intact (stippled area) and adrenalectomized monkeys receiving 5 mg of cortisol each day from 0800 to 0900 hrs EST (solid line); normal potassium rhythm was maintained. (*B*) Phase relationships of the potassium excretion (solid circles) and feeding (open circles) rhythms of monkeys before and after an 8-hr phase delay in the time of cortisol administration. (*C*) Response of adrenalectomized monkey to the continuous administration of adrenal steroids. For the first two days the daily dose of cortisol and aldosterone was administered between 0800 and 0900 hrs EST. For the remainder of the experiment the same daily dose was spread evenly over each 24 hrs. A persisting urinary potassium rhythm was observed with a shortened period. (*D*) Response of the potassium excretion rhythm of intact monkeys to a single infusion of 15 mg cortisol between 2000 hr and 2300 hr on day 2. There was little response when the infusion was given 180° out of phase with the endogenous peak of plasma cortisol concentration. (From Moore-Ede, Schmelzer, et al., 1977.)

Reproduction

The predominant rhythms of the mammalian reproductive system are infradian; that is, they have periods longer than 24 hours. The estrous cycles of rodents are 4–5 days long, and the menstrual cycles of the higher primates and humans are approximately 25 to 35 days. These periodicities are determined at least in part by the time required to ready the uterus and reproductive system for the possibility of sustaining fertilized ova. Superimposed on these rhythms in many mammals are seasonal patterns of breeding, so that mating and, more important, birth are scheduled for the time of year most advantageous for survival of the young.

The circadian system is integrally involved in the control of many of these infradian reproductive rhythms. The events on each day of the estrous cycle in rodents are precisely scheduled with respect to circadian time, and estrous rhythmicity ceases if the SCN are destroyed. Furthermore, the synchronization of circannual (approximately year-long) breeding cycles with the appropriate season relies on an accurate measurement of day length—and it is the circadian timing system that does the measuring. These findings have spurred much recent investigative effort, which we will outline below.

The involvement of the circadian timing system in the timing of

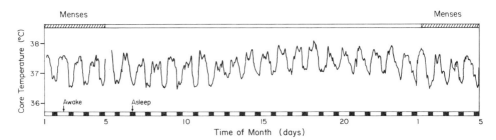

Fig. 5.30 The circadian rhythm of body temperature in a woman over the course of a menstrual cycle. The times of menstruation are marked by the shaded bars. (Kindly provided by J. Zimmerman, Department of Neurology, Montefiore Hospital, New York; copyright 1982 by Zimmerman.)

human reproductive cycles has as yet received little attention. There is a circadian rhythm in the time of spontaneous births, with more babies born between 3 and 4 A.M. than at any other time of day (Kaiser and Halberg, 1962). Furthermore, there is a modulation of the circadian body temperature rhythm over the course of the female menstrual cycle; in the latter half of the menstrual cycle after ovulation has occurred, the mean temperature becomes elevated (Fig. 5.30; Zimmerman et al., personal communication). A woman who wishes to become pregnant (or to avoid conception by the "rhythm method") must take her temperature every day at the same time to determine the timing of ovulation. The optimum time is the early morning since the temperature trough becomes elevated earlier in the menstrual cycle than does the daytime maximum.

The extent of circadian control over human reproductive processes is an inquiry long overdue if the recent findings of Steptoe and Edwards (Elliott, 1979) are borne out. In trying to achieve implantation of fertilized human ova ("test-tube babies") they have been successful so far in only 4 women out of 79. In all 4, the ovum was implanted between 10 P.M. and midnight—a 100% success at this single two-hour period of the day. Although this is a preliminary report, it points to the necessity of investigating further the circadian timing of human reproductive function.

RODENT ESTROUS CYCLICITY

Estrous cycles are several times longer than the circadian cycle, yet in many species the circadian timing system plays an important role in

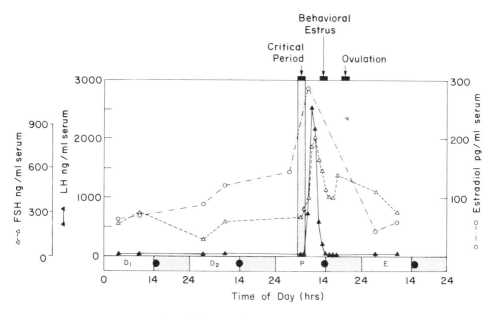

Fig. 5.31 Hormonal and behavioral events associated with the four-day (96-hr) hamster estrous cycle. P = proestrus; E = estrus; D_1 = diestrus 1; D_2 = diestrus 2. Solid circles = activity onsets on each day. (From Moline, 1981.)

determining their timing. This has been investigated mostly with hamsters and rats. Hamsters show a particularly precise 96-hour estrous cycle when maintained in a light–dark cycle with a 24-hour period (Fig. 5.31). The days of a four-day estrous cycle are referred to as estrus (day 1), diestrus 1 (day 2), diestrus 2 (day 3), and proestrus (day 4). Estrus is the state of sexual excitability during which the female will accept the male and is capable of conceiving. At this time the female shows a characteristic posture known as "lordosis" during the night preceding the day of estrus. Later in the same night, ovulation occurs, induced by a surge of luteinizing hormone (LH) 8–10 hours previously during proestrus. The day of estrus is also marked by changes in the appearance of the cells in the vaginal wall and in their mucoid secretion. This serves as a useful marker enabling investigators to confirm the animal's cyclic state.

A similar pattern of timing is seen in the rat. Rats typically show either a four- or five-day estrous cycle, but the exact phase of the events within the circadian day is maintained, so that each event is timed at precise 96- or 120-hour intervals. Individual animals may have either four- or five-day cycles, but it is possible to change four-day cycling rats to five-day cyclers by changing the length of the daily photoperiod (Hoffman, 1968). When the four-day rat and the five-day rat are compared, there are some small differences in the exact timing of estrous cycle events with respect to the circadian day, although there is good reproducibility in circadian timing from cycle to cycle in any individual animal.

Synchronization by light–dark cycles   It has long been known that the events of the female rodent estrous cycle are temporally related to the 24-hour light–dark cycle (Everett and Sawyer, 1950). However, only recently has it been recognized that the circadian system rather than the light–dark cycle directly is the major determinant of the timing of the estrous cycle. Thus, environmental light–dark cycles appear to act on estrous cycles only via their entrainment of the circadian system.

This can be most directly demonstrated by studying animals maintained in constant levels of illumination so that the circadian timing system will free-run with its own endogenous period. An example of such an experiment is shown in Figure 5.32. Fitzgerald and Zucker (1976) used as a marker for the circadian timing system the onset of wheel-running in a hamster maintained in constant dim illumination. Behavioral estrus was measured by timing the onset of lordosis. It can be seen that a consistent phase relationship was preserved between the estrous cycle marker and the circadian cycle, so that each estrous cycle still measured exactly four circadian days even though these days were lengthened in the free-running state. Thus, the period of the hamster's estrous cycle was a fourfold multiple of the circadian wheel-running rhythm both in free-running conditions and in the entrained state.

This result confirmed the earlier studies by Alleva and colleagues (1971), but Fitzgerald and Zucker (1976) went on to test one more feature of this temporal relationship. On the days marked by the stippled area in Figure 5.32, 50% deuterium oxide was provided instead of pure water for the animals to drink. As would be expected (see Chapter 2),

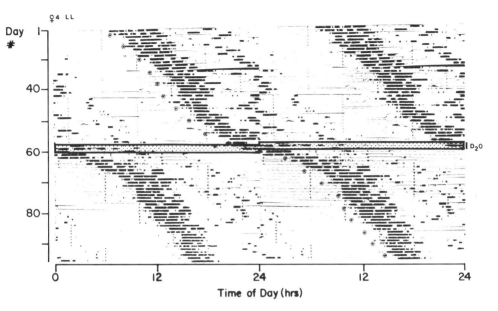

Fig. 5.32 Double-plotted free-running rhythms of estrous behavior onset (circled star) and wheel-running rhythm of a hamster maintained in constant dim illumination. In the middle of the experiment (stippled area) the animal was given water containing 50% D_2O. Estrous behavior maintained a stable phase relationship to every fourth circadian cycle. (From Fitzgerald and Zucker, 1976.)

deuteration slowed down the animal's circadian clock, lengthening the period of the circadian wheel-running rhythm. Even this did not uncouple the estrous cycle from the circadian system. Rather, the same phase relationships were conserved, so that the estrous cycle period was still four times as long as the lengthened circadian period. Similar findings have been obtained in rats. Although bright LL results in cessation of the estrous cycle, rats show free-running estrous cycles in constant dim illumination (McCormack and Sridaran, 1978).

Even in a light–dark cycle, the timing of estrous cycle events is not dominated by light and dark. Instead these events show a very tight correspondence under a wide variety of lighting regimens to the circadian timing system as indicated by the rhythm of wheel-running activity. Moline and co-workers (1981) demonstrated that the timing of the LH surge under various photoperiods was not reliably referenced to light onset, dark onset, or the time of mid-darkness or mid-light (Fig. 5.33). Rather, the most consistent relationship of the estrous cycle was with the endogenous circadian marker, the wheel-running rhythm.

These relationships were further examined by subjecting hamsters to acute phase shifts of the light–dark cycle (Albers, Moline, and Moore-Ede, 1981). It is well known, as we discussed in Chapter 2, that the circadian system takes several cycles to resynchronize with a new light–dark cycle after a phase shift. As expected, after a three-hour phase advance or phase delay of an LD cycle, it took approximately four days for the circadian wheel-running rhythm to shift. Similarly, it took several days for the estrous cycle to phase-shift (as indicated by the timing of the LH surge).

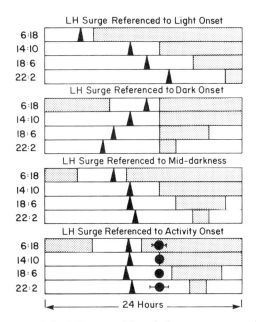

Fig. 5.33 Timing of the LH surge of female hamsters exposed to different photoperiods. The LH surge (dark triangle) is plotted (top to bottom) with respect to light onset; dark onset; mid-darkness; and activity onset (dark circles). The only consistent phase relationship across photoperiods was with the circadian activity rhythm, suggesting that the timing of the LH surge is regulated by the circadian system. (From Moline et al., 1981.)

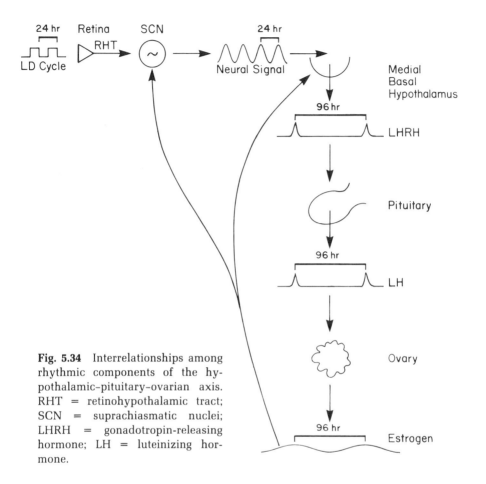

Fig. 5.34 Interrelationships among rhythmic components of the hypothalamic–pituitary–ovarian axis. RHT = retinohypothalamic tract; SCN = suprachiasmatic nuclei; LHRH = gonadotropin-releasing hormone; LH = luteinizing hormone.

Physiological basis   The female rodent estrous cycle is the product of an interaction between the ovary and the hypothalamus. As is schematically displayed in Figure 5.34, the various events are induced by surges of pituitary gonadotropin and ovarian hormone secretion. In the long "stimulatory" photoperiods of summer, when rats or hamsters are reproductively active, the secretion of estrogen from the ovary begins to increase on diestrus 2, causing a rise in plasma estrogen levels, which reach a peak during proestrus. The high levels of estrogen act on estrogen receptors in the preoptic and medial basal areas of the hypothalamus (Feder, Siegel, and Wade, 1974). As we will show, the estrogen signal interacts with a precisely timed neural signal originating from the circadian system to induce a surge of luteinizing hormone releasing hormone (LHRH) from the hypothalamus. The releasing hormone, a small peptide, is secreted into the blood traversing through the hypothalamo-hypophyseal portal circulation and perfusing the pituitary. This stimulates a rapid release of LH and FSH from the pituitary. The resultant surge in plasma LH levels is the key event triggering ovulation, approximately eight to ten hours later. There is also a secondary rise in FSH, initiating the next cycle of ovarian follicle maturation, during the early hours of estrus before FSH falls to baseline levels during diestrus 1 and diestrus 2.

Circadian timing   An increasing body of evidence now indicates that the circadian timing system provides the key signal that triggers the LH surge and subsequent ovulation. This role of the circadian system was not immediately appreciated, because it was difficult to imagine how an approximately 24-hour periodicity could drive a four- or five-day estrous cycle. However, it now appears that there is a neural signal at a precise time in every 24-hour cycle, but that the signal induces the release of gonadotropin-releasing factor and the LH surge only in the presence of the high circulating levels of estrogen found on the afternoon of proestrus.

The first hint of the daily potential of the estrous cycle was provided by the classic studies of Everett and Sawyer (1950). They demonstrated that there is a critical two-hour period during the afternoon of proestrus before which ovulation can be blocked by injection of a barbiturate. At any other time of day, the same dose is much less effective.

One of the most significant results of these studies was the finding that after an animal had been blocked, ovulation would spontaneously occur 24 hours later at precisely the same circadian time of day. The barbiturate injection could be repeated on several successive days (provided it was given during the critical period), and each time it would prevent ovulation occurring for 24 hours. More recently Stetson and Watson-Whitmyre (1977) have shown that barbiturate injection during the critical period blocks the surge of LH and FSH that normally occurs on the day of proestrus. Like Everett and Sawyer (1950), they showed that the hormone surges could be delayed by exactly 24, 48, or 72 hours, provided that the barbiturates were given during each critical period.

These studies suggest that the signal triggering the LH release and subsequent ovulation is provided by the circadian system, so that there is only a limited "window" each day during which these reproductive events can be initiated. The daily potential for an LH surge is apparently timed by the suprachiasmatic nuclei, since SCN lesions result in the elimination of estrous cycles (Stetson and Watson-Whitmyre, 1976; Nunez and Stephan, 1977; Raisman and Brown-Grant, 1977). Whether or not LH release is triggered by the signal appears to be determined by the levels of ovarian hormones in the blood. Legan, Coon, and Karsch (1975) have shown that after ovariectomy female rats will have daily LH surges if they are given replacement estrogens so that plasma estrogen levels are high every day instead of intermittently. Norman, Blake, and Sawyer (1973) and Buhl and colleagues (1978) have demonstrated the same phenomenon in hamsters maintained in stimulatory photoperiods after ovariectomy, again provided estradiol replacement was given.

In intact four-day cycling rats in stimulatory photoperiods, it is possible to advance the time of ovulation by exactly 24 hours by providing estrogen on diestrus 1 (Krey and Everett, 1973). Normally, estrogen levels do not rise to high levels until diestrus 2, but the elevated estrogen level on diestrus 1 enables the LH surge to occur a day earlier. However, in cycling animals it is not possible to induce LH release on every day of the cycle, because another ovarian hormone, progesterone, that is released following ovulation extinguishes the effect of estrogen on the LH release mechanism (Banks, Mick, and Freeman, 1980). Progesterone given to an animal on the day of estrus inhibits the estrogen from triggering LH release on proestrus, and thus the estrous cycle is delayed by exactly 24 hours to the next available window (Everett, 1948). However, there is some evidence to suggest that the effect of progesterone is biphasic. At some phases of the estrous cycle, such as the day of proestrus, progesterone will facilitate LH release, resulting in an earlier LH surge (Zeilmaker, 1966).

Effects of ovarian hormones   It has been recognized for many years that as rodents progress through the days of the estrous cycle there are modulations in the animals' circadian rhythms. A four- to five-day modulation in the total amount of wheel-running in rats was reported by Wang (1923), Slonaker (1924), and Richter (1927). Figure 5.35A shows a four-day oscillation in total activity of a female rat. Richter reported that most female rats run eight to ten miles on the night of ovulation, but only a fraction of a mile on the other days of the cycle.

High estrogen levels produce not only increases in activity but also phase advances in the timing of nocturnal activity. This results in the "scalloping" pattern shown in Figure 5.35B (Morin, Fitzgerald, and Zucker, 1977). The nightly wheel-running activity begins up to half an hour earlier on the evenings of the cycle when estrogen levels are highest, diestrus 2 and proestrus (Albers, Gerall, and Axelson, 1981). These effects of estrogen are antagonized by the progesterone released after ovulation, so that the daily onset of activity becomes delayed on the days of the cycle when plasma progesterone is elevated (Takahashi and Menaker, 1980; Axelson, Gerall, and Albers, 1981).

In addition to changes in the activity pattern, there are variations in feeding and drinking rhythms over the course of the estrous cycle (Tarttelin and Gorski, 1971). In rats maintained in LD cycles, on the day of estrus there are significant depressions in food and water intake. There is good evidence that circulating levels of estrogen are responsible for these modulations in the activity, feeding, and drinking rhythms. Wade and Zucker (1970) have shown that hypothalamic implants of estradiol can increase wheel-running activity and depress food intake. Tarttelin and Gorski (1971) have reported that estrogen depresses both feeding and drinking.

The estrous cycle also produces modulations in the core body temperature rhythms of female rats (Yochim and Spencer, 1976). The nocturnal peak of core temperature is highest during the night of ovulation and lowest the night before. Using ovariectomized rats, Marrone, Gentry, and Wade (1976) have shown that there is a dose-dependent rise in colonic temperature that is directly correlated with the levels of injected progesterone. Thus, there is good evidence that the estrous variations in ovarian hormonal levels cause these modulations in rhythm.

Estrogen not only modulates the expression of various circadian rhythms, it also affects the function of circadian pacemakers. When Morin and colleagues (1977) implanted silastic capsules containing es-

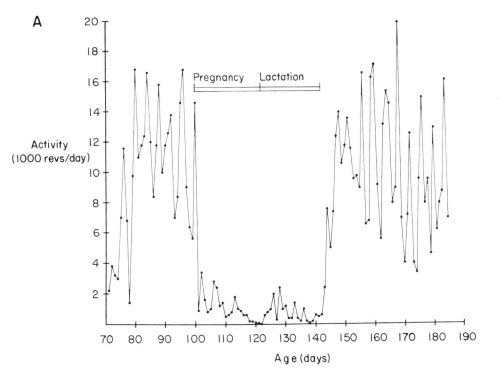

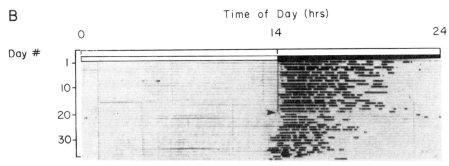

Fig. 5.35 Modulations of the circadian wheel-running rhythm over the course of the estrous cycle of (A) a female rat before and after pregnancy and lactation and (B) a female hamster. Maximum activity occurs on the night of estrus. The "scalloping" (see text) of wheel-running activity of the female hamster (B) is shown during exposure to an LD 14:10 cycle before and after blinding (at the time indicated by the arrow). A phase advance occurs on days 3 and 4 of the estrous cycle and is seen particularly clearly in the three estrous cycles that immediately preceded blinding. The scalloping continues while the circadian rhythm is free-running. The daily 10-hr dark period is indicated by the heavy horizontal line at the top of the figure. (A from Richter, 1965, reprinted courtesy of Charles C Thomas, publisher; B from Morin, Fitzgerald, and Zucker, 1977, copyright 1977 by the American Association for the Advancement of Science.)

tradiol benzoate into ovariectomized female hamsters, their studies demonstrated that estrogen shortens the free-running period in constant conditions. As we discussed in Chapter 2, the phase of the circadian system during entrainment is determined by the relationship between the periods of the pacemaker and the zeitgeber, with reductions in pacemaker period resulting in a phase advance of the circadian rhythm. Thus, the action of estrogen in shortening the period of the pacemaker may account for the earlier onset of wheel-running activity on the days when estrogen levels are high.

Some hint of the functional role of estrogen in the circadian timekeeping system and on the timing of the LH surge can be gained by some experiments of Albers (1981). He showed that in gonadectomized rats with free-running rhythms that were longer than 24 hours, estrogen shortened the period. The effect was quantitatively related to the length of the initial free-running period, so the greatest shortening occurred in animals with the longest original periods. The effect was particularly apparent in males and in neonatally androgenized females (who are not capable of spontaneous estrous cycles). However, in individual animals that had initial free-running periods somewhat shorter than 24 hours, estrogen lengthened rather than shortened the period. These results raise the intriguing possibility that estrogen acts to achieve homeostasis of the pacemaker's period.

PHOTOPERIODIC TIME MEASUREMENT AND SEASONAL BREEDING CYCLES

In the temperate zones of the earth, where there are significant seasonal alterations in the climate, most mammals have evolved annual reproductive cycles which ensure that the young are born at the time of year most promising for their survival (Van Tienhoven, 1968; Lodge and Salisbury, 1970). For many of these species, the seasonal variation in day length or "photoperiod" is the most important time cue ensuring the synchronization of the reproductive cycle with the appropriate season (Lodge and Salisbury, 1970; Negus and Berger, 1972). The detection of changes in day length, or "photoperiodic time measurement," is a function of the circadian system.

In the earliest studies of the effects of the daily photoperiod on mammalian reproduction, Bissonette (1932) showed with ferrets, and Baker and Ranson (1932) with field voles, that estrus could be induced in winter, before the normal breeding season, by lengthening the daily photoperiod to which the animals were exposed. In autumnal breeders such as the sheep, in contrast, estrus could be induced out of season by reducing the daily photoperiod (Yeates, 1949). This early work has been more recently extended to show that for both males and females in many mammalian species, the season of breeding is timed by the circannual change in photoperiod. For example, in autumn breeders such as the goat and the ram (Bissonette, 1941; Ortavant, Mauleon, and Thibault, 1964), short days are necessary for the induction and maintenance of spermatogenesis, whereas in spring breeders such as the hamster (Elliott, 1976), vole (Clarke and Kennedy, 1967), snowshoe hare (Davis and Meyer, 1972), and ferret (Bissonette, 1935; Hammond, 1954), testicular function is stimulated by long days.

Some animals show a remarkable sensitivity to small changes in photoperiod, making it apparent that the length of day is very precisely measured. Thus, in the hamster, shortening the photoperiod by just half an hour (12 hours instead of 12.5 hours of light per day) will induce testicular regression (Fig. 5.36). We will show in the next section that it is the circadian timing system which makes this precise measurement of environmental time.

It takes several weeks for the animal's reproductive system to adjust after a change in the photoperiod. Thus, if a male hamster is switched from a long photoperiod to a short, nonstimulatory photoperiod, the testes slowly shrink from 3 grams weight to half a gram over the subsequent eight to ten weeks. Similarly, the female hamster's estrous cycle persists for six to eight weeks after the animal is exposed to short photoperiods before the cycle ceases. In these regressed animals the LH surge becomes a daily event, emphasizing the circadian basis of the LH triggering mechanism when ovarian function is suppressed (Bridges and Goldman, 1975; Moline, 1981).

The gonadal regression in hamsters persists for approximately 20 weeks before reproductive function starts up again spontaneously, even if the photoperiod remains short. This process, which is called "recrudescence," represents an escape from the inhibition caused by the short environmental photoperiod. In nature it would appear that recrudescence enables the animal to anticipate the mating season and develop full reproductive function before the days become long, even if it is living in a dark burrow and getting no seasonal day length information. Thus, just as daily anticipation is a benefit conferred by the

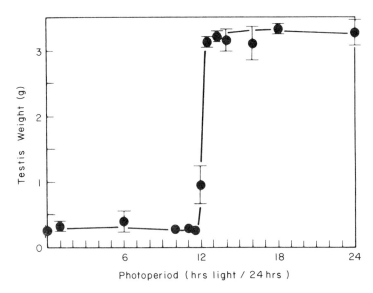

Fig. 5.36 Testicular response to photoperiods ranging in short steps from 0 to 24 hr of light per day. Each point represents the mean paired testis weight of a group of hamsters subjected to the indicated photoperiod for about 3 months. Photoperiods of at least 12.5 hr per day are required for maintenance of the testes. The critical day length is between 12 and 12.5 hrs. (From Elliott, 1976.)

circadian system, so is seasonal anticipation an advantage provided by these circannual timing mechanisms.

Circadian versus hourglass mechanisms   Two hypothetical mechanisms have been put forward for photoperiodic time measurement. The first assumes that the time measurement is based on some form of "hourglass," in which a physiological process requires a certain number of consecutive hours of light for the accumulation of a reaction product that is essential to reproduction function. This is essentially an exogenous hypothesis, since the reproductive state would not be a function of an internal oscillator, but instead a direct consequence of the characteristics of the light–dark cycle. In some insect species, hourglass time measurement does appear to be the predominant mechanism (Lees, 1966), but in mammals for which adequate studies have been performed, this mechanism has been ruled out.

The second hypothesis—that photoperiodic time is measured by the endogenous circadian system—was first advanced by Bünning in 1936. Bünning postulated that the circadian system was insensitive to light falling on most parts of the cycle, but that light falling on certain phases would bring about photoperiodically induced responses. On long days light would fall on a greater portion of the cycle and thus would fall on a critical phase that would stimulate the physiological response. This hypothesis is analogous to the principle of varying sensitivity to light over the circadian cycle (the phase response curve; see Chapter 2), which is responsible for entrainment of the animal to a 24-hour day.

Elliott (1976) has now shown in the hamster that the circadian system measures photoperiodic time by a mechanism that utilizes Bünning's proposed circadian variation in photoperiodic photosensitivity. Elliott, Stetson, and Menaker (1972) subjected male hamsters to light–dark cycles in which the duration of light was held constant at 6 hours but the length of darkness was varied. Figure 5.37 shows that in the LD 6:18 cycle light fell only during subjective day (when this nocturnal species is inactive), but in the LD 6:30 cycle alternate 6-hour photoperiods fell in the animals' subjective night. Similarly, in the LD 6:42 cycle (with twice the period length of the LD 6:18 cycle) light fell only in subjective day, whereas in the LD 6:54 cycle light fell in subjective night every fourth cycle.

The results were impossible to explain by an hourglass mechanism. The LD 6:18 and 6:42 cycles induced testicular regression, while the LD 6:30 and 6:54 cycles maintained testes weight. Neither the length of the light nor the length of the dark phase of the cycles could predict whether reproductive function would be stimulated. It was only the phase of the circadian system on which the light fell that was important, and as long as light fell in early subjective night every few days, testicular function was maintained.

In other experiments Elliott (1974) found he could entrain hamsters with one hour of light per day (a short photoperiod), and by altering the circadian period (the number of hours between successive light pulses) he could vary the phase of the animals' circadian systems on which the light pulse fell. As Bünning's hypothesis predicted, when the one-hour daily light pulse fell on certain phases of the circadian cycle, the testes were fully functional, and when it fell on other

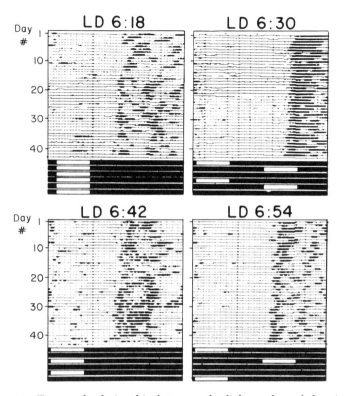

Fig. 5.37 Temporal relationship between the light cycle and the circadian wheel-running rhythm in hamsters exposed to four different LD cycles. Six consecutive days of the LD cycle are diagramed at the bottom of each record on a 24-hr time scale (solid bar = dark; open bar = light). On the LD 6:18 and 6:42 cycles the activity rhythm entrains in such a way that light is present only during the hamsters' subjective day and therefore fails to stimulate gonadal development. On the LD 6:30 and 6:54 cycles light is present both early and late in the subjective night and is photostimulatory to the reproductive system. (From Elliott, 1976.)

phases, they were not. These studies by Elliott and his colleagues have elegantly shown that in the male hamster the circadian timing system plays a critical role in inducing the seasonal switching-off and switching-on of the reproductive system. The differential sensitivity of the circadian system to light ensures that in long photoperiods reproductive function will be stimulated because light will impinge on a certain key phase of the hamster's circadian system.

Physiological basis   Although many of the elements of the physiological system which are responsible for mediating seasonal changes in reproductive function are known, the jigsaw puzzle has not yet been successfully assembled. In particular, the role of the pineal gland is poorly understood. However, it is known that the light–dark cycle is detected by the retina and the information conveyed via the retinohypothalamic tract to the SCN, which apparently interprets day length (the photoperiod). That is, animals with SCN lesions will not respond to short, nonstimulatory photoperiods and remain fully active reproductively (Rusak and Morin, 1976). Information about day length apparently travels from the SCN to the pineal gland by an indirect route. Neural fibers exit from the cranium to reach the superior cervical ganglion in the neck before reentering the brain to reach the pineal via sympathetic nerves coursing along the walls of blood vessels (Axelrod, 1974). Via this pathway the SCN are the source of circadian rhythmicity in several key aspects of pineal function. For example, Moore and Klein (1974) have shown that the circadian rhythm in pineal N-acetyltransferase activity and therefore in the synthesis of melatonin is abolished after destruction of the SCN or after ablation of the superior cervical ganglion.

The pineal gland appears to play a major role in transducing photoperiodic time information from the SCN into changes in gonadal function. Pinealectomy, like SCN lesions, has been shown to render animals incapable of responding appropriately to changes in the photoperiod (Reiter, 1974; Turek and Campbell, 1979). For example, pinealectomized ferrets, voles, and hamsters do not undergo testicular regression in nonstimulatory photoperiods, and their reproductive function continues through all seasons of the year (Turek and Campbell, 1979). These findings have led to the hypothesis that the pineal secretes an antigonadal substance. However, the story is not so simple. In some mammalian species (for example, the ferret and the Djungerian hamster), pinealectomy also blocks gonadal growth in animals exposed to long stimulatory photoperiods. Hence, this gland may have both antigonadal and progonadal functions.

The pineal's gonadal activity may be mediated by the hormone melatonin. Administering melatonin to an animal can stop testicular regression in short photoperiods. The mechanisms, however, are unclear because melatonin administered in long photoperiods can induce testicular regression (Turek and Campbell, 1979).

Part of the mechanism of seasonal switching on and off of the reproductive system involves a change of sensitivity of the hypothalamus to the feedback effects of reproductive steroids. When animals are exposed to nonstimulatory photoperiods, the hormones produced by the target organs—testosterone from the testis and estrogens from the ovary—have a marked inhibitory effect on the hypothalamus and the pituitary. Turek and his colleagues (Ellis and Turek, 1979; Ellis, Losee, and Turek, 1979) have shown that testosterone suppresses LH and FSH release in male hamsters in short photoperiods. The time course of this enhanced sensitivity is consistent with the change in reproductive function, developing within 3 weeks of exposure to a short photoperiod and then spontaneously recovering at about 20 weeks, when recrudescence of reproductive function begins. Like other aspects of the photoperiodic reproductive responses, this enhanced sensitivity is blocked in SCN-lesioned animals, indicating that the SCN plays an important role in mediating this response (Turek, Jacobson, and Gorski, 1980).

The enhanced sensitivity to gonadal steroids also appears to show a seasonal variation in female hamsters (Moline, 1981) and in sheep (Karsch, 1980). Sheep, for example, are autumnal breeders, becoming reproductively quiescent as the days lengthen. This is accompanied by an increased negative feedback sensitivity to estrogen (Karsch, 1980).

Thus, the circadian system plays as important a role in timing reproductive function as it does in timing other aspects of mammalian physiology. Even though the predominant rhythmicity in reproductive function is infradian (that is, estral, menstrual, or circannual), in species that have been carefully examined, reproductive events are switched on and off by circadian mechanisms. However, most studies to date have not defined the exact neural and endocrine pathways by which the circadian system exerts its influence. This would seem a profitable line for future inquiry.

The Human Circadian Timing System

Having examined the anatomy and physiology of the circadian timing system in the previous chapters, we will now examine its overall organization with the aid of a mathematical model. Because of the complex dynamics of coupled circadian oscillators, mathematical models have proved especially useful in investigating circadian organization. Some of the behaviors of coupled oscillator systems are not intuitively obvious and modeling has provided a way to formulate experimental questions and to examine experimental data.

A number of investigators, including Wever (1964, 1979), Pavlidis (1973), Pittendrigh and Daan (1976c), Daan and Berde (1978), Enright (1979), and Winfree (1980), have undertaken mathematical modeling of mammalian circadian systems. A review of these modeling approaches is beyond the scope of this book, but the interested reader may wish to consult a recent symposium volume (Moore-Ede and Czeisler, 1982). In this chapter we will focus on the organization of the human circadian system and will draw specifically upon a mathematical model recently developed by Kronauer and colleagues (1981).

Separate Circadian Pacemakers

Our knowledge of the human circadian system has been greatly aided by the study of its organization after internal desynchronization. As we discussed in Chapter 3, some rhythmic variables on occasion may maintain a different period from other rhythms in individuals who are isolated from time cues. Over the course of several weeks of study, one group of rhythms will totally lap another, with all 360° of internal phase relationships between them observed. The fact that each group of rhythms may lap the other several times over the course of an experiment indicates that there must be at least two separate pacemakers in the human circadian system. Each pacemaker maintains phase control over a number of diverse rhythmic variables.

Aschoff's and Wever's studies of subjects in temporal isolation (Aschoff, 1965b; Aschoff, Gerecke, and Wever, 1967; Wever, 1979) showed that when internal desynchronization occurred, the rest–activity cycle and the rhythm of urinary calcium excretion separated from the rhythms of body temperature and excretion of urinary potassium and water. Figure 6.1 shows the results of the spectral analysis of the data from a typical experiment. Both periodicities contribute to the rhythmicity of each variable, but the main spectral component follows either the temperature or the rest–activity cycle.

These findings have recently been confirmed and extended (Czeisler, 1978; Weitzman, Czeisler, and Moore-Ede, 1979; Czeisler, Weitzman, et al., 1980) in studies that used polygraphic EEG sleep recordings and blood sampling via an indwelling catheter. As is summarized in Figure 6.2, REM sleep propensity, plasma cortisol concentration, and urinary potassium excretion maintained a temporal relationship with the core body temperature rhythm during internal desynchronization. In comparison, the rhythms of skin temperature, plasma growth hormone concentration, and urinary calcium excretion tended to follow the rest–activity cycle and the circadian timing of slow-wave sleep. We have labeled the pacemaker driving the first group of variables as X and that driving the second group, which includes the rest–activity cycle, Y.

It is immediately apparent from Figure 6.2 that several physiological systems receive inputs from both pacemakers. For example, the thermoregulatory system receives a predominant input from the X pacemaker, which determines the core body temperature rhythm, but it also receives an input from the Y pacemaker, which governs the circadian rhythm of skin temperature (and therefore the timing of much of the body's heat loss). Similarly, the sleep and arousal centers receive inputs from the Y pacemaker, which determines the timing of slow-

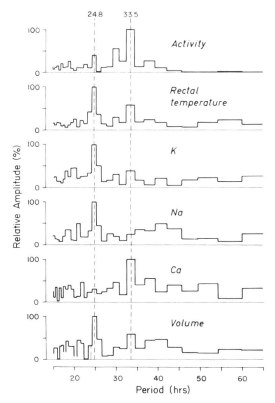

Fig. 6.1 Desynchronization of circadian rhythms in a human subject living in isolation without time cues. Period analysis of the rhythms shows the predominant spectral component in rectal temperature and in urinary potassium (K), sodium (Na), and water (volume) excretion was at 24.8 hours, whereas the circadian rhythms of activity and urinary calcium (Ca) excretion had their major spectral component at 33.5 hours. However, each rhythm showed a weaker component at the other period. (From Wever, 1979.)

wave sleep, but at the same time REM sleep is timed by the X pacemaker, so it is entrained with the circadian rhythm of core body temperature (Czeisler, Zimmerman, et al., 1980). Even in the kidney, certain rhythms, such as urinary potassium excretion, appear to be dominated by the X pacemaker, whereas the Y pacemaker appears to have the major influence on the rhythm of urinary calcium excretion.

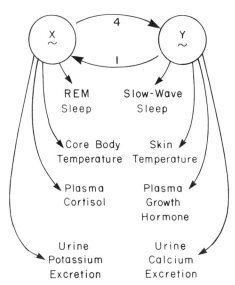

Fig. 6.2 The two groups of rhythms appear to be driven by separate pacemakers. Pacemaker X drives the rhythms of REM sleep, core body temperature, plasma cortisol concentration, and urinary potassium excretion. Pacemaker Y drives the rest–activity cycle and the rhythms of slow-wave sleep, skin temperature, plasma growth hormone concentration, and urinary calcium excretion. The coupling force exerted by X on Y is approximately four times greater than that of Y on X. (Copyright 1982 by Moore-Ede, Sulzman, and Fuller.)

Even if there are only two major circadian pacemakers in humans, this does not mean that there are no other oscillators in the circadian timing system. Indeed, as we discussed in Chapter 3, there is considerable evidence from other species that many tissues contain secondary oscillators which can generate rhythms but do not have the stability of the central pacemakers. These include the adrenal cortex, which generates the plasma cortisol rhythm, and the distal tubule of the kidney, which is apparently responsible for the rhythm of urinary potassium excretion. However, the various secondary oscillators normally appear to be phase controlled by the two central pacemakers, ensuring the internal synchronization of the entire circadian system so that interdependent body functions are appropriately timed with respect to one another.

Normally during entrainment to a 24-hour day–night cycle or during the first few weeks of free-run, all of an individual's rhythms are internally synchronized and demonstrate the same period. In the studies of Aschoff and Wever, only a quarter of their subjects demonstrated internal desynchronization during temporal isolation (Wever, 1979), but these experiments typically lasted no longer than a month. In longer-term studies virtually all subjects eventually appear to desynchronize. Czeisler, Weitzman, and co-workers (1980) have studied

12 free-running subjects in experiments lasting between 11 and 160 free-running days. During the first month, 8 out of 12 subjects maintained internal synchronization. However, the researchers found, in most subjects studied in free-running conditions for longer than two months, both in their studies and in the data they reanalyzed from the literature, that the circadian system eventually desynchronized into the two groups of rhythms defined in Figure 6.2. Figure 6.3 shows a subject who remained internally synchronized for over a month before the two pacemakers uncoupled. Internal desynchronization is not restricted to individuals with a particular psychological state (Lund, 1974) or of a particular age (Wever, 1979). If such factors play any role, it is apparently only to increase the rate of desynchronization.

INTERNAL COUPLING BETWEEN PACEMAKERS

In most human subjects studied in free-running conditions, the various circadian rhythms initially remain synchronized, indicating that there are internal coupling mechanisms between the two pacemakers. The coupling influence can be most simply demonstrated by examining Figure 6.3, which displays an output of pacemaker X, the body temperature rhythm, and an output of the Y pacemaker, the sleep–wake cycle. After desynchronization, the period of each circadian rhythm deviates from the free-running period, when they were synchronized. The subject in Figure 6.3 had an initial free-running period of 25.3 hours; after internal desynchronization the mean period of the sleep–wake cycle increased to 29.3 hours, and that of the body temperature rhythm decreased to 24.5 hours.

The fact that both rhythms change their period indicates that there must be a mutual coupling between the two pacemakers. The relative coupling strengths can be derived from the change in the periods of core body temperature rhythm and sleep–wake after internal desynchronization. The period of body temperature rhythm remained much closer to the internally synchronized period, so the X pacemaker must exert much more influence on the Y pacemaker than Y exerts on X. Wever (1979) calculated a coupling strength ratio of approximately 12:1 for X on Y versus Y on X, as measured by relative changes in the periods of the rhythms after desynchronization. However, the actual ratio of coupling strengths is probably nearer 4:1 (Kronauer et al., 1982) because the pacemakers cannot adopt their natural endogenous peri-

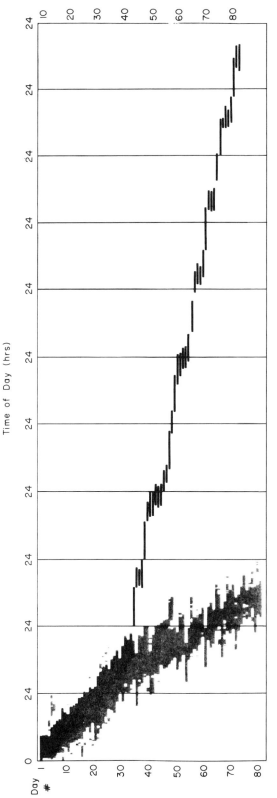

Fig. 6.3 Internal desynchronization between the rest-activity and body temperature rhythms in a human subject in temporal isolation. Bed rest episodes (horizontal bars) and body temperature (stippled areas indicate temperature below mean level) were entrained to a 24-hour day–night cycle on days 1–5, free-ran but maintained internal synchrony until day 35, then spontaneously internally desynchronized. After that point the period of the body temperature rhythm shortened and the rest-activity cycle increased its period (From Czeisler, Zimmerman, et al., 1980, reprinted by permission of Raven Press.)

ods, even after internal desynchronization, because of the continued interaction between them.

Conceptually, then, we can indicate the coupling between the two pacemakers as the two arrows in Figure 6.2. The coupling links between the pacemakers are presumably neural or hormonal (see Chapter 3), and act to ensure their mutual entrainment under normal conditions. The unequal coupling strengths are indicated in terms of their ratio. Neither coupling mechanism, however, is strong enough to maintain internal synchronization indefinitely in the absence of zeitgeber inputs from the environment.

Even after the two groups of rhythms have desynchronized from each other, there is still clear evidence of coupling interactions between the pacemakers (Aschoff, Gerecke, and Wever, 1967). For example, in Figure 6.3 the pattern of sleep after internal desynchronization shows a modulation corresponding to the phases of the core body temperature rhythm through which the independent sleep–wake rhythm is passing. There is thus *internal relative coordination* (Czeisler, 1978) between the two rhythms, analogous to the external relative coordination we discussed in Chapter 2 (Fig. 2.13), when a circadian pacemaker is subjected to a zeitgeber cycle that is not quite strong enough to entrain it. The sleep–wake cycle, and therefore the Y pacemaker, shows an analogous periodic modulation as it passes through the weak entraining signal provided by the X pacemaker.

When the data from such human studies is double- or triple-plotted, the interactions between the two groups of rhythms become readily apparent. Thus in Figure 6.4A (a triple-plotted version of Fig. 6.3) the short sleeps are clustered in the last half of the body temperature trough. The long sleeps, however, begin near the body temperature maximum, at quite a different phase relationship. This grouping of the phases of long sleeps and of short sleeps we will refer to as "phase clustering." It is interesting to note that the double-plotting technique, in which rhythmic data are reproduced side by side as well as vertically, has been used much less frequently for presenting human data than it has for animal experiments.

TRANSITION TO DESYNCHRONIZATION

In long-term studies, where internal desynchronization occurs after a month or so of study, the rhythms driven by pacemakers X and Y usually pass through a number of transition states. We will examine this transition in some detail, using the examples in Figure 6.4 because important information on the behavior and interactions of these pacemakers can be derived (Kronauer et al., 1982). The following stages can usually be recognized, although some individuals may jump immediately to stage 2 or 3.

Stage 1: internal phase drift (days 5–23 in Fig. 6.4A). After release from entrainment on day 6, the phases of the Y pacemaker rhythms (such as sleep) steadily drift apart from the X pacemaker rhythms (such as core temperature). Thus in Figure 6.4A the time of sleep occurred successively later in the trough of body temperature. In contrast to the phase relationships during entrainment (days 1–4) when body temperature started falling after the subject retired for the night and reached a minimum late in sleep, in free-running subjects the temperature rhythm reached its minimum close to the time when the subject fell asleep. This phase delay of the rest–activity cycle (or, in other terms, phase advance of the body temperature rhythm with respect to the rest–activity cycle) was first recognized by Wever (1973, 1979). The X pacemaker rhythms that tend to follow the body temperature rhythm, such as urine volume and potassium excretion (Wever, 1979) and REM sleep and cortisol (Czeisler, Weitzman, et al., 1980) similarly become phase advanced with respect to the sleep–wake cycle.

Stage 2: phase trapping (days 23–25 in Fig. 6.4A). When a certain phase relationship between the body temperature rhythm and the sleep–wake cycle is reached, an oscillation (or beating) occurs in the timing of sleep. This is an intermediate state between synchrony and desynchrony which we term "phase trapping" because while the phase relationship between sleep and body temperature is shifting there are constraints on the phase relationships tolerated. For the subject in Figure 6.4B (Jouvet et al., 1974), the phase-trapping stage is particularly pronounced on days 15–49, where abrupt jumps in the phasing of sleep can be seen. Phase trapping typically becomes more and more pronounced until internal desynchronization eventually occurs.

Stage 3: internal desynchronization (days 35–83 in Fig. 6.4A). When the two pacemakers eventually become desynchronized, the period of the sleep–wake cycle in most subjects becomes much longer and the period of the temperature rhythm a little shorter (Wever, 1974;

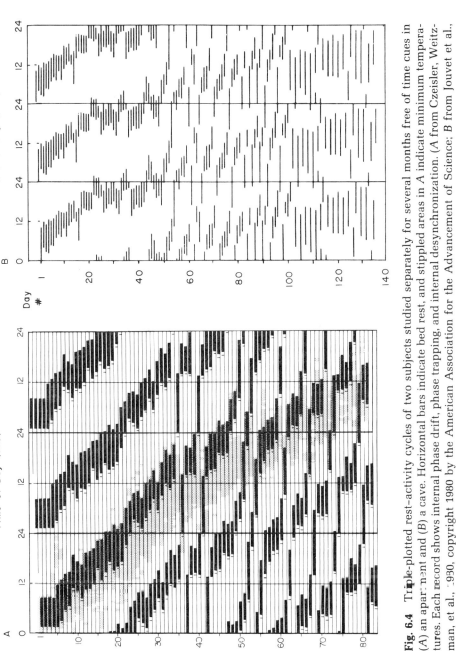

Fig. 6.4 Triple-plotted rest-activity cycles of two subjects studied separately for several months free of time cues in (A) an apartment and (B) a cave. Horizontal bars indicate bed rest, and stippled areas in A indicate minimum temperatures. Each record shows internal phase drift, phase trapping, and internal desynchronization. (A from Czeisler, Weitzman, et al., 1980, copyright 1980 by the American Association for the Advancement of Science; B from Jouvet et al., 1974.)

Czeisler, 1978), as we pointed out in Figure 6.3. However, the free-running sleep–wake cycle (and to a much lesser extent the body temperature rhythm) show internal relative coordination, so that the rate of advance of the sleep–wake cycle is much faster at certain phases of the body temperature rhythm and slower as it crosses through other phases and as phase clustering of the sleep periods occurs. The effects of this on sleep length (and wakefulness) are clearly visible in Figure 6.4A. At the point that desynchronization begins on day 35, the subject shows a delay in the onset of sleep (an extended period of wakefulness), then a long sleep that begins near the maximum of the body temperature rhythm. This is followed by several short sleeps and another long sleep, and this repeating cycle occurs throughout the record. The phase cluster line of these short sleeps adopts the shortened period of the body temperature rhythm.

LENGTHENING PERIOD OF THE Y PACEMAKER

One other feature of these records which gives an important clue about the underlying process, is that the period of the sleep–wake cycle tends to gradually lengthen during the course of internal desynchronization. This is most dramatically illustrated in Figure 6.4B, where the sleep–wake cycle eventually attained a period length of over 40 hours. Jouvet and colleagues (1974) referred to this as the "bi-circadian day," although there is little evidence to suggest that the period of the sleep–wake cycle (and the Y pacemaker) ever maintains a period that is twice that of the body temperature rhythm (and X pacemaker).

This gradual change in the natural period of pacemaker Y in fact starts before internal desynchronization has occurred. Phase trapping appears to represent beats (just like two musical notes that are close in frequency) between pacemakers X and Y. From the period of the beating can be calculated the period of pacemaker Y that must be responsible. Such calculations indicate that the natural period of Y is also steadily increasing during the phase-trapping regime and this accounts for the increase in the beat frequency that is often seen. Even in Stage 1 (internal phase drift) one of the simplest explanations for the progressive phase delay of the sleep–wake cycle with respect to the body temperature rhythm is a gradual increase in the period of pacemaker Y. As we discussed in Chapter 2, the phase relationship between two oscillators (or a zeitgeber and an oscillator) is dependent on the difference between their periods.

Wever (1979) has reported that some of his subjects adopted a shortened, rather than a lengthened, sleep–wake cycle with a period of 16–17 hours when their circadian systems became internally desynchronized. Does this mean that the Y pacemaker was gradually shortening its period? This is not necessarily the case since it is possible that the Y pacemaker lengthened its period as usual but the sleep–wake cycle split—a circadian phenomenon discussed in Chapter 3 (Fig. 3.5). This interpretation is supported by the observation that the short sleep–wake cycles have a mean period that is approximately half the mean period typically seen when the sleep–wake cycle lengthens in internal desynchronization. Furthermore, Czeisler, Weitzman, and colleagues (1980), who discourage their subjects from taking naps in the middle of the day, have never observed a subject to internally desynchronize with a short-period sleep–wake cycle.

Thus it appears that a drift in a single variable, the natural period of the Y pacemaker, could account for the apparent complexities of internal phase drift, phase trapping, and internal desynchronization which are seen in the sleep–wake patterns of human subjects isolated from environmental zeitgebers. This conclusion is confirmed in the next section, where we will show that a simple, mutually coupled oscillator model (Kronauer et al., 1982) can replicate all these behaviors of the human circadian system when the period of Y is steadily increased, an increase that may be analogous to the aftereffects we discussed in Chapter 2.

Mathematical Modeling of Coupled Oscillators

To examine some of these behaviors of the human circadian system, Kronauer and colleagues (1982) designed a model using two mutually coupled oscillators. They found that the free-running behavior of the human circadian system could be readily modeled if the two oscillators were coupled so as to provide amplitude information to each other. Until recently, amplitude has not been utilized as an important variable in the modeling of the circadian system's behavior, but it turns out to be a key parameter if the human system is to be understood.

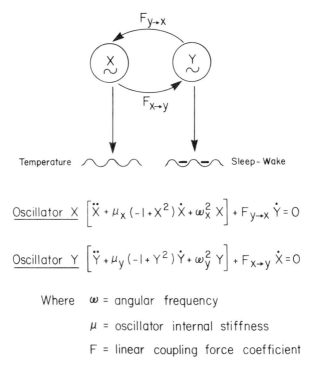

Oscillator X $\left[\ddot{X} + \mu_x\,(-1+X^2)\,\dot{X} + \omega_x^2\,X\right] + F_{y \to x}\,\dot{Y} = 0$

Oscillator Y $\left[\ddot{Y} + \mu_y\,(-1+Y^2)\,\dot{Y} + \omega_y^2\,Y\right] + F_{x \to y}\,\dot{X} = 0$

Where ω = angular frequency

μ = oscillator internal stiffness

F = linear coupling force coefficient

Fig. 6.5 The human circadian timing system modeled by two coupled van der Pol oscillators. Oscillator X drives the body temperature and related rhythms, and oscillator Y drives the sleep–wake cycle (solid bar = time of sleep) and the rhythms associated with it. The equations of each oscillator are shown. The mutual coupling is applied by adding a linear coupling force coefficient to each equation. (From Kronauer et al., 1982.)

To model these relationships, Kronauer and his colleagues (1982) used van der Pol oscillators, one of the simplest types of self-sustained oscillator equations, as shown in Figure 6.5. In keeping with our previous nomenclature, the output of the X oscillator was taken to be temperature, and the output of the Y oscillator was a rhythm which when it reached its minimum two-thirds corresponded to the timing of sleep. This provided sleep for one-third of every cycle (8 hrs in 24 hrs). The values of the various parameters in the oscillator equations were obtained from analysis of physiological data from human subjects.

INTERNAL DESYNCHRONIZATION

With this very simple model, all the stages of internal desynchronization could be readily replicated by steadily increasing only one variable, the period of the Y pacemaker; internal phase drift, phase trapping, and the phase clustering of long and short sleep periods were all simulated (Fig. 6.6A). This simulation, with the period of the body temperature pacemaker maintained throughout at 24.37 hours, shows the internal phase drift of the rest–activity cycle with respect to the temperature minimum (days 5–53), phase trapping (days 53–75), and the ultimate desynchronization (days 75–100). Phase clustering of short sleep periods occurred near the end of the temperature minimum, whereas the long sleeps occurred when sleep began closer to the maximum of body temperature. In Figure 6.6B the parameters of the simulation are shown.

The data from a number of different individuals studied for extended periods of time in temporal isolation have been successfully simulated using this model. These simulations have suggested a number of important features of the human circadian system. First, there are at least two mutually coupled pacemakers. Second, the coupling is much stronger from the X to the Y oscillator than in the reverse direction. Third, all changes can be explained by a steady increase in the period of pacemaker Y.

Perhaps one of the most important lessons from these studies is that the transition from the synchronized to the desynchronized state involves a very minor change in a single parameter. Indeed, some human subjects drift in and out of desynchronization over the course of such an experiment. There is nothing fundamentally different between the synchronized and the desynchronized state, although the consequences for physiological function are considerable.

ENTRAINMENT TO ENVIRONMENTAL ZEITGEBERS

The mathematical model also makes it possible to examine how zeitgeber inputs are coupled to the two pacemakers. The zeitgeber inputs could impinge predominantly on oscillator X or predominantly on oscillator Y. The model provides quite distinct predictions for these two possibilities, as shown by the results of the two simulations in Figure 6.7. If the zeitgeber impinges predominantly on X, the timing of

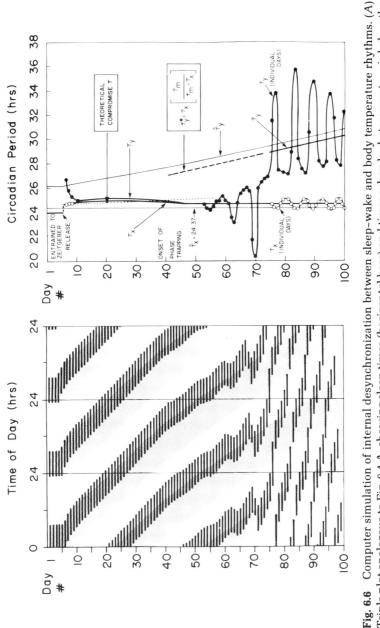

Fig. 6.6 Computer simulation of internal desynchronization between sleep–wake and body temperature rhythms. (A) Triple plot analogous to Fig. 6.4 A, showing sleep times (horizontal bars) and times when body temperature is below the mean (stippled). (B) Periods of the X pacemaker ($\hat{\tau}_x = 24.37$) and the steadily increasing period of the Y pacemaker ($\hat{\tau}_y$) are plotted over the "days" of the simulation, together with the observed periods of the body temperature rhythm (τ_x) and the sleep–wake cycle (τ_y). The period of the Y oscillator during phase trapping (τ_y^*) can be computed from the beat frequency and is shown by the dashed line. Internal phase drift, phase trapping, and internal desynchronization can all be simulated by steadily increasing $\hat{\tau}_y$. (From Kronauer et al., 1982.)

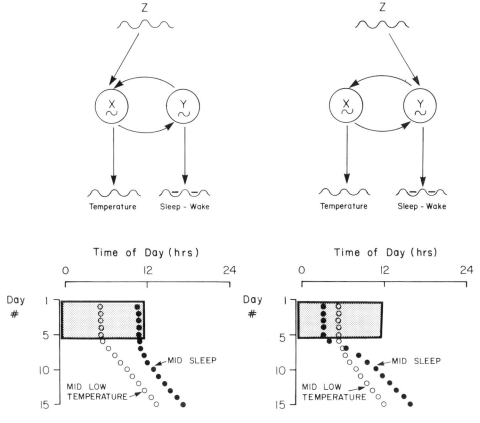

Fig. 6.7 Predicted effects of a zeitgeber, Z, impinging predominantly on the X oscillator (*left*) or on the Y oscillator (*right*). Lower panels show the phase of mid-low temperature and mid-sleep during 5 days of entrainment and then during the first 10 days of free-run in the absence of zeitgeber cues. If Z impinged on X, mid-sleep would phase-advance with respect to mid-low temperature on release into free-run, while if Z impinged on Y, mid-sleep would phase-delay with respect to mid-low temperature, as is the case in human experiments (see Fig. 6.4A). (From Kronauer et al., 1982.)

sleep will phase-advance with respect to the body temperature rhythm on release into free-running conditions, whereas if the zeitgeber impinges predominantly on Y, sleep will be phase-delayed with respect to the body temperature rhythm. The data from human subjects who

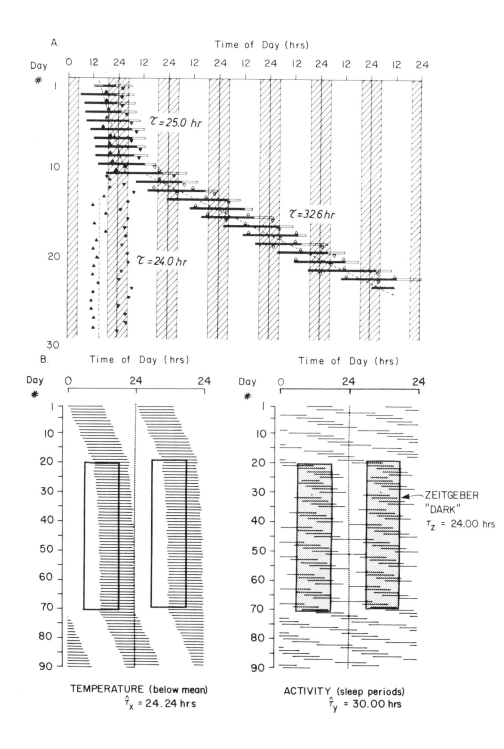

A.

Time of Day (hrs)

τ = 25.0 hr

τ = 32.6 hr

τ = 24.0 hr

B.

Time of Day (hrs)

Time of Day (hrs)

ZEITGEBER
"DARK"
τ_z = 24.00 hrs

TEMPERATURE (below mean)
$\hat{\tau}_x$ = 24.24 hrs

ACTIVITY (sleep periods)
$\hat{\tau}_y$ = 30.00 hrs

Fig. 6.8 Partial entrainment of the circadian system by a weak 24-hour zeitgeber cycle. (*A*) Data from a human subject who maintained internal synchrony until day 11, when the body temperature rhythm (upright triangle = temperature maximum; inverted triangle = temperature minimum) became desynchronized from the rest–activity cycle (black bars = activity; white bars = bed rest). The temperature cycle became entrained to the 24-hour zeitgeber, but the rest–activity cycle free-ran. (*B*) Computer model simulation indicating that a weak zeitgeber with a 24-hour period can entrain the body temperature rhythm (*left*) but not the sleep–wake cycle (*right*), even though the zeitgeber impinges on the Y pacemaker. Times of zeitgeber application indicated by boxes. (*A* from Wever, 1979; *B* from Kronauer et al., 1982.)

have been released from entrainment clearly show (as Fig. 6.4A illustrated) an immediate phase delay in sleep with respect to the body temperature rhythm, clearly suggesting that the zeitgeber input falls predominantly on Y.

There are some rather important consequences if zeitgebers predominantly entrain the human circadian system via the Y pacemaker, particularly since the coupling drive of oscillator X onto oscillator Y is so much stronger. For example, Wever (1979) has demonstrated that a weak zeitgeber (light–dark cycle with reading lights available to subjects at night) with a 24-hour period can entrain the body temperature rhythm (Fig. 6.8A) but not the rest–activity cycle. At first this might seem to suggest that the zeitgeber acts on the X rather than on the Y pacemaker. However, simulations again show that this need not be the case. In Figure 6.8B a zeitgeber with a 24-hour period was applied only on the Y pacemaker; although it failed to entrain the Y pacemaker with its much longer period, it did entrain the X pacemaker. The zeitgeber influenced the waveform of the Y pacemaker but did not entrain the pacemaker because the zeitgeber period was outside the range of entrainment. However, the modulations in the waveform of Y were conveyed through the coupling link to X, which had a period much closer to 24 hours. These weak modulations in the mutual coupling were sufficient to ensure entrainment of the body temperature rhythm to a 24-hour period.

A reverse situation is seen when the zeitgeber is applied with a longer period, as in the example of Wever's (1979) data plotted in Figure 6.9A. The rest–activity cycle was entrained but the body tempera-

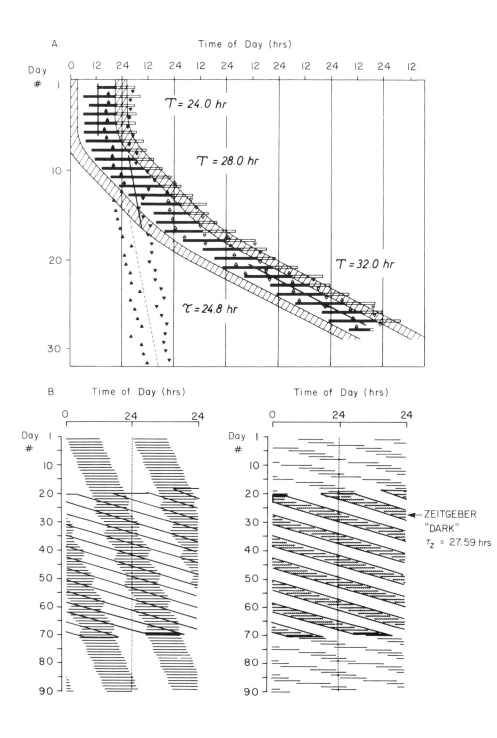

Fig. 6.9 Partial entrainment of the circadian system by a long-period zeitgeber. Symbols as in Fig. 6.8. (A) Data from a human subject whose rest–activity cycle entrained to 28- and 32-hour zeitgeber periods, while the body temperature rhythm free-ran with a period of 24.8 hours. (B) Computer simulation, similarly indicating that a long-period zeitgeber (T_z = 27.59 hrs) can entrain the rest–activity cycle (right) but does not entrain the body temperature rhythm (left), although some relative coordination is seen. (A from Wever, 1979; B from Kronauer et al., 1982.)

ture rhythm was not. Again, this was easily simulated by the model (Fig. 6.9B), which showed that the sleep–wake cycle was entrained by the zeitgeber (which was close to the natural period of Y) but the body temperature rhythm was not, since the zeitgeber period was outside the range of entrainment of X. The coupling links, however, do result in internal relative coordination in the pattern of the body temperature rhythm. An examination of Wever's data also suggests that internal relative coordination is occurring in the body temperature pattern. It should be emphasized that all these simulations were achieved without altering any of the model's parameters. Thus, this model of two simple oscillators mutually coupled together can mimic a wide variety of behaviors of the human circadian timing system.

Anatomical and Physiological Basis

So far in this chapter we have undertaken a black-box analysis of the human circadian timing system, considering only zeitgeber inputs and overt rhythmic outputs. We have deduced that there are two main pacemakers, but their structure and many details of their function remain undefined. We are faced with a problem, though, in trying to peer into the black box because of the ethical and practical limits of human experimentation. We must resort to animal experimentation but must be cautious in interpretation because we cannot be certain that the organization is the same in humans as in other mammals, even the primates.

However, to attempt to conceptualize the human circadian system from animal experimentation is not an unreasonable goal. After all, there are considerable similarities in the anatomical structures and physiological systems of most mammals. For example, all mammals have a pituitary-adrenal axis with a pituitary that secretes ACTH and an adrenal gland that secretes a corticosteroid. It is reasonable to talk about the mammalian pituitary-adrenal axis because the general design of the system is the same even if there are interspecies differences. Thus, the exact chemical nature of the predominant adrenal steroid differs between species (for example, corticosterone in rats and cortisol in humans), and the rate of secretion varies (cortisol secretion = 20 mg/kg/day in the squirrel monkey and 0.3 mg/kg/day in man). The example of the pituitary-adrenal axis is repeated in virtually every physiological system, so that while the details may vary, the "big picture" does not.

There is no reason to suspect that the circadian timing system is any different. Clearly there are interspecies differences; some animals (most rodents) are nocturnal and others (most primates) are diurnal, but there are nocturnal primates (the owl monkey), and diurnal rodents (chipmunks). Some animals (humans and nonhuman primates) have highly prominent circadian rhythms, whereas others (cats and dogs) are only weakly rhythmic. However, all mammals so far studied have an SCN that receives input via an RHT from the retina, and the characteristics of the measurable endogenous circadian rhythms are very similar.

Some of the supposed interspecies differences may in reality be a result of the experimental design. For practical reasons, quite different experimental designs have been used when studying different species. For example, many studies of humans in free-running conditions have used self-selected light–dark cycles where the subject has control of the light switch, whereas most animal studies have used constant levels of illumination not controlled by the subject. The differences between these two experimental designs is considerable, since the self-selected light–dark cycle provides a feedback loop between an output of the system (the timing of wakefulness) and the input (the light–dark cycle).

A comparable problem exists because of differences in the physiological variables that have been measured in the various mammalian species. Variables are often selected because of their ease of experimental measurement in a particular species. Many circadian studies of rodents have used wheel-running activity, a highly specific circadian variable with no human counterpart. Furthermore, in studies of rhythms in blood constituents in rodents, because of their small blood volume, each animal may contribute only a single sample as it is sacrificed. The rhythm reported is thus composed of group-mean data from a large group of animals rather than a series of samples from a single subject, as can easily be obtained from man.

IDENTITY OF THE X AND Y PACEMAKERS

All mammalian species that have been examined have a pair of suprachiasmatic nuclei located dorsal to the optic chiasm and bilateral to the anterior tip of the third ventricle. Until recently, there was some doubt about whether humans even had suprachiasmatic nuclei. However, as we discussed in Chapter 4, Lydic and colleagues (1980) have recently demonstrated that a pair of homologous neural clusters appear to exist in the human brain. The human SCN are smaller and more diffusely organized than those of other mammalian species, but there is no reason to believe that they differ in their functional role in the circadian system.

In Chapter 4 we reviewed how lesions of the SCN in various nocturnal and diurnal mammals disrupted circadian rhythmicity in a wide variety of physiological variables. Therefore the SCN might contain either the X or the Y pacemaker or both. However, recent data, obtained mostly from the squirrel monkey, suggest that the SCN contain (or "are") the Y pacemaker and that the X pacemaker is located outside the SCN (Fuller et al., 1981). After total bilateral SCN destruction, the rest–activity, feeding, and drinking rhythms were disrupted, but the circadian rhythm of body temperature persisted. Furthermore, after SCN destruction in rodents, some evidence suggests that although the rest–activity rhythm is disrupted, the rhythms that persist include body temperature (Dunn, Castro, and McNulty, 1977; Powell et al., 1977; Nakayama, Arai, and Yamamoto, 1979), plasma cortisol concentration (Krieger, Hauser, and Krey, 1977; Szafarczyk et al., 1979; Reppert et al., 1901) and REM sleep (Stephan and Nunez, 1977; Mouret et al., 1978; Yamaoka, 1978). It is noteworthy that these are the very rhythms which the human data from internally desynchronized subjects suggest are driven by the X pacemaker.

The weight of evidence, albeit some of it preliminary, suggests that the Y pacemaker corresponds to the SCN and that the X pacemaker is located elsewhere. The location of the X pacemaker is as yet unknown, although evidence reviewed in Chapter 4 suggests the ventromedial or lateral hypothalamus as possible sites.

ENTRAINMENT PATHWAYS

In Chapter 2 we discussed how the light–dark cycle (Czeisler, Richardson, Zimmerman, et al., 1981) and social cues (Wever, 1979) appear to be the major environmental zeitgebers of the human circadian timing system. In all other mammalian species the RHT conveys photic information directly from the retina to the SCN. As yet it has not been feasible to confirm the presence of an RHT in humans, but there is no reason to suppose that one does not exist.

The coupled oscillator simulation of Kronauer and co-workers (1982) indicated that in humans the predominant target of environmental zeitgeber information is pacemaker Y. Since that pacemaker appears to correspond to the SCN, and the SCN in other mammals receive a major input of light–dark cycle information via the RHT, the model conforms very well to the physiological evidence.

INTERSPECIES DIFFERENCES

While the model of the human circadian system and the physiological data from animal experiments give some credence to our proposed organization of the human circadian timing system, there are some clear differences between human and nonhuman species. The circadian timing system of other mammalian species has not, for example, been reported to show internal desynchronization of the rest–activity and body temperature rhythms. This does not mean that two separate pacemakers do not exist; instead the two pacemakers may be much more tightly coupled, or pacemaker Y may not have such a labile period as it does in man. Similarly, in other mammalian species food availability is an effective zeitgeber, but there is no clear evidence as yet that this is true for humans, perhaps because the appropriate experiments have not yet been undertaken.

The problems that remain represent more than a lifetime or two of research objectives, but with the increasing numbers of investigators becoming intrigued with the problem of physiological time, enormous strides are already being made in determining the anatomical and physiological basis of timekeeping. We are only now beginning, 250 years after deMairan's original discovery of biological clocks, to extend our considerable knowledge of biology in three dimensions into an understanding of biological organization in the fourth dimension—time.

Medical Implications of Circadian Rhythmicity

In this chapter we will examine the implications of circadian rhythmicity for human health, considering first the health consequences of manipulating the day–night cycle. We will then deal with the normal rhythmic variations in measurements used for diagnosis and with circadian variations in the effectiveness of drugs and other forms of therapy. Finally, we will discuss malfunctions of the circadian timing system and develop a classification of the various disorders.

Alterations in Environmental Time

Within the past hundred years, a mere instant on the evolutionary time scale, there have been dramatic changes in the temporal environments to which man is exposed. *Homo sapiens,* like most other species on earth, evolved in a regular 24-hour light–dark cycle. Although the earth's daily spin on its axis has been slowing down over the one million years that it took man to evolve (Coale, 1974), the day length when humanlike forms first appeared was only 20 seconds shorter than it is now (Rosenberg and Runcorn, 1975). In comparison, the environmental changes of the last hundred years have been immense; the past century has seen Edison's invention of the light bulb and the development of the airplane, changes that have enabled rapid shifts to be made in the timing of environmental light–dark cycles. Travel in an easterly or westerly direction used to be sufficiently slow that the subjective daily light–dark cycle rarely deviated by more than a few minutes from 24.0 hours—well within the range of entrainment. The time zones below the Arctic Circle are 500 to 1,000 miles wide—much more than a day's travel in earlier times. Nowadays, however, millions of people are subjected to abrupt shifts in environmental time cues when they fly across time zones or work on shift work schedules. Still others, such as patients in an intensive care ward, are subjected to environments where they are isolated from daily time cues. What are the consequences of these alterations of our temporal environment?

The most widely recognized disorders relate to abrupt environmental phase shifts either because of a change of work shift or because of a rapid flight across multiple time zones. Because of the variation in the degree of stress imposed by different kinds of work schedules, we will discuss the effects of a single acute shift in environmental time separately from the chronic effects of repeated schedule shifts.

ACUTE SHIFTS—JET LAG

Since the 1950s and the advent of widespread commercial jet travel, increasing numbers of people have been exposed to the effects of moving rapidly across time zones. Each year several hundred million people fly on trips that cross one or more time zones. The majority of international air routes are in fact in an east–west rather than a north–south direction, because so many of the main centers of international commerce in Europe, the United States, and Japan are at similar latitudes. The age range of those exposed varies from the newborn to the elderly. There is minimal selection on the basis of health, and little attention has been paid to the consequences of such shifts on people with specific clinical pathology.

What is jet lag? Although individuals vary widely in their responses to travel across multiple time zones, the general symptoms include disruption of sleep, gastrointestinal disturbances, decreased vigilance and attention span, and a general feeling of malaise. The extent of the symptomatology depends on the number of time zones crossed; most people can cope with a phase shift across a single time zone, but virtually everybody has a problem flying across 12 zones.

Anecdotes abound about the consequences of jet lag on human performance. Because it takes a significant amount of time for people to readjust, particularly after crossing six or more time zones, significant deterioration in performance is frequently the immediate consequence. One of the earliest anecdotes is of the time in the 1950s when Secretary of State John Foster Dulles flew to Egypt to negotiate the Aswan Dam treaty. The sensitive negotiations were conducted very shortly after his flight, and apparently as a consequence the project was lost to the USSR. Because this initiated a decade of Soviet influence in that country, he henceforth advised the diplomatic corps to be very cautious about conducting meetings soon after such travel. The potential effects of jet lag are enormous, not only for diplomats but also for businessmen and anyone else who must make difficult decisions shortly after arrival.

Twice a year the entire population of many countries in the temperate latitudes is subjected to a one-hour time shift. We adjust our work–rest schedules forward in the spring for daylight saving time and backward in the autumn for standard time. This might be thought a trivial adjustment for the circadian system, but it takes several days to make a complete adjustment (Monk and Folkard, 1976; Monk and Alpin, 1980). Furthermore, there is evidence that the traffic accident rate increases significantly in the week after the time shift (Monk, 1980; Hicks, Lindseth, and Hawkins, 1980). Presumably even a small change in alertness and psychomotor performance can affect the statistical outcome of a large number of potential highway accidents.

Pathophysiology Three factors appear to underlie the body's response to these timing conflicts. We will call Factor 1 that portion of the fatigue and/or sleep loss which is directly consequent to travel and is independent of the timing conflict. For example, although flying from north to south does not result in a time zone shift, the stress of the flight itself may influence an individual's performance. This factor is particularly exacerbated by the airlines' decision to make virtually all eastward transatlantic flights at night. This enables them to use their expensive aircraft twice per 24 hours but places economic considerations at a higher priority than the sleep disruption of passengers and crew.

The other two factors result from the shift between external and internal time. Factor 2 is the direct effect of external desynchronization, the disparity between external and internal body time. For example, an individual's minimum daily psychomotor performance capability may be at 4:00 A.M., a time when there are normally no external challenges (Klein, Wegmann, and Bruner, 1968). However, after a flight across five time zones from the United States to Europe (a phase advance of five hours), the circadian timing system takes several days to resynchronize. Before it starts to adjust, the person's body time will be at 4:00 A.M. when the local environmental time is 9:00 A.M., a time of day when the person may be expected to operate with maximum effectiveness at an important business meeting or may have to combat rush-hour traffic.

Factor 3 is more subtle. As we discussed in Chapter 3, the circadian timing system is composed of several separate oscillators or "clocks," each of which resets to the new environmental time at a different rate. Therefore there is internal desynchronization between the rhythmic functions of the various physiological systems, which further reduces a person's optimum abilities. Several reports suggest that psychomotor performance is influenced by the internal phase relationships between the different oscillators of the circadian timing system (Winget, 1974; Wever, 1979).

How fast does a person adapt to a time zone shift? This depends on the variable being studied, because some physiological rhythms resynchronize faster than others. It also depends on the number of hours of phase shift that must occur and the strength of the zeitgeber signal (see Chapter 2). Figure 7.1 shows that people exposed only to weak time cues after they arrive in a new time zone, because they stay in their hotel room where food and a bed are continually available, readjust much more slowly than those who go out during the local daytime and hence are exposed to illumination, feeding cues, social cues, and other environmental zeitgebers (Gerritzen, 1962; Klein and Wegmann, 1974).

Another factor that determines the rate of resynchronization is the direction of travel. As Figure 7.2 shows, westward travel, which phase-delays the circadian system, is followed by more rapid adjustment than eastward travel. Estimates by Klein and Wegmann (1979) suggest that the rate of resynchronization is approximately 50% faster with westbound flight. A series of careful studies by Wegmann and co-workers (1973) have shown that the effect of travel direction is independent of the time of day during the flight (whether it is a day or a night flight), the subjects' place of origin (providing they were fully

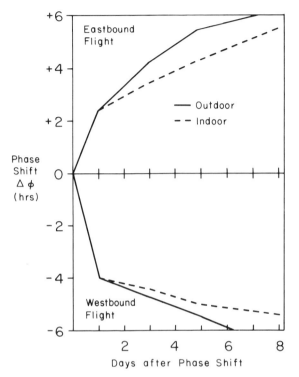

Fig. 7.1 Phase shift of the maximum of the performance rhythm after transmeridian air travel, phase-advancing (*upper panel*) or phase-delaying (*lower panel*) across six time zones. Subjects who remained indoors upon arrival took longer to resynchronize than subjects who were exposed to multiple environmental zeitgebers because they were outdoors. (After Klein and Wegmann, 1974.)

synchronized before travel), or the amount of sleep deprivation caused by the time of flight. The average rate of resynchronization was calculated for a combination of psychophysiological variables, including heart rate, body temperature, catecholamine, 17-hydroxycorticosteroids (17-OHCS), and various mental performance tasks. Reentrainment was shown to occur at a rate of about 88 minutes per day after westbound travel, but at only about 56 minutes per day after eastbound travel (Klein and Wegmann, 1979). These figures, however, have to be treated with caution, because the mean rate of resynchronization

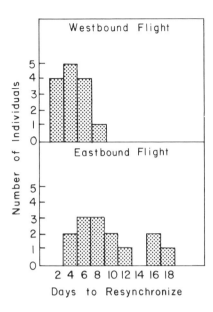

Fig. 7.2 Frequency distribution of the length of time required to resynchronize the psychomotor performance rhythm after transmeridian air travel across six time zones westbound (phase delay) or eastbound (phase advance). The average rate of resynchronization was faster for phase delays than for phase advances, with some individuals taking up to 18 days to resynchronize after a 6-hour phase advance. (After Klein and Wegmann, 1979, AGARD Lecture Series no. 105; data from Beljan et al., 1972.)

of rhythmic functions is less important than the time it takes for all of them to resynchronize, so that internal synchrony is regained.

Figure 7.2 also illustrates that there are marked intraindividual differences in the rate of resynchronization after a phase shift. One factor contributing to this variability was identified by Mills, Minors, and Waterhouse (1978a), who showed that many individuals subjected to a phase advance of environmental time do not phase-advance their body temperature cycle, but instead phase-delay it the "wrong" way around the clock, taking many more days to achieve the same eventual phase relationship. Thus, as the data of Klein and Wegmann (1979) show in Figure 7.3, some individuals after an eastward flight across nine time zones shift their body temperature cycle with a 9-hour phase advance, while others resynchronize with a 15-hour phase delay. In fact, the studies of Mills, Minors, and Waterhouse (1978a) even suggest that the majority may shift the "wrong" way around the clock in eastward travel and the "right" way around the clock in westward travel.

Not all rhythms necessarily shift in the same direction. Figure 7.3 shows that in all subjects after an eastward flight the rhythm of urinary 17-OHCS excretion shifted via a phase advance, even in those subjects whose body temperature rhythm phase-delayed. Aschoff

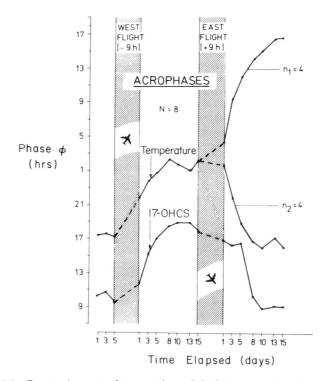

Fig. 7.3 Reentrainment after westbound flight across nine time zones (phase delay) and, two weeks later, "reentrainment by partition" after a return flight eastbound (9-hour phase advance). The mean acrophases of body temperature and urinary 17-OHCS excretion of eight individuals are shown. All subjects adjusted to the westbound flight by phase-delaying their circadian rhythms, but after the eastbound flight, the body temperature rhythm of half the subjects achieved the net phase advance of 9 hours by phase-delaying 15 hours. The 17-OHCS rhythm in all subjects, however, phase-advanced after the eastbound flight. (From Klein and Wegmann, 1979, AGARD Lecture Series no. 105.)

(1978b) has termed this "reentrainment by partition" since the different oscillators resynchronize by independent routes.

Which individuals will shift rapidly and which more slowly can best be predicted by the amplitude of the individual's body temperature rhythm (Fig. 7.4). This relationship, first proposed by Aschoff (1978c), has been tested by Reinberg and colleagues (1978a, b), who have con-

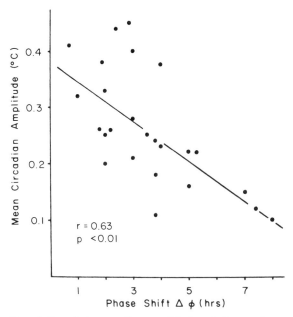

Fig. 7.4 Correlation between the amplitude of the oral temperature rhythm and the phase shift of the temperature rhythm acrophase on the first day after transfer from day to night shift. Individuals with low-amplitude circadian temperature rhythms shifted much more rapidly than those with a high-amplitude rhythm. (After Reinberg et al., 1978b.)

firmed that there is a significant correlation between the rate of adjustment after an environmental phase shift and the previous amplitude of the rhythms of body temperature and urinary 17-OHCS (although not other urinary rhythms). Subjects with the lowest-amplitude rhythms shifted most readily. Aschoff (1978c) has suggested that the amplitude of an individual's circadian rhythm indicates the "degree of persistence" of his circadian oscillators and therefore their inertia to phase-shifting stimuli. Certainly it is a matter of common experience that some individuals cope with shifts in environmental time fairly well, while others suffer insomnia and malaise for much longer periods. This has implications for the selection of individuals who can tolerate shift work schedules, as we will discuss in a later section.

Finally, we should recognize that a person's performance is a complex function. Although Kleitman (1963) suggested that performance is correlated with body temperature rhythm, more recent analysis has shown that this is an oversimplification. Folkard and colleagues (1976) have demonstrated that while the ability to do tasks requiring little memory is greatest at the maximum of the body temperature rhythm, tasks requiring memorization are best accomplished near the minimum of the body temperature rhythm. After a time zone shift, memorization may suddenly improve during the new daytime but then deteriorate again as the circadian system becomes reentrained.

A cure? How then to minimize the effects of jet lag? First, it is most important to decide whether complete adjustment to the new local environmental time is advisable. If an extended stay is planned, then methods which increase the rate of resynchronization are required. If, however, return to the original time zone is imminent, it is preferable not to shift one's circadian clocks at all; where possible one should remain on one's original schedule. Ideally, the schedules imposed by the local environment, whether business meetings or visits to the opera, should be scheduled to take into account the circadian phases of the guest as well as the host. The least stress will be endured if the home schedule of sleeping and waking can be maintained.

For those who will stay in a new time zone for several weeks, it is possible to speed up the rate of adjustment to the local day–night schedule by maximizing the strength of environmental zeitgebers such as social cues (Wever, 1979; Vernikos-Danellis and Winget, 1979) and the light–dark cycle (Czeisler, Richardson, Zimmerman, et al., 1981) by maintaining the greatest contrast between the day and night portions of the cycle (see Chapter 2). The experiments of Klein and Wegmann (1974), which were illustrated in Figure 7.1, indicated that exposure to the social activities and light–dark schedule of everyday life speeded up the rate of resynchronization as compared to staying in a hotel room and avoiding social contact and bright daylight. Although meal timing has not been definitively examined as a potential zeitgeber in humans, its potential synchronizing impact should not be ignored. Certainly the diurnal variation in a number of body functions is determined by the timing of meals (Goetz et al., 1976).

Thus the best advice for the traveler who wants to adjust to a new time zone is to immediately be active during the new daytime and to sleep during the new night, eat meals at the local times, and spend the day out and about in well-lit environments with maximum social contact. Even with such an exposure to environmental schedules, one may feel below par for several days before complete adjustment occurs. There has, therefore, been some attempt to develop special therapeutic methods for hastening the resynchronization process, which would be particularly useful for people who have problems in readjusting to a new schedule. For example, Figure 7.2 showed that some people take up to 18 days to readjust after traveling across only six time zones.

In recent years it has become apparent that certain pharmacological agents will reset biological clocks. At certain phases of the cycle these agents induce phase advances, at other phases they induce phase delays, and at other times there may be no effect. Some of the compounds produce shifts in one direction only. Much of the work to date has been done on simpler organisms, but recent work suggests that the effects may be generalized to higher animals. Such pharmacological agents offer a potentially very effective way to rapidly reset our biological clocks after crossing multiple time zones. If this therapeutic principle can be applied in man, such drugs could form the basis of a "jet-lag pill."

Compounds which are effective at resetting circadian clocks in lower organisms include the methyl xanthines, theophylline (Mayer, Gruner, and Strubel, 1975), and caffeine (Mayer and Schere, 1975); certain protein synthesis inhibitors, such as chloramphenicol (Frelinger, Matulsky, and Woodward, 1976), and puromycin (Feldman, 1967; Applewhite, Satter, and Galston, 1973; Rothman and Strumwasser, 1976; Karakashian and Schweiger, 1976), the ionophore valinomycin (Bünning and Moser, 1972; Sweeney, 1974), and ethanol (Enright, 1971; Bünning and Moser, 1973; Sweeney, 1974). It is interesting that some of these agents are present in the beverages we commonly drink while flying. Caffeine and theophylline are found in coffee and tea, and of course ethanol is the alcoholic constituent of wines, beer, and liquor. However, we normally take these with no regard to the time of day and do not consider their potential effect on our biological clocks.

Only a few of the agents listed above can be used in mammals. However, Ehret, Potter, and Dobra (1975) have shown that theophylline and pentobarbital will phase-shift the body temperature rhythm of rats. Phase-response curves were developed for the clock-resetting effect; there were either phase advances or phase delays of varying

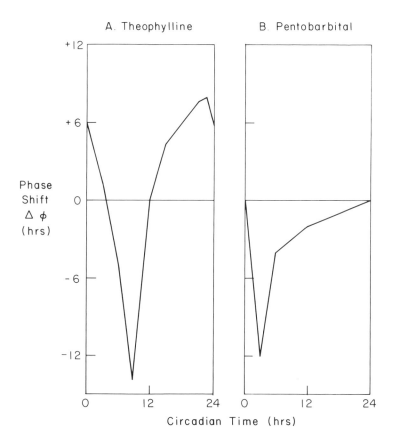

Fig. 7.5 Phase-response curves of rats to (A) theophylline and (B) pentobarbital. Injection of a 75 mg/kg dose of theophylline produces either phase advances or phase delays of varying magnitudes in the body temperature rhythm, depending on the circadian phase when it is administered. These shifts can be plotted as a phase-response curve analogous to those for the phase-resetting action of light pulses (see Fig. 2.23). Pentobarbital also has phase-shifting action, but produces only phase delays, which vary in magnitude depending on the circadian phase. (After Ehret, Potter, and Dobra, 1975, copyright 1975 by the American Association for the Advancement of Science.)

magnitude, depending on the time of day the drug was given (Fig. 7.5). They have also investigated the clock-resetting properties of other agents, including L-dopa (Ehret, Meinert, and Groh, 1979) and dexamethasone (Horseman, Meinert, and Ehret, 1979). In studies using hamsters, Moline and Moore-Ede (unpublished observations) have shown that haloperidol has some phase-resetting properties, and they have defined a phase-response curve. However, the problem with all these studies to date has been that the doses that induce phase shifts also have other side effects. Furthermore, complex interactions among the various oscillators of the circadian timing system in mammals may interfere with the phase-resetting efficacy of a drug. Hence, it will be necessary to search for pharmacological compounds with much more specific actions and fewer side effects before the dream of obtaining a jet-lag pill can be realized.

CHRONICALLY SHIFTING SCHEDULES

While most people can tolerate an occasional adjustment to a new time zone without too much discomfort, repeated shifts in rest–activity schedules present a much greater stress. Wiley Post, the first person to fly around the globe, in his book *Around the World in Eight Days* (1931) reported that he was one of the first to discover the negative effects of crossing time zones repeatedly. Nowadays millions of people experience repeated conflicts between body time and environmental time (Johnson et al., 1980). These include not only frequent travelers such as commercial airline pilots, but also workers on shift work schedules.

The number of people involved in some kind of shift work is staggering. For example, 21.9% of the French working population, 20.4% in the Netherlands, 16% in the United States, 23.3% in Japan, and 18.2% in the United Kingdom are shift workers (Rutenfranz, Knauth, and Colquhoun, 1976; Shift Work Committee, 1979). The definitions of shift work vary, and some of the surveys are old, but it is safe to say that 15–25% of the working population of the industrialized countries are shift workers—a total of perhaps 60 million people.

Shift work has developed because of the economic demands for expensive equipment to be used around the clock; the need for continuous attention to technological processes in the chemical, steel, and energy industries; and the demand for 24-hour service in hospitals, transportation, and emergency services. Shift work schedules vary greatly. The working shifts may vary from 6 to 12 hours in length; some workers may be permanently assigned to a given shift, others may rotate every one or two days or every week or month (Rutenfranz et al., 1977). Some crews rotate by successive phase advances and others by phase delays. Even if a person always works on the night shift, he still may be subject to shifting environmental schedules if he is active during the daytime on weekends and vacations.

Airline pilots and flight attendants are exposed to particularly irregular schedules. Figure 7.6 shows the work, sleep, and meal schedules of two airline pilots; neither has a consistent pattern of sleep and wake times. Because of work demands, the external time of the environment is rapidly and randomly shifted on a day-to-day basis from the pilot's home time. The timing of meals, which may also influence body time, is also shifted, but not always to the same degree as the work schedule. Finally, the social time frame of his home life, which is interspersed with his schedule, may disrupt the environmental schedule even further. Those who are most likely to be exposed to highly disruptive schedules are often the most senior pilots, who select these schedules because vacation allowances are highly favorable for those who fly those routes. However, it should be a matter of some concern that the older pilots, with first pick of the schedules, choose the most disruptive work routines, for they may be less able to cope with them.

Unfortunately, pilot duty schedules in the United States are currently governed by Federal Aviation Administration (FAA) regulations in which rest time is computed solely on the basis of accumulated duty time with no consideration of the individual's circadian or body time. Such regulations suppose that the human body is a machine running at a constant level through night and day. The fallacy in such an assumption will be obvious to anybody who has tried to sleep at an unaccustomed time. It is not as easy to achieve an adequate amount of sleep when going to bed at 10 A.M. as compared to 10 P.M., yet the FAA regulations assume that it is. It is essential, therefore, that the FAA's regulations be revised to take into consideration the body times of the air crew.

Health consequences   There are major differences in the tolerances of different individuals to chronically shifting schedules. When schedules involving night work were first adopted in the armaments industry in World War I, an especially large number of stomach diseases occurred among night shift workers (Vernon, 1921). In 1929 Duesberg

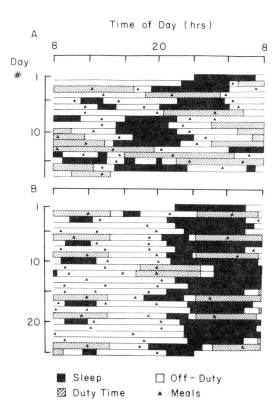

Fig. 7.6 Single plot of sleep, duty time, and mealtime schedules of two commercial airline pilots. Sleep (black bars), duty time (hatched bars), and meals (triangles) show no consistent scheduling with respect to circadian time, and there are frequent alternations between subjective daytime and subjective nighttime flying. The data were obtained from daily logs kept by the pilots. The scheduling algorithm used by the airlines to generate these work schedules calculated off-duty time solely on the basis of work time, with no regard for circadian factors. (From Fuller, Sulzman, and Moore-Ede, 1981.)

and Weiss calculated that the risk of stomach ulcer was eight times greater for shift workers than for people in regular daytime work. However, since World War II, comparisons of the health of shift workers with that of the general population have not revealed such striking medical problems. This appears to be because a process of self-selection is continually taking place among shift workers. Unlike the situation in wartime, no one is forced into shift work, and in almost all countries shift workers undergo regular medical examinations. Those who find they cannot cope with shift work schedules usually apply for other jobs. Thus, statistical comparisons between shift workers and nonshift workers are biased against finding any difference.

That chronically shifting schedules induce serious health impairment in some people is indicated by the study of workers who have requested transfer to nonshift work schedules. Aanonsen (1959), in a study of 128 people who had left shift work, found that 16% had had problems with sleep before starting shift work, whereas 90% complained of chronic tiredness and insomnia on shift work schedules. Furthermore, the incidence of gastrointestinal disorders such as peptic ulcer was two- to threefold greater than that of a general population of workers who had never been on shift work. Ten to twenty percent of those who began shift work reacted with gastrointestinal diseases. Hence, it is clear that such schedules can be tolerated only by a certain segment of the population.

The major health consequences of shift work are disorders of digestion and of sleep (Rutenfranz, Knauth, and Colquhoun, 1976). The digestive disorders include gastric and peptic ulcers, gastritis, and various intestinal disorders (Shift Work Committee, 1979). The disturbances of sleep consist of changes in the length and in the quality of sleep. There are also some reports of respiratory problems and lower back pain (Shift Work Committee, 1979). The incidence of cardiovascular disease and neurological disorders, however, does not seem to be greater than in the population at large (Rutenfranz et al., 1977).

The hazards of shift work may affect a much larger population than the workers themselves. For people who have not fully synchronized to the new schedule—and most shift workers on rapidly changing schedules do not become synchronized—the incidence of work errors is much higher in the early morning hours (say, 3 to 5 A.M.) than at any other time of day. For example, the delay before answering a telephone (Browne, 1949), the number of errors in reading meters (Bjerner, Holm, and Swensson, 1955) or in answering warning signals (Hildebrandt, Rohmert, and Rutenfranz, 1974), all increase during the early morning hours. This is in keeping with laboratory tests of human performance, which show that psychomotor ability and the capability for mental arithmetic fall to a minimum between 3 and 5 A.M.

Disrupted schedules may affect the health of only the individual concerned. For example, truck drivers are 200% more likely to have a single-vehicle accident at 5 A.M. than during regular daytime hours (Harris, 1977). But the consequences may be much more widespread. Human errors by pilots or air traffic controllers contribute to many aircraft accidents, and part of the cause undoubtedly is incomplete adjustment to their shift work schedule. The exact contribution of a disrupted duty–rest schedule is hard to ascertain, because the extensive accident reports which the National Transportation Safety Board prepares after an aircraft accident provide no evaluation of the prior work–rest schedules of the crew or the air traffic controllers (Moore-Ede and Fuller, 1980).

A number of documented aviation incidents illustrate the dramatic loss in vigilance that may occur in individuals not fully adapted to a night shift. For example, air traffic controllers at Los Angeles International Airport were shocked one night when a Boeing 707 that had filed a flight plan to land at the airport overshot Los Angeles at 32,000 feet and headed west over the Pacific. The entire crew had fallen asleep, and the plane was cruising on automatic pilot; it was not until the jet was 100 miles out over the Pacific Ocean that the controllers were able to awaken one of the crew by triggering a series of alarms in the cockpit. Fortunately, the plane carried enough fuel to make the return journey to Los Angeles (Anonymous, 1978). Similarly, a jetliner with about 150 passengers aboard was coming in early one morning after a night flight from Honolulu. Just before touchdown, with the plane only 200 feet above ground, the pilot fell asleep, even though he was supposed to be flying the plane. Fortunately the co-pilot realized in time what had happened, and with the pilot's help he uneventfully landed the Boeing 720 (Aviation Safety Institute File: RS, HAZARD 770923001).

Some shift work schedules are designed without considering the limits of the human circadian system. For example, the crew of a 707 that crashed in Bali in the Pacific in 1974, killing all 96 passengers and 11 crew, had just flown an exhausting schedule (here corrected to San Francisco time). They departed from San Francisco on April 17 at 7:44 P.M. and flew to Honolulu, arriving at 1:32 A.M. The next day they left at 3:39 A.M. and flew to Sydney, Australia, arriving at 2:35 P.M. The following day they left Sydney at 6:21 P.M. and landed at Jakarta, Indonesia, at 1:30 A.M., then left the same night at 2:18 A.M. and arrived in

Hong Kong at 6:40 A.M. The flight that ended in a fatal crash departed
from Hong Kong the next day at 4:00 A.M. and crashed at Bali at 8:30
A.M. (Gologoski, 1979). The combination of crossing at least 12 time
zones and mixing day and night flying, with little time allowed for rest,
may well have been a major contributor to the "pilot error" in this accident. Such schedules reinforce our recommendation (Moore-Ede,
Gander, and Czeisler, 1980) that the FAA should not be calculating
duty time and rest time solely from the number of prior hours of work.
They should consider the phase of the pilot's circadian system and
where possible avoid flights which impinge on the air crew's subjective night. While scheduling rest time and duty time is undoubtedly
easier if one doesn't have to take into consideration circadian phase,
the overriding influence of circadian phase on the duration and quality
of sleep (Czeisler, Weitzman, et al., 1980) makes it essential to pay attention to circadian factors when designing such schedules.

Much greater dangers exist. The accident at the Three Mile Island
nuclear power plant in 1979 occurred at 4 A.M. in the middle of the
night shift (11 P.M.–7 A.M.) with a crew that had been on night duty for
only a few days and had been rotating shifts around the clock on a
weekly basis for the previous six weeks. Ehret (1981) has pointed out
that the operator in the control room of a nuclear power plant is con-
fronted by a complex array of instruments and instrument panels. In
this monotonous environment, which demands a high quality of vigilance and psychomotor tasks such as reading meters and responding
to emergency calls, shift work schedules may have contributed to the
sequence of events that precipitated the Three Mile Island accident.

Pathophysiology   Disruptions of mealtimes and rest periods are the
most obvious causes of gastrointestinal and sleep disorders. Food is
provided at times of day when the endogenous circadian rhythmicity
of the gastrointestinal tract does not anticipate it (Suda and Saito,
1979). Similarly, it is usually not feasible for shift workers to sleep at
the normal phase of the circadian cycle. Instead, they may try to sleep
near the maximum of their endogenous rhythm of alertness and thus
fail to obtain the needed rest. Because they are out of synchrony with
their family and friends, their schedules cause stresses in social and
family life (Shift Work Committee, 1979), especially when shift workers want to sleep during the daytime, when their children are awake
and making noise.

Cycles in the environment such as light–dark, social cues, and food availability, which normally synchronize the circadian system, are therefore disrupted and may even provide conflicting phase information. Furthermore, with continual shifts, the body is constantly in a state of transient internal desynchronization. Hence the normal predictability of internal physiological events is continually disrupted—in other words, "the timing is off."

No long-term studies have as yet been conducted in humans to investigate the effect of extended exposure to shift work. That these are badly needed is indicated by animal studies of longevity. Blowflies show a 20% decrease in lifespan when they are subjected every week to a 6-hour phase shift of the light–dark cycle (Aschoff, Saint-Paul, and Wever, 1971), and Halberg, Nelson, and Cadotte (1977) found that mice subjected to a complete 180° inversion of the light–dark cycle every week showed a 6% decrease in longevity. Translated into human terms, this might mean a 4- to 5-year shortening of the lifespan of people who all their lives are subjected to repeatedly shifting schedules. As a cautionary note, there is an essential difference between the animal experiments and the human situation. The animals cannot select whether they want to remain on the shifting schedules, whereas humans can. In light of the evidence that some people who cannot tolerate shift work avoid it, the chronic effects in humans may not be as great as in experimental animals. To resolve these questions, long-term studies are badly needed, perhaps similar to the Framingham study (Dawber, 1980) in which extensive populations are studied over many years to determine the long-term consequences of life style (Moore-Ede, 1976b).

Treatment and prevention   The simplest treatment for those who have medical problems resulting from repeatedly shifting schedules is to avoid shift work altogether and transfer to regular daytime work. The evidence is clear that medical problems are more frequent when people who cannot tolerate such schedules are forced to, as in wartime (Duesberg and Weiss, 1939; Aanonsen, 1959). However, those who tolerate shift work schedules can probably do so without major risk.

In scheduling shift workers, managers should decide whether total readjustment to each shift is required or no adjustment at all. The schedules that change once a week seem to be the most pernicious, because the individual is always in a state of readjustment to a new schedule. It may take most of the week to adjust to the new times of sleep and work, and when adjustment is complete, it is time to change work and sleep times all over again. Many workers prefer fast-rotating schedules with only one or two days on each shift, which does not allow enough time to resynchronize to any new schedule. However, on the night shift the circadian rhythms of performance will show the typical dip between 3 and 5 A.M., and extra care must be taken, for this is the time when accidents or mishaps are most likely to occur.

The ideal shift work arrangement for one's circadian clocks is for each crew to be permanently assigned to a particular shift so that they have ample time to readjust and be at their best during their work period each day. The problem is that only a few people will agree to work at night permanently, and even those who do may revert to daytime activity on vacations and weekends and hence spontaneously adopt a type of rotating shift. While some sort of rotating schedule is often necessary in industry, further work is needed to determine the best shift work schedules. This is an important area of applied circadian research with potential benefits for millions of people.

ABNORMAL PERIOD ENVIRONMENTS

Some shift workers are exposed to an average day length that is radically different from 24 hours. A particularly extreme example is that of the men in the U.S. Navy who operate nuclear submarines (Schaeffer et al., 1979). They are typically assigned to 6 hours of duty and 12 hours of rest, in other words, an 18-hour day–night cycle. The Navy's rationale is that an 8-hour duty span is too long for maintaining peak vigilance, and there is only space on the submarine for three shifts of men—hence the 18-hour (3 × 6) schedule. During the early 1960s the first Polaris submarine crews used a 4-hours-on, 8-hours-off schedule, which could be kept synchronized to a 24-hour day. However, because other duties have to be performed during the 8 hours off, the crew slept only 5–6 hours per day and became progressively sleep deprived.

The long-term consequences of human exposure to an 18-hour day–night cycle have not been adequately characterized; however, shorter-term studies have demonstrated problems with insomnia, emotional disturbance, and impaired coordination (Andrezsyuk, 1968; Dushkov and Komolinskii, 1968). Sleep is highly fragmented in the naval watch schedules at sea, as compared to those of shore-based naval personnel (Johnson, 1979). Perhaps the best indicator of the difficulties is the enormously high turnover of enlisted men in U.S. submarine crews—as high as 33 to 50% per voyage—and only a small number of men undertake more than two or three of the 90-day submarine missions. This has serious consequences for training, because on each trip a third or half of the crew are trainees, and the first months they spend learning their tasks. The whole process is not helped by the fact that each submarine, even within the same class, is mechanically somewhat different. However, it is interesting to note that officers, who usually maintain a 24-hour day–night schedule, tolerate submarine duty much longer and may spend years on active duty. As Schaeffer and colleagues (1979) pointed out, there would be considerable advantages to adopting 24-hour rest–work schedules for the whole crew; in this way the men could function optimally and the drop-out rate might be less. There should be some global concern about the health and performance of these men, since they are the ones with their fingers directly on the nuclear button!

Pathophysiology Circadian oscillators are capable of synchronizing only to environmental cycles with periods close to 24 hours, as we discussed in Chapter 2. This range of entrainment means that human circadian rhythms will free-run when exposed to periods outside a limited range. With an 18-hour day–night cycle, most circadian pacemakers cannot synchronize (Kleitman, 1963), and the shift workers' performance rhythm will be out of synchrony with the demands of their environment much of the time. For submarine crews the problem is complicated further because social activities and meal times are on a 24-hour day, and the crew, therefore, receives conflicting zeitgebers. The studies of Schaeffer and co-workers (1979) of 11 submariners on patrol demonstrated that some of the men showed no adaptation at all to the 18-hour watch, whereas others showed both 18-hour and 24-hour components in each of their rhythms.

We have conducted a simulation of an 18-hour day–night cycle using a squirrel monkey on an LD 9:9 schedule. Figure 7.7 shows the complex pattern of the body temperature rhythm, with two different components (Fuller, Sulzman, and Moore-Ede, 1981). The first component is a circadian rhythm free-running with a near-24-hour period,

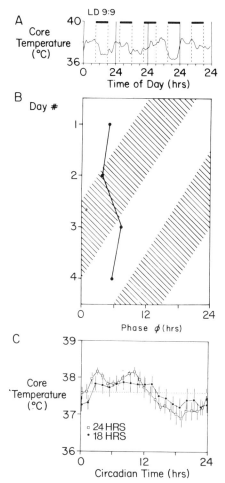

Fig. 7.7 Core body temperature rhythm in a monkey exposed to an 18-hour light–dark cycle (LD 9:9). (*A*) Body temperature (°C) plotted against time of day for four days. Times of darkness are shown by dark bars. (*B*) Plot of core body temperature rhythm daily acrophases, indicating that the rhythm does not entrain to the 18-hour light–dark cycle (hatched areas indicate darkness). (*C*) Educed waveforms of body temperature at a 24-hour period (open squares) and an 18-hour period (closed circles). The major component is an approximately 24-hour rhythm, free-running independently of the light–dark cycle. However, the light–dark cycle has passive effects on body temperature (see Fig. 5.14), so a smaller component with an 18-hour period is also seen. (Copyright 1982 by Fuller, Sulzman, and Moore-Ede.)

because an 18-hour light–dark cycle is outside the monkey's range of entrainment. On day 1 the maximum of the body temperature rhythm occurred during lights-on, on day 2 after lights-off, on day 3 just after lights-on, and on day 4 at about the same time as on day 1. This circadian component is thus showing four cycles superimposed on the five 18-hour LD cycles. Light additionally has a direct influence on the body temperature rhythm, causing it to be higher on average when the lights are on. Thus the autocorrelation analysis of the body temperature pattern in Figure 7.7C shows rhythmic components with both 18- and 24-hour periods. The internal circadian component (the 24-hour cycle) shows the largest amplitudes, but there is still a significant rhythm at the 18-hour period as a result of the concurrent LD cycle.

Studies in lower organisms have demonstrated the deleterious effects of long-term exposure to day–night cycles that deviate from the organism's natural free-running period (Went, 1959). For example, when Pittendrigh and Minis (1972) exposed fruit flies to light–dark cycles with periods of 21 and 27 hours, they demonstrated that lifespan was considerably reduced for flies maintained on any day–night cycle outside the range of 24–25 hours. Similar findings have been reported by Saint-Paul and Aschoff (1978). The periodicity of food intake is also important. Rats fed at 20- and 28-hour intervals consume less food and gain less weight than rats fed once every 24 hours (Saito and Noma, 1980). As yet there has been no long-term evaluation of human populations exposed to environmental schedules that deviate from 24 hours, but clearly this should be undertaken. The only treatment available is the avoidance of work–rest schedules outside the human circadian range of entrainment.

INSUFFICIENT ENVIRONMENTAL CUES

Anthropologists tell us that *Homo sapiens* evolved in temperate or subtropical latitudes which had a 24-hour day–night cycle. As the human race spread out over the globe, some populations finally settled near or above the Arctic Circle. There, during midwinter and midsummer, they are subjected to either continuous night or continuous daylight; only during spring and autumn are they exposed to a 24-hour day–night cycle. In its absence, in midsummer and especially midwinter adults and children who live above the Arctic Circle are active at all times of day and night and show highly disrupted circadian systems, sometimes with very low-amplitude rhythms (Lobban, 1960). Lewis and Masterton (1955) showed that members of the British North Greenland Expedition maintained normal circadian rhythmicity throughout most of the two years of study but showed a disrupted sleep–wake pattern during the winter darkness. There are similar reports from other expeditions. Yoshimura (1973) reports a pattern during midwinter darkness of short naps taken at various times of day rather than the consolidated single long sleep taken during a light–dark cycle. However, there have been quite significant differences reported between the rhythms of native Eskimos and those of visitors who have previously lived in a highly regular daylight cycle. The visitors continued to show circadian rhythms in midsummer and midwinter with significant amplitudes, whereas the Eskimos had low amplitude and sometimes undetectable rhythms (Lobban, 1960).

In a more recent report, however, Lobban (1976) noted that modern technology had influenced the circadian rhythmicity of the population by providing electric light and imposing regular schooltime schedules for the children. Now significant amplitudes are found in circadian rhythms at all four seasons of the year. Thus it appears that the diminished rhythmicity reported in Eskimo populations was not a genetic phenomenon but mostly a result of the lack of exposure to light–dark cycles. This conclusion is supported by comparisons between the circadian rhythm amplitudes of new visitors to the Antarctic and of expedition members who had become acclimatized by a year of Antarctic living. At the same season of the year under identical conditions the acclimatized individuals showed lower-amplitude circadian rhythms, even though they were not natives (Yoshimura, 1973).

While the introduction of modern technology has promoted circadian rhythmicity in Eskimo populations, it has also led to the loss of environmental time cues for certain individuals at lower latitudes. In technologically advanced societies it is possible to advertently or inadvertently isolate people from the time cues of the natural environment. This may occur, for example, in the intensive care units of modern hospitals, where patients are monitored very closely and where the room lighting is often maintained at a constant level at all times. If they are conscious, the patients' social contacts, with nurses, for example, occur equally during night and day. Even meals may have no circadian schedule if all metabolic requirements are given continuously by vein. These patients, who are already highly stressed, are thereby deprived of any external temporal cues.

Health consequences When Byrd went to the Antarctic in the 1920s he is said to have taken along, in addition to two coffins, a dozen straitjackets (Mullin, 1960), a precaution that proved overly pessimistic. However, there are a number of important health consequences of deprivation of 24-hour time cues, especially during midwinter's continuous darkness. The most common is insomnia, often referred to as "big eye" (Mullin, 1960). Visitors to the Arctic and the Antarctic (Lewis and Masterton, 1955; Yoshimura, 1973) as well as native residents above the Arctic Circle have considerable problems with disturbances of the sleep–wake cycle. In a recent survey of residents of Tromsö, Norway, which has two months of midwinter continual darkness, 24% reported insomnia (Lingjaerde and Bratlid, personal communication). More serious are the major changes in mood that occur during midwinter darkness, commonly referred to as the Morketid syndrome or Arctic hysterias (Foulks, 1972). The incidence of mental illness, suicide, and violence is higher in populations living at high latitudes than at low latitudes (Kraus and Buffler, 1979) but many other factors, particularly the harsh environment and the disruption of native societies in recent years, contribute importantly to these statistics. Suicidal behavior is more common during midwinter darkness, but there are also increases in suicidal behavior in non-Arctic populations in December and January.

The other health consequences of environmental time cue deprivation are harder to define. However, evidence from plant and animal studies points to the importance of providing a periodic environment. Plants and animals fail to function optimally when deprived of environmental time cues. In constant light tomato plants fail to thrive (Hillman, 1956) and fruit flies show significant reduction in lifespan (Pittendrigh and Minis, 1972). In studies of squirrel monkeys in constant light, we have found that the response of the thermoregulatory system to environmental challenges becomes remarkably impaired (Fuller, Sulzman, and Moore-Ede, 1978b). As Figure 7.8A shows, animals synchronized to environmental time cues were able to maintain body temperature without difficulty during exposures to mild cold (a fall of 8° C in ambient temperature), but when time cues were re-

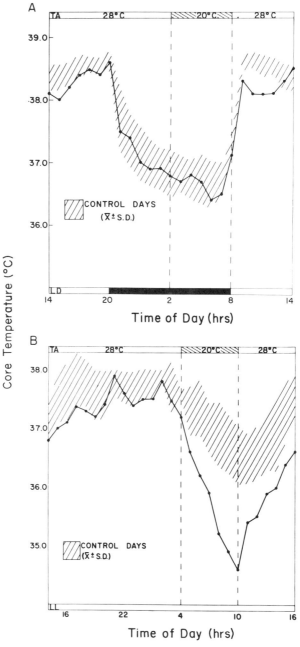

Fig. 7.8 Effects of 6-hour exposures to cold (20 °C ambient temperature) on body temperature in monkeys normally maintained at 28 °C when (*A*) entrained to an LD 12:12 cycle, and (*B*) free-running in constant light (lighting regimen indicated at the base of each graph). The shaded areas show the body temperature rhythm (x̄ ± SD) for the three previous control days, and the solid line is the body temperature on the day of the cold pulse. The cold pulse had virtually no physiological effect in LD, but produced a significant fall in core body temperature when the animals were isolated from environmental time cues. (From Fuller, Sulzman and Moore-Ede, 1978a, copyright 1978 by the American Association for the Advancement of Science.)

moved (Fig. 7.8B) body temperature fell when the animal was exposed to the same mild cold stress. This effect appears to be the result of internal circadian dissociation of heat production and heat loss mechanisms, which often occurs when no time cues are present (Fuller, Sulzman, and Moore-Ede, 1978b, 1979b). Thus, effective thermoregulation appears to require the proper temporal synchronization of the various physiological systems responsible for maintaining body temperature.

It is possible that circadian dissociation of thermoregulatory mechanisms may contribute to the susceptibility of elderly people to hypothermia when environmental temperature is reduced (Moore-Ede, 1981). Elderly people appear to be more subject to internal desynchronization (Wever, 1979) and circadian dissociation of various rhythms has been reported in animals with advanced age (Sacher and Duffy, 1978; Albers, Gerall, and Axelson, 1981). Hypothermia is a serious medical problem (Fox et al., 1973), and the possibility that a circadian disorder may contribute to it deserves further attention.

One solution to aperiodic environments such as the intensive care ward is to provide temporal cues in whatever ways are possible. Instead of nutrients being provided by vein continuously throughout day and night, as is the normal practice, they should be provided periodically. In squirrel monkeys totally maintained by an intravenous infusion of an amino acid, glucose, and lipid mixture, weight maintenance was enhanced when the nutrients were provided periodically at double the rate for half (12 hours) of the day instead of a continuous infusion throughout each 24 hours (Finn et al., 1982). Periodic intravenous nutrition has also been reported to reduce the incidence of the liver pathology which is often associated with prolonged continuous feeding by vein (Maini et al., 1976). Much needs to be done, however, to evaluate other ways of providing circadian periodicity in intensive care environments.

Medical Consequences of Normal Circadian Timekeeping

Traditionally, the concept of homeostasis has been a central theme in introductory physiology courses in medical school. The idea that the body's internal environment is maintained at a constant level was first put forward by Claude Bernard (1878) and subsequently refined by

Walter B. Cannon (1921). Bernard's and Cannon's teachings that the integrity of higher forms of life relies on maintenance of a constant internal milieu have been central to the development of modern physiological theory.

Unfortunately, the teaching of this concept to successive generations of medical students has led to an overly simple perception being embedded in the collective medical consciousness. Cannon never suggested that every physiological variable is tightly regulated within narrow limits, nor did he indicate that even the most well-regulated variables were maintained at an absolutely constant level. Rather, it should be obvious that many physiological variables are components of effector systems or are directly controlled by them and thus may change their level dramatically from moment to moment. Take as an example skin temperature. Although the regulated variable of the thermoregulatory system—core body temperature—is relatively constant, this constancy is achieved at the expense of large fluctuations in skin temperature, which enable the animal to increase or decrease heat loss.

Even highly regulated variables such as core body temperature show reproducible rhythmic changes. We discussed in Chapter 5 the circadian rhythms in core temperature as well as the temperature changes that occur over the course of the menstrual cycle. The reader of this book will recognize the myriad of other physiological variables which show these rhythmic variations.

It is interesting to note that Cannon (1929) in the article in which he first outlined his concept of homeostasis, specifically pointed out that even the most tightly regulated variables may oscillate. He accordingly defined homeostasis as the process which regulates a physiological variable within certain limits, but that the variable may oscillate between those limits, and the limits themselves may change in response to some special demand.

IMPLICATIONS FOR DIAGNOSIS

Almost every physiological variable that might be measured as part of a diagnostic procedure has a circadian rhythm. Probably the most comprehensive documentation of circadian rhythms in humans is provided by Conroy and Mills (1970) in their book *Human Circadian Rhythms*. Indeed, we may have now reached the point where it is more interesting to demonstrate that a certain variable has no circadian variation than to show that it is rhythmic.

The consequences for diagnosis are obvious; every laboratory result must be evaluated in terms of the time of day when the sample or measurement was taken. For some variables the circadian variation is relatively small, maybe less than 5%, but for others the changes can be large. The concerned but naive intern managing a critically ill patient may stay up all night taking repeated urine samples, sending them to the lab for "stat" electrolytes at hourly intervals, and altering the IV infusion accordingly, without ever realizing that the normal nocturnal rates of urinary electrolyte excretion may be five times less than daytime levels (Moore-Ede, Brennan, and Ball, 1975). This is not to suggest that such close attention is necessarily inappropriate. However, the circadian variation in a variable must always be taken into account.

To illustrate this point, consider the example shown in Figure 7.9. We have plotted the range of daily variation of a hypothetical parameter, with the horizontal dashed lines defining the upper and lower range of values seen in normal individuals. Two additional points are plotted; the one on the right is outside of the usual range of values and would be recognized as abnormal. The point on the left is clearly within the bounds of normal values and would probably not cause any concern. However, it is significantly outside the range of normal values for that time of day. Recognition of the circadian variation in the results of such tests may enable physicians to apply diagnostic criteria more precisely.

A circadian rhythm that is now quite regularly taken into account in clinical practice is that of plasma cortisol concentration, which typically reaches a peak just before a person awakens and drops to a trough in the few hours before he goes to sleep. In a normal person the early morning samples are consistently higher than those taken in the evening. Thus, the diagnostic criteria for Addison's disease (adrenal hypofunction) or Cushing's disease (adrenal hyperfunction) have to take into account the time of day when the plasma cortisol sample was obtained from the patient. If the normal maximum levels seen during the morning are maintained throughout the 24 hours, a patient may have Cushing's disease. Similarly, an Addisonian patient might demonstrate the low levels typical of evening throughout day and night.

Unfortunately, the maxim "life is never simple" is most appropri-

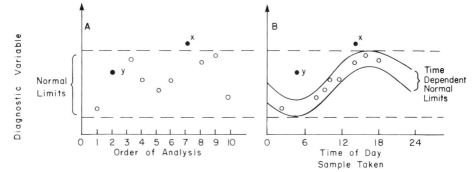

Fig. 7.9 Normal limits for a diagnostically useful variable plotted (A) without and (B) with regard to the time of day. When the time of day is taken into account, the detection of abnormal values is improved, so that not only value x but also value y can be identified as being outside the normal range. (Copyright 1982 by Moore-Ede, Sulzman, and Fuller.)

ately applied to the practice of medicine. We do not intend to suggest that the mere knowledge of the normal circadian variation in the diagnostic variable does any more than minimize one more source of error. As an example, let us take the diagnosis of adrenal malfunction based on sampling plasma cortisol, recognizing that not only is there circadian variation but also a highly episodic pattern of secretion, so that levels may change severalfold over the course of an hour, depending on whether the sample is taken during a secretory episode or between such episodes. Additionally, one has to be aware that the patient may not show a normal circadian variation because of some factor in the temporal environment in which he is studied (hospitals are not the most normal of temporal environments, with noise, light, and activity maintained throughout day and night). Alternatively, the patient may have some disorder of his circadian timekeeping system so that the cortisol rhythm is abnormally phase-related to the 24-hour environmental time scale (Martin, Mintz, and Tamagaki, 1963). These two issues we deal with later in this chapter. In addition, the clinician must deal with other factors that influence cortisol levels, such as stress and errors in laboratory measurement. However, when all is said and done, circadian rhythmicity is a significant source of variance in many important diagnostically tested variables. Taking this variation into account can only improve the precision of any test.

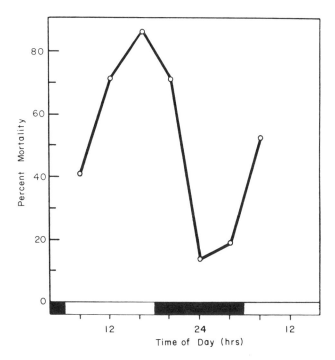

Fig. 7.10 Susceptibility rhythm of mice to intraperitoneal injections of *E. coli* endotoxin. A dose compatible with survival for most animals at one time of day is highly lethal when injected into comparable mice at a different circadian phase. (After Halberg, 1960.)

SUSCEPTIBILITY TO TRAUMA AND TOXINS

Another consequence of circadian rhythmicity is that the body is more susceptible to challenges at certain phases of the circadian cycle than at others. Varying susceptibility has been demonstrated for a wide range of exogenous challenges, from bacterial toxin administration to auditory trauma (Halberg et al., 1960; Reinberg, 1967). Figure 7.10 presents an example of such a susceptibility rhythm, which greatly influences an animal's survival.

The effects demonstrated by these studies are quite dramatic. However, to evaluate the experiments, one must understand their design. Most of these studies were done around either the Lethal Dose 50 (the dose at which 50% of the animals are killed) or the Effective Dose 50 (the dose at which 50% of the animals show the defined effect). Animal studies that show major circadian changes, such as 20% dying at one time of day and 80% at another time with identical treatments, are this dramatic because the studies are being conducted near the Lethal Dose 50. The circadian differences in lethality may be comparable to changes in lethal effects produced by altering the dose by a relatively small amount if the dose-response curve is steep at the dose being used in the study.

This is not to minimize the findings of circadian rhythms in susceptibility. Indeed, they may be highly relevant to human clinical medicine, particularly in touch-and-go situations. A number of authors have documented circadian rhythmicity in the timing of death, with the highest death rate at night in both surgical (Frey, 1929) and nonsurgical patients (Jenny, 1933; Jusatz and Eckhardt, 1934). Of course, factors such as the alertness of the nursing and medical staff at that time must be taken into account, but the findings still appear to be real. Parenthetically, the timing of human births also reaches a peak in the early hours of the morning, leading more than one wit to point out that we tend to live an exact number of days.

What lessons can be learned from these observations? Probably the most significant application is in situations where the chances of a patient's recovery are marginal, and circadian rhythmicity could tip the balance. It might be advantageous, for example, to monitor an individual more carefully at certain times of day. However, all these circadian rhythms of susceptibility are themselves susceptible to the precautionary statements in the previous section, because they can be altered if there has been a shift in environmental time cues or if the patient has a disorder of the circadian timekeeping system itself (Moore-Ede, 1976a).

Recent evidence indicates that patients with specific disorders may have abnormal susceptibility rhythms. For example, while there is a circadian rhythm in lung airway resistance in normal individuals, it is usually of such low amplitude that it does not significantly limit our respiratory performance. However, Hetzel and Clark (1979) have shown that patients with respiratory failure due to bronchial asthma had a greatly exaggerated rhythm of bronchial constriction, with maximum constriction seen around 6 A.M. In fact, the amplitude of the rhythm was so significant that respiratory arrests tended to occur at this time, and several of their patients died at this time of day. Such a dramatic rhythm in bronchoconstriction pointed to the possibility of concentrating therapy and providing particularly careful observation in the early morning for this group of patients.

DRUG EFFECTIVENESS AND TOXICITY

Many drugs have been shown to have circadian variations in therapeutic effectiveness or toxicity, and this subject has been the topic of a number of reviews (Halberg, 1969; Reinberg and Halberg, 1971; Moore-Ede, 1973). One of the most exciting prospects of this research is that the therapeutic benefits may be maximized and the toxic side effects minimized by administering a drug at the appropriate time of day. Of course, not all drugs show sufficiently different effects to warrant such care in the timing of therapy, but there are a number of important instances where the clinician may be operating in a narrow range between the effective dose and the toxic dose, so that such care with timing becomes important.

Many circadian rhythms in drug effectiveness have been discovered by chance. Carlsson and Serin (1950a), for example, were carrying out extensive tests on nikethamide toxicity but found a disconcertingly wide variation in the Lethal Dose 50 in different experiments. Careful examination of their records finally revealed that the time of day the dose was administered was the only factor that could account for the variation. On further systematic study (Carlsson and Serin, 1950b), they found that 67% of a group of mice died after a 0.3 gm dose of nikethamide at 2 P.M., whereas only 33% of a similar group died when the same dose was given at 2 A.M. It is interesting that similar anecdotes underlie the discoveries of circadian rhythmicity in the actions of many other drugs, and in a number of instances the investigator was unaware that the phenomenon had been reported previously.

The effectiveness of a drug is a function, in part, of the rates of absorption, metabolism, and excretion; the degree of dilution in the body fluid compartments where it is distributed; and the susceptibility of the target tissue. As summarized diagrammatically in Figure 7.11, the circadian rhythmicity in each of these factors contributes to a circadian variation in the therapeutic response.

The rate of absorption of any orally administered drug will depend partly on the amount of food in the stomach (Goodman and Gillman, 1975). Since diurnal animals, such as man, tend to eat during the day,

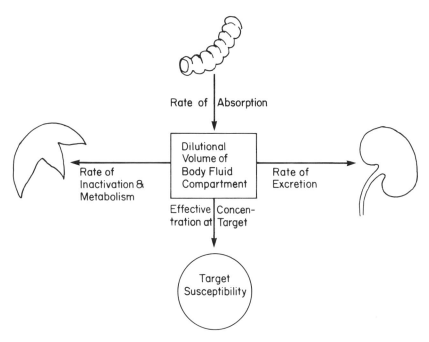

Rate of Absorption

Dilutional Volume of Body Fluid Compartment

Rate of Inactivation & Metabolism

Rate of Excretion

Effective Concentration at Target

Target Susceptibility

Fig. 7.11 Body functions with demonstrated circadian rhythms that determine the overall circadian rhythm of a drug's effectiveness and toxicity. (From Moore-Ede, 1976a.)

and nocturnal animals, such as rats, tend to eat by night, this factor may contribute to circadian variations in the response to drugs. But beyond the day–night variation in stomach contents, there are prominent circadian rhythms in intestinal function which persist in fasted animals. These include rhythms in intestinal enzyme activity (Suda and Saito, 1979), gastric acidity (Johnston and Washiem, 1924), and the rate of glucose uptake into the circulation (Jarrett, 1979). Although there have been very few studies of circadian variation in the uptake of orally administered drugs, it is reasonable to assume that such rhythms exist. Similarly for drugs administered intramuscularly, there may be rhythmicity in the rate of uptake.

Circadian variations in the rates of metabolism and inactivation are well documented. Radzialowski and Bousquet (1967, 1968) examined the activities of the rat liver enzymes that are responsible for the inactivation of hexobarbital, aminopyrine, p-nitroanisole, and 4-dimethylaminoazobenzene (4-DAB). Each enzyme activity varied with the time of day, reaching a peak at 2 A.M. and falling to a minimum at 2 P.M. Most of these rhythms are apparently driven by adrenal steroid rhythms, since they were abolished in animals who had been adrenalectomized or who were maintained with constant corticosterone levels by exogenous steroid treatments. Other investigators have obtained similar results. Vesell (1968) and Nair and Casper (1969) independently have confirmed the rhythm of liver hexobarbital oxidase activity. Jori, DiSalle, and Santini (1971) have also examined the activity of imipramine N-demethylase in rat liver homogenates from animals killed at different times of day. They found a rhythmic activity with a maximum at the beginning of the dark period.

Rates of excretion are also subject to circadian variation. For example, Reinberg and co-workers (1967) have examined the urinary excretion of salicylate in healthy humans and shown that the slowest excretion is in the morning. Beckett and Rowland (1964) have reported that methylamphetamine shows high rates of excretion in the morning and low rates later in the day. The urinary excretion of sulfonamides has also been shown to vary with the time of day. Dettli and Spring (1967) reported that the plasma half-life of sulfasymazine was three times longer at night than during the day. These variations in urinary excretion rates are probably related to the rhythm in urinary acid excretion (Mills and Stanbury, 1952). Urine excreted during the night is more acid, while that produced during the daytime is more alkaline. Since only nonionized forms are thought to be reabsorbed in the tubules, the excretory rates for most drugs depend on their acid dissociation constant (pKa). Thus the excretion of salicylate (with a low pKa) is reduced when the urine is acidic, whereas methylamphetamine and sulfonamide (which have higher pKa's) are excreted more rapidly in the early-morning acid urine.

Circadian variations also occur in blood volume (Cranston and Brown, 1963) and extracellular fluid volume (Moore-Ede, Brennan, and Ball, 1975), and this may have some small additional effect on the degree of dilution of a drug. Furthermore, there may be circadian variations in plasma protein concentration, which would influence the binding of drugs carried by plasma proteins.

All the aforementioned factors will have a net effect on the circadian variation in the concentration of a drug as it diffuses to the target tissue. In turn, however, there may be a circadian rhythm in the suscepti-bility of the target. In response to a circadian variation in a trophic hormone, for example, there may be a change in the number of recep-tor sites on a cell or the inducibility of intracellular enzyme systems. Relatively little work has yet been done on such changes in target sus-ceptibility, but they are probably important.

The above discussion of the multitude of factors influencing drug ef-fectiveness, each of which may show independent circadian rhythms, might lead the reader to conclude that surely all these factors cancel each other out so that there are no significant rhythms in drug effec-tiveness and toxicity. This may be true for some drugs, but clearly there are others whose rhythmic components are in a phase relation-ship that amplifies the net effect. The proof, then, is in direct compari-son of the therapeutic effect of administering a particular dose of a drug at different times of day. We will discuss here some drugs for which circadian variations may be clinically important.

Cancer chemotherapy   There is increasing interest in utilizing the circadian rhythm of drug effectiveness and toxicity in the treatment of cancer. The drugs used to treat malignant growth are toxic not only to the tumor but also to the host, which is one of the major problems in cancer chemotherapy. Since many of these drugs show a rhythm in toxicity, some investigators have examined whether there might be an optimum schedule for administering chemotherapeutic agents, utiliz-ing the rhythms of sensitivity and of toxicity. In other words, a higher dose can be tolerated by the body at its time of least sensitivity, and this increased dosage could tip the scale in favor of a cure.

The patient's sensitivity to the drug may be due to the body's various normal cellular and biochemical rhythms. For example, a potentially lethal dose of cytosine arabinoside, a drug that inhibits DNA synthesis and is widely used in the treatment of human leukemia, kills as many as 74% of the treated mice when administered at one phase, while as few as 15% die when treated at another time (Cardoso, Scheving, and Halberg, 1970). The synthesis of DNA in the bone marrow and intes-tines of mice also varies with the time of day. The maximum toxicity of cytosine arabinoside occurs at the beginning of the DNA synthesis phase of the cell division cycle. This temporal relationship suggests that the maximum of the rhythm of susceptibility may be due to the timing of the maximum sensitivity of bone marrow and gut to inhibition of DNA synthesis.

Another chemotherapeutic agent, cyclophosphamide, also has a toxicity rhythm, but it is somewhat out of phase with that of cytosine arabinoside. Scheving and co-workers (1977) have determined that giving these two drugs in combination to leukemic mice at the optimal times brings about an increased survival time and cure rate. In one study these authors injected mice with leukemia cells and then examined the effects of giving these drugs in a rhythmically varying "sinusoidal" regime or a "homeostatic" (continuous level) regime. About 95% of the mice on the sinusoidal schedule survived, compared to only 70% of the mice on the homeostatic schedule, even though both groups were given the same total amounts of each drug.

With different sinusoidal schedules, phased to scan the 24-hour day, a very interesting result was obtained. The cure rate (the percentage of healthy mice at the end of the study) also showed a pronounced circadian rhythm. If the peak dose of the sinusoidal treatment was given in the early part of the light period, only 44% of the mice were cured. But if the peak dose was given at the end of the light period, 94% of the mice were cured (Fig. 7.12). These data suggest that the tumor cells, as well as the host, have a circadian rhythm of resistance and susceptibility to the drug.

In addition to the sensitivity rhythms to various chemical agents, the body shows a differential sensitivity to certain physical agents, at least one of which—sensitivity to X rays—is relevant to cancer therapy. Some of these responses are quite striking. In one study, Haus and colleagues (1974) showed that mice exposed to 550 roentgens of whole-body X-irradiation at different times of the day had very different mortality rates. Eight days after exposure, all mice irradiated at the midpoint of the activity time (mid-dark) were dead, while all those exposed during the second part of the rest span were still alive (Fig. 7.13). If these findings are confirmed and extended, then the timing of therapy for maximum resistance of the host and maximum sensitivity to the target could offer exciting possibilities for future cancer chemotherapy.

Anesthetics and analeptics Anesthetic agents have been shown to have especially marked rhythms of effectiveness and toxicity. The

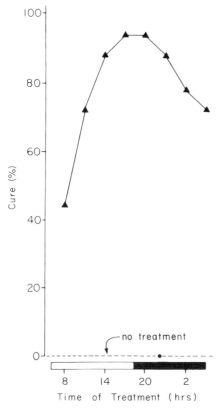

Fig. 7.12 The percentage of cures (% mice alive at 75 days after inoculation with leukemia cells) plotted against time of day of cyclophosphamide treatment. (After Scheving et al., 1977.)

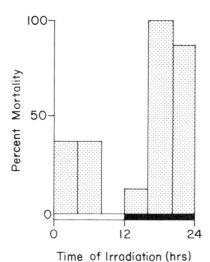

Fig. 7.13 The percentage mortality of mice irradiated with 550 roentgens of X rays at different times of day. (After Haus et al., 1974.)

most important finding from animal studies has been that the susceptibility to anesthesia-producing doses may have a circadian rhythm
which is 12 hours out of phase with the circadian variation in the lethal effects of the same drug. Thus the therapeutic index (Lethal Dose
50/Effective Dose 50) varies widely over the 24-hour day.

Munson, Martucci, and Smith (1970) have demonstrated a circadian
rhythm in the minimum alveolar concentration of halothane (Fluothane) needed to keep rats anesthetized. The lowest minimum concentration is needed at noon and the highest minimum concentration at 8
P.M. However, at 8 P.M. rodents show the greatest susceptibility to
doses of halothane in the lethal range (Matthews, Marte, and Halberg,
1964), whereas they are least susceptible at 8 A.M.

Barbiturate anesthetics show the same phenomenon. The largest
doses of pentobarbital are needed at night to anesthetize rodents
(Davis, 1962; Emlen and Kem, 1963), and they are most susceptible to
potentially lethal doses at this time (Pauly and Scheving, 1964). Thus,
the therapeutic index for anesthetics is highest (that is, the margin of
safety is widest) during the rodents' resting period and lowest during
the active period. Whether such a relationship exists in man remains
to be investigated. It is worth noting that most elective surgery in
humans is done during our active period.

Although the anesthetics studied to date all show similar circadian
rhythms in duration of anesthesia, the time of peak effectiveness
varies with different anesthetics studied under similar conditions in
the same animal. These rhythms of anesthesia and lethality may be
partly due to the rhythms of drug-metabolizing enzyme activity. For
example, there is evidence that the rhythm of hexobarbital-induced
anesthesia duration is inversely related to the rhythm of liver hexobarbital oxidase activity (Vesell, 1968; Nair and Casper, 1969).
Rhythms of urinary excretion and drug redistribution rate may also be
involved in anesthesia and lethality rhythms, and the rhythm of the
brain's susceptibility to anesthetics may play a role (Roberts, Turnbull,
and Winterburn, 1970).

Drugs having stimulant effects on the central nervous system also
show circadian rhythms of effectiveness. In what was probably the
first description of a drug circadian rhythm, Agren, Wilander, and
Jorpes (1931) showed that mice were more resistant to insulin-induced
convulsions if the insulin was given in the evening instead of at noon.

To cause the same percentage of convulsions in a group of mice in the evening required twice the dose that was effective at noon. Since then, other drugs producing convulsions, such as tremorine (Pauly and Scheving, 1964), hexafluorodiethylether (Davis and Webb, 1963; Webb and Russell, 1966), and lignocaine (Lutsch and Morris, 1967), have been shown to have circadian rhythms of effectiveness in nocturnal rodents. However, unlike insulin, these drugs were most effective when the animal was active (at night), and the animal was least susceptible to the drug while it was asleep (during the light period). Thus, although a similar response is produced by insulin and the other convulsants, the different mode of action of these drugs results in a different circadian rhythm of effectiveness.

This point is important, for some authors (Morris and Lutsch, 1967) have used the timing of a drug's circadian rhythm as evidence that it has a stimulant or depressant effect on the central nervous system. Figure 7.14 illustrates the timing of the maxima and minima of various drugs acting on the central nervous system of nocturnal rodents. It can be seen that depressants such as pentobarbital (Davis, 1962; Emlen and Kem, 1963) have their greatest anesthesia-producing effect in the middle of the day, whereas stimulants such as tremorine (Pauly and Scheving, 1964) and hexafluorodiethylether (Davis and Webb, 1963; Webb and Russell, 1966) are most effective during the night. However, it should also be noted that there are nocturnal maxima in rodent susceptibility to lethal doses of amphetamine (Scheving, Vedral, and Pauly, 1968), a stimulant, and chlordiazepoxide (Marte and Halberg, 1961; Marte, 1961), a central nervous system depressant. Nikethamide (Carlsson and Serin, 1950a, b), another stimulant, has a maximum effect near the time of minimum responsiveness to amphetamine (Scheving, Vedral, and Pauly, 1968).

Most of the drugs discussed here have been studied only in rodents. These are nocturnal animals, which sleep predominantly during the light period. Thus, the circadian rhythms of man and the rodent might be expected to be similar but out of phase by 12 hours. Unfortunately, this is an oversimplification, for many cardiovascular and metabolic rhythms which reach a maximum toward the end of the daytime active period in man (Conroy and Mills, 1970) in rodents rise to the maximum at the beginning of the nocturnal active period. Rectal temperature, for example, is highest at 8 P.M. in rats living in a dark (active)

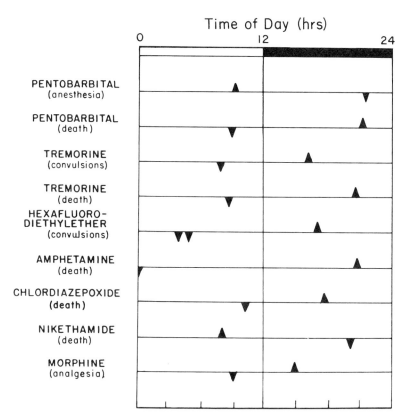

Fig. 7.14 Maxima (upright triangles) and minima (inverted triangles) of rodent susceptibility to various drugs acting on the central nervous system. (From Moore-Ede, 1973.)

period from 6 P.M. to 6 A.M. (Halberg et al., 1954). In man, this variable reaches the maximum typically in late afternoon (Conroy and Mills, 1970). There is therefore no substitute for repeating these studies in man for clinical applicability. Although the basic principle may be studied in rodents, care must be taken in interpreting the findings.

Corticosteroids In man, corticosteroid secretion by the adrenal cortex undergoes a marked circadian rhythm, with a peak value just before the time of awakening (Bartter, Delea, and Halberg, 1962), as we discussed in Chapter 5. The secretory rate falls throughout the day, with minimal values reached about midnight (DiRaimondo and Forsham, 1956). Between midnight and 9 A.M., 70% of the 24-hour output of corticosteroid is secreted (Nichols and Tyler, 1967).

The therapeutic implications of this rhythm were first recognized by DiRaimondo and Forsham in 1956. They suggested that suppression of the adrenal cortex during corticosteroid therapy could best be avoided by giving the daily therapeutic requirement as a single dose in the morning between 8 and 10. In this way the administered glucocorticoid would supplement the low daytime plasma corticosteroid concentration without suppressing ACTH secretion by the pituitary, which is least sensitive at this time of day to circulating corticosteroid levels. Conversely, when inhibition of adrenal cortical activity is desired, as in the treatment of androgenic adrenal hyperplasia, they predicted that a dose of hydrocortisone would prove nearly twice as effective at night as in the morning.

Experimental support for these suggestions has now been amply provided. Nichols, Nugent, and Tyler (1965) found that an oral dose of 0.5 mg dexamethasone given to human subjects at 8 A.M. reduced the adrenal corticosteroid output from 19.2 mg per 24 hours to 7.2 mg per 24 hours, but that the same dose given at midnight caused much more suppression, reducing the endogenous 24-hour output to 1.9 mg. Others (DiRaimondo and Forsham, 1958; Segre and Klaiber, 1966) have shown that urinary 17-hydrocorticosteroid excretion is suppressed to a much greater degree by doses spread evenly throughout the day or confined to late afternoon and evening than by a single dose at 8 A.M. Schulz and Retiene (1971) have shown that these findings hold true for patients treated with steroids for longer periods. Similar findings have been reported in rodents (Retiene, Frohns, and Schulz, 1967; Schulz and Retiene, 1971), although when interpreting the results, allowance must be made for their reversed activity pattern, compared to man's. Retiene, Frohns, and Schulz (1967) and Tarquini, Fantini, and Cagnoni (1967) found a marked reduction in adrenal weight in rats treated with dexamethasone before the plasma corticosteroid peak, but very little reduction in the group given dexamethasone just after the endogenous peak.

The timing of human corticosteroid therapy to obtain minimal adrenal suppression has been investigated in clinical practice. Behl and Garg (1967) have studied the effects of giving triamcinolone for dermatological conditions either as a single morning dose at 8 A.M. or as divided doses spread over the day. They judged that triamcinolone given as a single dose in the morning was marginally more effective than divided doses and that side effects were fewer in the group on the morning dosage. Dubois and Adler (1963) found that patients taking a single dose of steroid in the morning could get the same therapeutic effect from a lower dosage than they could on the usual divided dosage scheme. These investigators also reported an apparent reduced incidence of peptic ulcer activation in the group on the single regimen. Harter, Reddy, and Thorn (1963) claimed that a further advantage of a single-dose regimen is that it can be stopped abruptly with rarely any side effects, a remarkable finding when one considers the risk of precipitating an adrenal crisis after abrupt cessation of a 6-hourly steroid regimen (Goodman and Gillman, 1965). A single dose of corticosteroid each morning is also more convenient, and Dubois and Adler (1963) found that their patients much preferred it to split-dosage regimens.

A knowledge of the circadian rhythm of adrenal corticosteroid secretion can therefore be of advantage in selecting the most beneficial dosage regimen for a patient. Caution, however, must be exercised in treating patients who also have certain central nervous system disorders, for in such conditions the adrenal circadian rhythms may be absent (Eik-Nes and Clark, 1958; Perkoff et al., 1959) or altered (Mills, 1966). Obesity and thyroid disorders may also affect the timing of these rhythms (Martin, Mintz, and Tamagaki, 1963). If any doubt exists, it may be helpful to determine the patient's plasma corticosteroid circadian rhythm to establish the optimal time for corticosteroid therapy.

Anabolic steroids   The circadian variation of antidemineralization effectiveness of an anabolic steroid, oxymetholone, was studied by Moore Ede and Burr (1973) in patients with traumatic paraplegia. These subjects, during the months of immobilization that follow their injury, excrete an increased amount of calcium in their urine (Freeman, 1949). It was found that most of this incremental calcium was excreted during a few hours in the middle of the day in immobilized subjects (Moore-Ede, Faulkner, and Tredre, 1972; Moore-Ede and Burr, 1973). To discover whether therapy would be most effective if it was timed to cover the midday peak, oxymetholone was administered to similar groups of patients as a single dose at different times of day. A

10 mg dose at 6 A.M. produced a significant reduction in urinary calcium excretion, but a similar dose at other times of day was not effective. The rhythm of oxymetholone effectiveness appeared to be related
to a rhythm in bone salt mobilization, which peaked in the early
morning.

Histamine and antihistamines Reinberg, Sidi, and Ghata (1965)
tested the erythemal response to intradermal injection of histamine at
various times of day. They found the greatest response when the injection was given at 11 P.M. and the smallest response between 7 A.M.
and 11 A.M. The reactions to penicillin and to house dust showed a similar circadian variation (Reinberg et al., 1969), with maximum and
minimum at the same times as in the histamine test. Cormia (1952)
studied the same phenomenon in a different manner. Having observed
that pruritus seemed to be more severe at night, he investigated the
itch threshold by injecting increasing concentrations of histamine in-
tradermally. He found that the histamine dose needed to produce an
itch sensation at 2 P.M. was 100-fold higher than that which induced an
itch sensation at midnight. Further studies by Reinberg and Sidi (1966)
have shown that the effectiveness of an antihistaminic drug, cyproheptadine, varied with the time of day. The antihistaminic effect was
judged by noting the response to an intradermal injection of histamine
at intervals after the administration of cyproheptadine. When the antihistamine was given at 7 A.M. the effect lasted 15 to 17 hours, but at 7
P.M., near the time of maximum response to histamine, the antihistaminic effect lasted only 6 to 8 hours.

Disorders of Internal Timekeeping

The circadian timing system may on occasion malfunction, even in
individuals living in a regular 24-hour day–night environment. Just as
disease processes may interfere with any other physiological system,
they may disrupt circadian timekeeping by altering zeitgeber reception, pacemaker function, or the mechanisms which couple oscillators.
In this section we will discuss the specific disorders of the circadian
timing system that have been identified and propose a method of classification.

So far, most of the known disorders of circadian timekeeping are those that interfere with the sleep–wake cycle. The reason for this is obvious. We are not subjectively aware of most rhythmic functions of the body, such as hormone levels and body temperature. Disruptions in these rhythms are not readily apparent and are not easily detected in current diagnostic tests, although as these tests become more sophisticated, rhythmic disorders in other variables will be recognized. The physician does not normally take more than one blood sample or blood pressure measurement in an office visit and therefore can have no idea of an abnormal circadian rhythmicity in these variables. In contrast, changes in the sleep–wake cycle are readily apparent to the patient. If he cannot sleep at night or falls asleep during the daytime, this represents a major disruption of the normal routine and may interfere with the ability to concentrate at work or enjoy social activities. Patients with these disorders may present with a primary complaint of a sleep disorder or may report it as a problem secondary to some other condition.

DIAGNOSIS AND EVALUATION

The simplest and least expensive way to evaluate circadian sleep–wake disorders is to have the patient record every day in a diary the times of going to sleep and waking and the timing of any naps for at least one month (Kokkoris et al., 1978; Czeisler, Richardson, Coleman, et al., 1981; Weitzman et al., 1981). This can be done either by writing down the times of each bedtime and awakening (although patients' records often become confused if they sleep twice in any one day), or by filling out a chart like the ones in Figure 7.15. When this chart is duplicated and the copy placed beside the original, a circadian double plot can be obtained, like those used earlier in this book. Examples of sleep diaries from a normal individual and from individuals with disorders of the circadian timing system, which we will discuss below, are shown in Figure 7.15. These show highly characteristic patterns which can be easily recognized. This circadian actogram should be part of the diagnostic workup of any patient in whom a malfunction of circadian timekeeping is suspected.

The diaries cannot, of course, replace the polygraphic recording of sleep–wake cycles. Patients frequently overestimate the time that they lie awake at night, and other errors may creep into subjective estimates of sleep and wakefulness. However, polygraphic recording of

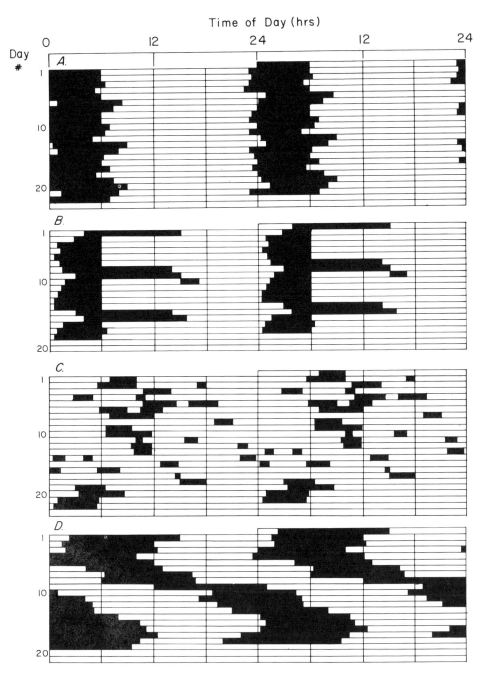

Fig. 7.15 Circadian actograms from patients with sleep disorders. Daily logs on which sleep time is recorded are double-plotted to show records of patients with (A) normal sleep–wake cycle; (B) delayed sleep phase insomnia; (C) disrupted sleep–wake cycle; and (D) period disorder with failure to entrain to 24-hour day. (Copyright 1982 by Moore-Ede, Sulzman, and Fuller; B after Czeisler, Richardson, Coleman, et al., 1981a, reprinted by permission of Raven Press; C plotted from data supplied by Q. Regestein.)

sleep is very expensive, requiring admission to a sleep disorder clinic, expensive recording equipment, and the constant vigilance of a technician all night long. It is rarely feasible except in research environments to record more than one or two nights of the sleep–wake cycle. This length of observation may not be sufficient to diagnose circadian sleep–wake disorders in which the pattern of sleep and wakefulness is the essential feature. Thus the patient's diary recorded over a much longer period of time may be essential in the differential diagnosis.

Another feature of the workup is to determine the phase relationship of the patient's sleep–wake and alertness rhythms to environmental time cues. It is a matter of common experience that some people like to stay up late at night, whereas others are at their best early in the morning. The extreme cases are easiest to identify. The extreme "lark" arises at 4 or 5 A.M. and does his best work during the morning, becoming tired and less able to concentrate by 8 or 9 at night. In contrast, the extreme "owl" has an enormously hard time getting up in the morning, prefers to sleep until noon if possible, then slowly gains momentum throughout the day and is often at top form at night, maybe staying up until 2 or 3 A.M.

The distribution of these behavior patterns has been quantified, using a questionnaire developed by Horne and Ostberg (1977). By a series of questions aimed to evaluate a person's relative performance at different times of day, it is possible to determine where he is on the owl–lark spectrum. The important thing is not the time of day when the person sleeps and awakens because, for example, he could be on a shift work schedule and be required to work at an unusual time of day. Rather it is the relative performance and ability to concentrate at different times in the cycle that gives the best indicator of the phase relationships of the circadian system. For example, if there is little time to prepare an important lecture or business presentation, the owl will stay up late to complete the task; in contrast, the lark will get up early. Each is using the time when he feels at his best. Although a wide range of behaviors can be considered normal, extreme lark or owl behavior is a useful diagnostic pointer.

Owls and larks have always put considerable social pressure on each other. Often larks view owls as slothful and, indeed, there are records going back through history of people who prefer to lie in in the morning complaining bitterly about the self-righteousness of early risers. Samuel Johnson pointed out his own hypocrisy on this subject: "I have all my life long, been lying till noon; yet I tell all young men, and tell them with great sincerity, that nobody who does not rise early will ever do any good" (Boswell, 1785). Similarly, owls may consider larks "party-poopers" because they nod off to sleep or are socially moribund at evening gatherings.

CLASSIFICATION

As disorders of circadian timekeeping are increasingly recognized, it is useful to consider how to classify them. They could be classified by the affected component of the circadian system. For example, we could differentiate between disorders of zeitgeber transduction, pacemaker function, and internal coupling mechanisms. However, in many cases the mechanism that accounts for a clinical disorder is as yet undetermined, and at least two possibilities exist. It therefore seems safer at present to classify circadian pathologies by their predominant influence on the patient's observed rhythms. We will therefore identify phase disorders, period disorders, and amplitude disorders, depending on the major effects observed when the patient is living in a regular 24-hour day–night environment. For accurate diagnosis, the nature of the environmental time cues to which the patient is subjected must be confirmed, because many disorders of circadian function are the result of disrupted environmental time cues rather than internal disorders of the circadian timing system.

PHASE DISORDER: DELAYED SLEEP PHASE INSOMNIA

The most common circadian sleep disorder is characterized by the complaint that the patient has considerable trouble falling asleep at night, followed by corresponding difficulty awakening in the morning. This disorder, termed "delayed sleep phase insomnia," or DSPI (Czeisler, Richardson, Coleman, et al., 1981; Weitzman et al., 1981), has long been recognized. The patients are on the extreme owl end of the owl–lark spectrum but are incapacitated by chronic sleep loss. Robert MacNish (1836) provided an early description of these patients: "They lie awake for perhaps 2 or 3 hours, after going to bed, and do not fall into slumber till toward morning. Persons of this description often lie long and are reputed lazy by early risers, although it is probable, they actually sleep less than these early risers themselves."

The characteristic diagnostic features of patients with delayed sleep phase insomnia are (1) reported sleep onsets and wake times intractably later than desired; (2) actual sleep times at nearly the same clock hour daily; and (3) essentially normal EEG recording except for the delayed sleep onset. Indeed, the most marked feature, which distinguishes these patients from those with other forms of insomnia, is that they have no difficulty maintaining sleep once it has been initiated.

The disorder is common. In a study of 100 patients complaining of insomnia at the Stanford University Sleep Disorders Center, 10% of the patients were given a diagnosis of "circadian rhythm disruption," most of whom had DSPI (Dement, Guilleminault, and Zarcone, 1975). In a retrospective analysis of a series of 450 insomnia patients at the Montefiore Hospital Sleep–Wake Disorders Center, approximately 7% were found to have the specific diagnosis of DSPI (Weitzman et al., 1981). However, the prevalence of DSPI insomniacs is probably higher. Since sleep, once achieved, is usually maintained without disruption, many people with DSPI would probably not consult a sleep clinic or even a physician for diagnosis. Some of them even alter their daytime work schedules to accommodate their sleep time (say, by taking a 3 P.M.–11 P.M. job) without ever seeking medical treatment. Kleitman (1963) speculated that delayed sleep onset was "undoubtedly the most common" type of insomnia. In fact, in a recent epidemiological survey of a representative sample of the adult population, Bixler and co-workers (1979) reported that 23.4% of the general population complained of difficulty with falling asleep.

A common feature of those who suffer from this disorder is that they have tried many different medications over the years without success. Indeed, their insomnia may be compounded by the effects of the medication. The indiscriminate use of sleep medications such as benzodiazepines and barbiturates (25.6 million prescriptions in 1977) interferes with their daytime performance and may cause rebound insomnia leading to drug dependency (Institute of Medicine, 1979). In fact, most patients with DSPI have been taking hypnotics and/or other drugs (both prescription and nonprescription) for many years without sustained relief (Weitzman et al., 1981).

The sleep–wake record of a patient with DSPI is shown in Figure 7.16. In the first part of the record, the subject, a medical student, had to get up each morning at 7 A.M. to attend lectures. However, when he

Time of Day (hrs)

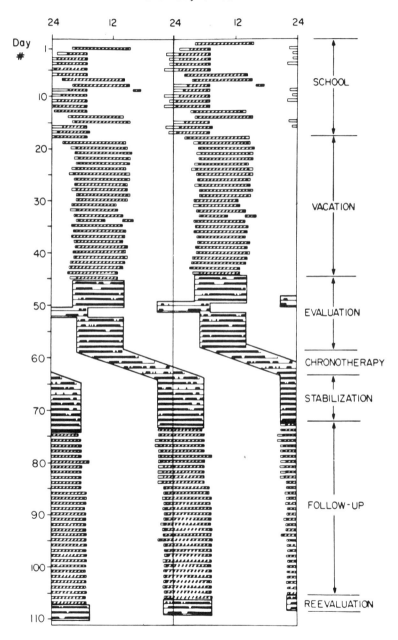

Fig. 7.16 Sleep times (striped horizontal bars) from a diary like that in Fig. 7.15 and EEG recording of sleep (dark horizontal bars) in a patient with delayed sleep phase insomnia. When attending school five days a week, day 1–day 18, he lay awake in bed for several hours (open horizontal bars) before falling asleep but could sleep at a delayed phase with no difficulty on weekends (days 1, 7, 8, 14, 15) and on vacation (days 19–45). On days 46–59, he was evaluated in the laboratory and on days 51 and 52 was subjected to an acute phase advance, which confirmed his inability to sleep at the desired time (2200–0600 hrs). He was treated by chronotherapy on days 58–65 by phase-delaying his bedtime 3 hours each day until the desired phase relationship with the 24-hour clock was obtained. (From Czeisler, Richardson, Coleman, et al., 1981a, reprinted by permission of Raven Press.)

tried to go to bed at 10 or 11 at night, he did not fall asleep until 2 or 3 A.M. During weekends, when he had no imposed schedule, he fell asleep at 3 or 4 A.M. and then slept until the afternoon. During the weekend his sleep–wake cycle would start to free-run and then, during the week, he would gradually phase advance the cycle but still could not fall asleep until after midnight, despite considerable sleep deprivation by the end of the week. However, when the subject went on vacation and could self-select the time of going to bed and waking, he stayed synchronized to a 24-hour routine, slept 8 hours every night, and had no problem with going to sleep or with sleep loss. Thus DSPI is not a problem of sleep once it is attained; it is rather a problem with falling asleep at the desired time.

The exact cause of delayed sleep phase insomnia is unknown, but based on our discussions of the properties of the circadian system in Chapter 2, we can examine possible mechanisms. The most obvious cause would be a change in the period of the pacemaker driving the sleep–wake cycle. As was demonstrated in Figure 2.21, if the period of the pacemaker changes (as with the addition of D_2O to an animal's drinking water; Richter, 1977), then the phase of the rhythms driven by that pacemaker—for example the sleep–wake cycle—would show a delayed phase relationship with respect to the zeitgeber.

This mechanism cannot adequately explain DSPI, because humans self-select their light–dark cycle. When allowed to sleep on his own schedule while on vacation, the subject in Figure 7.16 did not show delayed sleep onset with respect to his self-selected light–dark cycle. If his pacemaker had had a lengthened period, his sleep–wake cycle would have adopted a phase relationship that was delayed with respect to any light–dark schedule he chose, and he would always have returned to the pattern of lying awake in bed in the dark for several hours before falling asleep.

The basis of DSPI instead appears to be a reduced phase-advance capability such as might be caused by a low-amplitude advance portion of the PRC. Although a daily shortening of the endogenous circadian period can be achieved so as to keep the subject entrained to a 24-hour period, there is a very limited ability to achieve larger phase advances, so that once a subject has a retarded sleep onset, it is difficult for him to move his sleep onset time earlier. In short, DSPI patients cannot readily recover from sleeping late over the weekend and have an exaggerated version of the Monday morning blues all week long.

Treatment Sleep medications such as the benzodiazepines and barbiturates are ineffective in the long-term management of this disorder. Many patients have tried such medications (prescribed and nonprescribed) with little success. However, drugs which can reset the phase of the circadian pacemakers might prove useful once they become available.

The most exciting advance, however, has been in a drug-free form of therapy (Czeisler, Richardson, Coleman, et al., 1981). Since these patients can maintain synchrony with a 24-hour zeitgeber period, their problem is in phase-advancing their sleep–wake cycle to the desired phase relationship with the 24-hour day–night cycle. Czeisler, Richardson, Coleman, and co-workers (1981) therefore postulated that it should be possible to phase-delay their sleep–wake cycle around the clock until the desired phase relationship was reached.

This form of therapy has proved remarkably successful, as the lower half of Figure 7.16 shows. The subject was admitted into the hospital, and the pattern of falling asleep between 3 and 4 A.M. and waking up in the afternoon was confirmed objectively by polygraphic EEG recording. When subjected to an acute phase advance of the day–night cycle to his desired time of going to bed (9 P.M.) for two nights, he was virtually unable to sleep. However, when the timing of the light–dark cycle and his bedtimes were progressively delayed by three hours per day, he was eventually able to reach a phase relationship so that he fell asleep at 9 P.M. and awoke at 6 A.M. He was able to maintain this normal schedule after release from the hospital, and although he slipped back one more hour, he was able to maintain a sleeping time from 10 P.M. to 7:30 A.M. each night. His symptoms of chronic tiredness and sleep onset insomnia disappeared, and for the first time in years he felt sleepy when it was time to go to bed and had no more difficulty in arising at 7 A.M. than he used to have in awakening at 2 to 3 P.M.

An interesting footnote is that the subject in Figure 7.16, a medical student, was later required to go on night call duty. As a result of staying up late, his sleep slipped back to its previous phase relationship to the 24-hour clock. He was again unable to phase-advance his sleep–wake cycle and again developed the symptoms of DSPI. However, having once learned the technique of phase-delaying his sleep–wake cycle around the clock, he was able to self-schedule this treatment and regain the desired timing of sleep and wakefulness.

The important feature of this rescheduling regimen is that the phase delays are achieved with a zeitgeber period on the order of 27 hours (24 + 3) that is within the human circadian range of entrainment. Treatment at home can be accomplished under the very close supervision of the physician and the use of a telephone answering machine, for example, to make sure that telephone calls do not wake the patient during the daytime. It is important to rigidly control the sleep and wake times during phase readjustment therapies. If this is not done, patients may become very confused and get into a worse situation than they were in previously. We strongly recommend, therefore, that patients with DSPI should not try to readjust their sleep–wake cycle themselves without the supervision of a sleep disorder specialist.

PHASE DISORDERS IN PSYCHIATRIC ILLNESS

Changes in the phase of the circadian sleep–wake cycle have long been recognized in certain mental disorders. One of the classic symptoms of depression is early morning waking, in which the subject complains of waking up early in the morning and being unable to fall back asleep (Mendels and Cochrane, 1968).

A study of 114 chronic psychiatric patients (Morgan and Drew, 1970) showed that their average bedtime was 8 P.M., even though the staff continually encouraged them to stay up later. These patients also woke up and got up earlier than they had to. Furthermore, the schizophrenic patients went to bed an average of an hour earlier than the others. Subsequent studies demonstrated that the daily body temperature rhythms of schizophrenic patients were significantly phase-advanced with respect to controls (Morgan and Cheadle, 1976). Similarly, studies of depressive patients have demonstrated that REM sleep is significantly phase-advanced in nocturnal sleep (Kupfer et al., 1978).

Phase shifts in the circadian sleep–wake cycle have been most extensively studied in patients who undergo manic-depressive cycles. It has long been known that certain patients will oscillate between a manic phase, in which they are hyperactive, and a depressed phase (Richter, 1965). The cycle length may vary from 48 hours to a year or more. In some detailed studies of the circadian rhythms in these patients, using a wrist activity monitor to record the rest–activity cycle, Wehr and colleagues (1979) have shown some characteristic phase shifts in the timing of the rest–activity cycle and the body temperature rhythm. As Figure 7.17 shows, during the manic phase the peak of the body temperature rhythm gradually advanced 3 or 4 hours earlier than during the depressive phase, and similarly the bedtime of these patients moved to an earlier time during mania, although to a lesser extent.

These reports of altered phase relationships of circadian rhythms in certain psychiatric disorders raise a number of interesting etiological questions. Most important, of course, is whether these shifts in internal phase relationships between circadian rhythms are a secondary consequence of the psychiatric illness or actually play a role in inducing the mental disorder.

Several lines of evidence now suggest that circadian phase disorders may contribute to the psychiatric disturbance. First, experimental phase shifts of the sleep–wake cycle of human subjects can result in symptoms ranging from emotional and psychosomatic disturbances (Taub and Berger, 1974) to depressive reactions and hostility (Rockwell et al., 1976). Second, Wehr and colleagues (1979) report that acute six-hour phase advances of the sleep–wake cycle can temporarily correct the abnormal timing of a depressed patient's circadian rhythms and thereby induce a two-week remission in the depression. Third, patients with psychiatric conditions associated with phase advances of the circadian system (Atkinson, Kripke, and Wolf, 1975; Mills et al.,

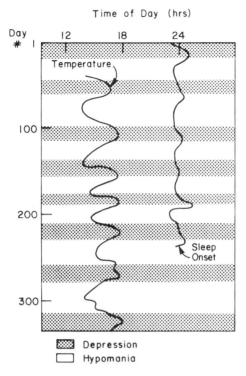

Fig. 7.17 Phase shifts in the timing of body temperature acrophase and sleep onset in a patient with a cyclic manic-depressive disorder. Shaded areas indicate days with clinical depression, and white areas indicate days of manic behavior. (Data kindly provided by Dr. T. Wehr, National Institute of Mental Health, copyright 1982 by Wehr et al.)

1977) may have free-running circadian rhythms with abnormally short periods (Kripke et al., 1978). Fourth, the clinical symptoms of some of these conditions are effectively treated by lithium salts, which lengthen the free-running period of circadian rhythms in plants and animals (Engelmann, 1973; Hoffmann et al., 1978, Kripke, Wyborney, and McEarchron, 1979) as well as in man (Johnsson et al., 1979; Johnsson and Engelmann, 1980).

As yet, most of these data are only suggestive, and it is difficult to separate out the underlying mechanisms when both schizophrenic and depressive illnesses appear to be associated with phase advances of circadian rhythms and shortening of the free-running period. However, a closer examination of one condition, cyclic manic-depressive illness, suggests some fruitful lines of inquiry.

It appears possible that a shortening of the period of the X pacemaker driving the body temperature rhythm may underlie cyclic manic-depressive phenomena. As we have mentioned, in the transitions between mania and depression, the temperature rhythm shows larger phase shifts than does the sleep–wake cycle (Wehr, personal communication). Furthermore, lithium, which is often clinically effective in manic-depressive disorders, has been reported to lengthen the period of the body temperature rhythm in healthy subjects up to a period of 27 hours (Johnsson et al., 1979; Johnsson and Engelmann, 1980). These studies suggest that the period of X can be lengthened more by lithium treatment than by just increasing the period of the sleep–wake Y pacemaker (Wever, 1979). Thus lithium may act on the pacemaker driving the body temperature rhythm, and it is possible that lengthening the period of this pacemaker may account for the beneficial effect of lithium therapy.

The etiological importance of a shortened temperature pacemaker period in manic-depressive illness has been further supported by modeling studies. Kronauer and co-workers (unpublished observations), using the coupled oscillator model discussed in Chapter 6, have shown that if the period of oscillator X is shortened, a phenomenon resembling manic-depressive cycling will occur. Both the sleep–wake and the body temperature rhythms cycled with a 20-day periodicity, with several days on which they were phase advanced alternating with several days when they were phase delayed. The similarity of this pattern to the manic-depressive cycles reported by Wehr and colleagues (1979) lends some credence to the idea that a shortening of the X pacemaker period may underlie sleep–wake phase disorders in manic-depressive patients, and maybe the manic-depressive syndrome itself.

A door may thus have been opened to the study and treatment of certain psychiatric illness. These studies suggest that phase disorders of the circadian system may play an important contributory role and that a change in the period of one of the circadian pacemakers may underlie these phenomena. This lead deserves to be followed and may produce some exciting and beneficial consequences for the management of some psychiatric illnesses. Besides lithium therapy, other techniques that change the period of circadian pacemakers may be worth trying. It is even possible that driving the X pacemaker either electronically or with phase-resetting drugs may at some time in the future prove to be beneficial therapeutic techniques.

PERIOD DISORDERS

Although there are major changes in circadian phase in the disorders discussed above, most patients remain synchronized to the 24-hour day–night cycle. In other conditions the patient fails to entrain to normal 24-hour cues in the environment. These patients fall into two groups: those who are blind and those who are sighted and yet still free-run.

Many blind people are capable of remaining well synchronized to the day–night cycle, even though they cannot visually perceive light and dark. Others show considerable problems in maintaining synchrony (Miles, Raynal, and Wilson, 1977), and their circadian timing system may free-run with a period different from 24 hours. Initially, there was some controversy as to whether blind people had circadian rhythms at all. However, it is now clear that they almost always have prominent circadian rhythms but that they may be free-running. The occasional failure to observe rhythms appears to have occurred when data were averaged from a group of blind subjects who had free-running rhythms with widely varying phases at the instant of measurement. As we discussed in Chapter 1, no mean circadian rhythm may be documented under such circumstances.

Figure 7.18A shows the sleep–wake record of a blind man who complained of intermittent insomnia (Miles, Raynal, and Wilson, 1977). For two or three weeks at a time he would find it hard to sleep at night and would have excessive daytime sleepiness. An examination of the circadian actogram shows a free-running circadian rhythm with a period of approximately 25 hours. Miles, Raynal, and Wilson (1977) showed that the rhythms of body temperature, alertness, performance, cortisol secretion, and urinary electrolyte excretion, as well as sleep and wake, all free-ran in this subject. Despite strenuous attempts to adjust to normal society, including the cyclic administration of hypnotic and stimulant drugs, he was unable to entrain to a 24-hour day–night cycle.

There are a number of reports of normally sighted people who have

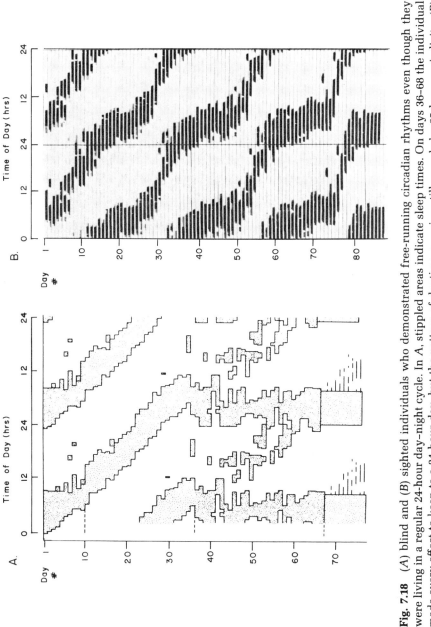

Fig. 7.18 (A) blind and (B) sighted individuals who demonstrated free-running circadian rhythms even though they were living in a regular 24-hour day-night cycle. In A, stippled areas indicate sleep times. On days 36–68 the individual made every effort to keep to a 24-hour day, but the pattern of daytime napping still revealed a 25-hour periodicity. (B) The sighted individual showed a similar relative coordination to the 24-hour clock as was shown in the animal experiments in Fig. 2.13. (A from Miles, Raynal, and Wilson, 1977, copyright 1977 by the American Association for the Advancement of Science; B from Kokkoris et al., 1978, reprinted by permission of Raven Press.)

free-running circadian rhythms. Elliott and co-workers (1971) investigated a student whose circadian sleep–wake cycle free-ran even though he was living in a 24-hour environment. Over several months Kokkoris and colleagues (1978) studied a patient with what they termed "a hypernyctohemeral sleep–wake cycle disturbance." The data from this patient are shown in Figure 7.18B. The mean period of the sleep–wake cycle and the body temperature rhythms were 24.8 hours, although the sleep–wake cycle showed a relative coordination with respect to the day–night cycle. The rate of delay of the sleep–wake cycle thus slowed as it passed through the time of day (midnight to 6 A.M.) normally associated with sleep. Weber and co-workers (1980) reported on four years of data from a man who consistently showed a non-24-hour sleep–wake cycle in a normal environment. His girlfriend attempted to maintain a corresponding schedule but with less success. Her period change was apparently not an endogenous disorder but an attempt to adapt to her boyfriend's schedule.

The failure of normal individuals to entrain to a 24-hour day may be more common than has been previously recognized. Those who attempt to maintain a 24-hour schedule may suffer from intermittent insomnia. In other words, depending on the phase of their endogenous circadian pacemaker, they have two to three weeks of insomnia at night and excessive sleepiness during the day, followed by two to three weeks of good nighttime sleep and full wakefulness during the day. The individual may feel fine whenever the imposed rest–activity cycle coincides with his endogenous circadian rhythms. Thus, he may not seek medical advice since the insomnia always temporally "cures itself." Such intermittent insomnia is also hard to characterize unless the person keeps a precise sleep diary and the data are plotted in a format such as the circadian actograms we demonstrate here. The insomnia may be missed unless the physician is aware of the necessity of obtaining a long-term sleep–wake records.

Failure to entrain to 24-hour environmental time cues may be caused either by a disorder of zeitgeber reception or a change in pacemaker function so that the endogenous period and range of entrainment diverge too much from 24.0 hours. Zeitgeber reception disorders are undoubtedly the major cause of altered circadian periods in blind subjects and may also be an important cause for the failure of sighted individuals to entrain.

Animal studies have shown that the SCN are innervated via a separate neural pathway from the retina (the RHT). If an RHT exists in humans, this may explain in part why some blind people can entrain to 24-hour light–dark cycles while others cannot and why some sighted people may fail to entrain. Animal studies have shown (see Chapter 4) that transection of the optic pathways behind the optic chiasm without damage to the RHT may interfere with visually guided behavior but will not compromise circadian entrainment (Moore, 1978). Similarly, lesions of the RHT would be expected to compromise circadian entrainment but not interfere with vision.

Furthermore, even if both the primary optic tracts and the RHT were destroyed, social cues, the other major zeitgeber for humans, may sometimes permit synchronization to a 24-hour period. This could explain how some blind people are able to remain entrained. It is also possible that some sighted patients may have a disorder in the perception of social cues. Some of the sighted people who free-run appear to be reclusive or have other psychological disorders which may isolate them from normal social cues. In the absence of effective social entrainment, they may develop free-running circadian rhythms. The causal contribution of such psychopathology is, however, difficult to evaluate since psychological disturbances (as we discussed earlier) may be induced by disordered phase relationships in the circadian system.

Changes in pacemaker period might also lead to failure to entrain to a 24-hour day–night cycle. For example, it is possible that such people may have an extreme case of delayed sleep phase insomnia, so that their sleep–wake cycle pacemaker has an extremely long endogenous period. In this case, although a 24-hour time cue is sufficient to provide some relative coordination, the period of the pacemaker would be too long for entrainment to occur. The subject would thus show free-running circadian rhythms.

At present, other than strengthening all possible zeitgeber cues, there is no specific therapy available for patients who fail to entrain to the environmental 24-hour day. However, as we indicated earlier, electronic stimulation techniques and drugs which induce circadian phase shifts by resetting the pacemaker may offer some hope for the future.

AMPLITUDE DISORDERS AND ARRHYTHMIAS

The third general type of disorder of circadian timekeeping involves reductions in the amplitude of rhythms or, in the extreme case, the loss of rhythmicity. There are various possible causes of such reductions in amplitude, including removal of environmental zeitgeber influences, failure of the pacemaker, breakdown of the transmission process, and pathology of the tissues in which the overt rhythmicity is seen.

Some amplitude disorders are a long-term response to arrhythmic environments. We have already discussed the influence of continual midwinter dark on the circadian rhythmicity of Eskimos. Rhythms in many variables are markedly reduced in amplitude; the urinary electrolyte rhythms are best studied (Lobban, 1960). This reduction appears to have been a response to the lack of environmental zeitgebers, because in recent years, with the imposition of regular day–night schedules, circadian rhythms in Eskimos have attained normal amplitude (Lobban, 1976). The causes of loss of amplitude in overt rhythms appear to be twofold. First, the removal of a regular light–dark cycle also removes the masking effects of light, which normally contribute to the amplitude of the overt rhythm; secondly, there may be a greater tendency for the secondary oscillators of the circadian system to dissociate and thus reduce the amplitude of the overt rhythm.

Alternatively, a loss in rhythm amplitude may occur because of damage to the central pacemakers. In Chapter 4 we discussed suggestions that tumors destroying the SCN region may disrupt the sleep–wake cycle (Fulton and Bailey, 1929; Gillespie, 1930). However, the reports were published long before the SCN were identified or the properties of the circadian timing system characterized. More recently, Krieger and Krieger (1967) have shown disruptions in urinary electrolyte rhythms in patients with hypothalamic disease, and Page, Galicich, and Grunt (1973) have shown that increases in brain third ventricle pressure disrupt the body temperature rhythm. Furthermore, it is possible that patients such as the one whose record is shown in Figure 7.15C may have damage to the pacemaker. However, a combination of careful recording of sleep–wake patterns and other normally rhythmic variables, and ultimate postmortem examination of the hypothalamus is needed before firm correlations can be made.

Finally, the loss of an overt rhythm can be due to a failure in the neural or endocrine transmission processes or in the target tissue itself. For example, in Addison's disease in which there is adrenal failure, the rhythm in plasma cortisol is lost and the rhythms in urinary electrolyte excretion are considerably damped (Levy, Power, and Kepler, 1946; Finkenstaedt et al., 1954). The loss of rhythmicity in a mediator, such as plasma cortisol, whether because of disease or because of masking due to stress, can lead to the desynchronization of the secondary oscillators which that mediator normally entrains (Moore-Ede, Czeisler, et al., 1977). Thus, adrenalectomized patients receiving replacement corticosteroids evenly distributed over each 24 hours show a free-running rhythm in urinary potassium excretion because the circadian rhythmicity in plasma cortisol concentration is lost (Moore-Ede, unpublished observations).

THE STUDY of the circadian timing system is thus at a fascinating stage. The general properties are mostly defined, and possible malfunctions can be predicted. An important task ahead is to identify clinical syndromes that are a result of circadian clock malfunction. In the last few years several syndromes have been identified, and undoubtedly more remain to be described. The ultimate benefit is to the patient, especially if research into the pathology of these timing disorders yields clues as to how they may be rationally treated.

Glossary

Acrophase

Phase angle of the crest (maximal value) of a sine function fitted to the raw data of a rhythm.

Active element

Component in a biological system which is capable of generating self-sustained oscillations.

Aftereffects

Long-term transients, often lasting for 100 days or more, after a rhythm is released into constant conditions where it can adopt its spontaneous free-running period.

Amplitude

Difference between maximum (or minimum) and mean value in a sinusoidal oscillation; often used in a looser sense for other oscillations.

Circadian rhythm

Self-sustained biological rhythm which in the organism's natural environment is normally entrained to a 24-hour period.

Circalunar rhythm

Self-sustained biological rhythm which in the organism's natural environment is normally entrained to the period of the moon (28 days).

Circannual rhythm

Self-sustained biological rhythm which in the organism's natural environment is normally entrained to the period of the 365.25-day seasonal variation in the environment.

Circa-rhythms
Classes of rhythms that are capable of free-running in constant conditions with a period approximating that of the environmental cycle to which they are normally synchronized, and that are entrainable by zeitgebers. (*See* Circadian, Circatidal, Circalunar, Circannual rhythms.)

Circatidal rhythm
Self-sustained biological rhythm which in the organism's natural environment is normally entrained to the period of the ocean tides (typically 12.4 hours).

Desynchronization
Loss of synchronization between two or more rhythms so that they show independent periods.

Endogenous rhythm
Self-sustained rhythm generated within an organism.

Entraining agent
Synonymous with zeitgeber.

Entrainment
Synchronization of a self-sustaining rhythm by a forcing oscillation (zeitgeber). During entrainment the frequencies of the two oscillations are the same or integral multiples of each other (entrainment by frequency demultiplication; see below).

Exogenous rhythm
Rhythm generated by the influences of an environmental periodicity on an organism.

External desynchronization
Loss of synchronization between rhythm and zeitgeber.

Forced internal desynchronization
Internal desynchronization (see below) produced by subjecting an organism to one or more zeitgeber cycles which entrain only some of the circa-rhythmic variables within an organism.

Free-run
State of a circa-rhythm in constant conditions, that is, in the absence of entraining agents (zeitgebers).

Frequency
Reciprocal of period.

Frequency demultiplication
Entrainment of a biological rhythm by a zeitgeber which has a period that is an integral fraction of the rhythm's period.

Infradian
Biological rhythm with a period longer than that of circadian rhythms.

Internal desynchronization
Loss of synchronization between two or more rhythms so that they free-run with different periods within the same organism.

Mean value
Arithmetic mean of all instantaneous values of an oscillating variable within one cycle.

Mediator
Neural or endocrine function which through its oscillations can transmit period and phase information so as to synchronize the rhythms in a target tissue.

Pacemaker
Functional entity capable of self-sustaining oscillations which synchronize other rhythms.

Passive element
Component in a biological system which is not capable of generating self-sustained oscillations and is rhythmic only if forced by another oscillation.

Period
Time interval between recurrences of a defined phase of the rhythm.

Phase
Instantaneous state of an oscillation within a period.

Phase angle
Value of the abscissa corresponding to a phase of the oscillation, usually given in degrees, where the whole period is defined as 360 degrees and the zero point is arbitrary. It can be given in units of time if the length of the period is stated.

Phase angle difference
Difference between corresponding phase angles in two coupled oscillations, given either in degrees or in units of time. (Often "corresponding" phase angles have to be defined arbitrarily.)

Phase control
Control of the period and phase relationship of a rhythm by a zeitgeber.

Phase relationship
Synonymous with phase angle difference.

Phase response curve
Graphical plot indicating how the amount and the direction of a phase shift, induced by a single stimulus, depend on the phase at which the stimulus is applied.

Phase shift
 Single displacement of an oscillation along the time axis; may occur instantaneously or after several transient cycles.

Photoperiod
 Duration of light in a light–dark cycle.

Range of entrainment
 Range of periods within which a self-sustaining oscillation can be entrained by a zeitgeber.

Range of oscillation
 Difference between maximum and minimum value (independent of shape of oscillation).

Relative coordination
 Modulations in the period of a rhythm which result when the organism is exposed to a zeitgeber cycle too weak to entrain the rhythm.

Secondary oscillator
 Oscillator within an organism capable of generating oscillations but which usually has less stability and persistence than a pacemaker, is not entrained directly by zeitgebers, and may not necessarily synchronize other oscillators.

Transducer
 Component in a biological system which detects environmental zeitgeber cycles and converts the temporal information into a form (typically neural or endocrine rhythms) which can synchronize biological oscillators.

Transient
 Temporary oscillatory state between two steady states.

Transient internal desynchronization
 Internal desynchronization occurring between circa-rhythms over only a few cycles. Typically this may occur after a phase shift of a zeitgeber when some rhythms resynchronize more rapidly than others.

Ultradian
 Biological rhythm with a period shorter than that of circadian rhythms.

Zeitgeber
 Forcing environmental oscillation which entrains a biological self-sustaining rhythm.

References

Aanonsen, A. (1959). Medical problems of shift work. *Industr. Med. Surg.* 28:422–427.

Abe, K., Kroning, J., Greer, M. A., and Critchlow, V. (1979). Effects of destruction of the suprachiasmatic nuclei on the circadian rhythms in plasma corticosterone, body temperature, feeding and plasma thyrotropin. *Neuroendocrinology* 29:119–131.

Abrams, R. and Hammel, H. T. (1965). Cyclic variations in hypothalamic temperature in unanesthetized rats. *Am. J. Physiol.* 208:698–702.

Adair, E. R. (1974). Hypothalamic control of thermoregulatory behavior. Preoptic-posterior hypothalamic interaction. In *Recent Studies of Hypothalamic Function* (Ledesis, K., and Cooper, K. E., eds.). Basel: Karger. 341–358.

Aghajanian, G. K., Bloom, F. E., and Sheard, M. H. (1969). Electronmicroscopy of degeneration within the serotonin pathway of rat brain. *Brain Res.* 13:266–273.

Agren, G., Wilander, O., and Jorpes, E. (1931). Cyclic changes in the glycogen content of the liver and muscles of rats and mice. *Biochem. J.* 25:777–785.

Akerstedt, T. and Gillberg, M. (1981). The circadian pattern of unrestricted sleep and its relation to body temperature, hormones and alertness. In *The Twenty-Four Hour Workday: Proceedings of a Symposium on Variations in Work-Sleep Schedules* (Johnson, L. C., Tepas, D. I., Colquhoun, W. P., and Colligan, M. J., eds.). Cincinnati: U.S. Department of Health and Human Services (National Institute for Occupational Safety and Health) publication 81-127:605–624.

Albers, H. E. (1981). Gonadal hormones organize and modulate the circadian system of the rat. *Am. J. Physiol.*, 241:R62–R66.

Albers, H. E., Gerall, A. A., and Axelson, J. F. (1981). Effect of reproductive state on circadian periodicity in the rat. *Physiol. Behav.* 26:21–25.

Albers, H. E., Gerall, A. A., and Axelson, J. F. (1981). Circadian rhythm dissociation: effects of long-term constant illumination. Neurosci. Lett.,
25:89–94.

Albers, H. E., Lydic, R., Gander, P., and Moore-Ede, M. C. (1982). Gradual
decay of circadian drinking organization following lesions of the suprachiasmatic nuclei in primates. Neurosci Lett. in press.

Albers, H. E., Moline, M. L., and Moore-Ede, M. C. (1980). Entrainment of the
preovulatory LH surge and locomotor activity following an acute shift in
the timing of the light–dark cycle. Paper presented at the 12th Eastern
Conference on Reproductive Behavior, Rockefeller University, New
York.

Allen, C. F., Kendall, J. W., and Greer, M. A. (1972). The effect of surgical isolation of the basal hypothalamus on the nychthemeral rhythm of plasma
corticosterone concentration in rats with heterotopic pituitaries. Endocrinology 91:873–876.

Alleva, J. J., Waleski, M. V., and Alleva, F. R. (1971). A biological clock controlling the estrous cycle of the hamster. Endocrinology 88:1368–1379.

Alvis, J. D., Davis, F. C., and Menaker, M. (1978). Sexual dimorphism of the biological clock in golden hamsters. Physiologist 21:2.

Amand, B. K. and Brobeck, J. R. (1951). Hypothalamic control of food intake in
rats and cats. Yale J. Biol. Med. 24:123–140.

Anand, B. K., Chhina, G. S., Sharma, K. N., Dua, S., and Singh, B. (1964). Activity of single neurons in the hypothalamic feeding centers: effect of glucose.
Am. J. Physiol. 207:1146–1154.

Andjus, R. K. and Smith, A. U. (1955). Reanimation of adult rats from body
temperatures between 0 and +2°C. J. Physiol. (London) 128:446–472.

Andrews, R. V. (1968). Temporal secretory responses of cultured hamster adrenals. Comp. Biochem. Physiol. 26:179–193.

Andrews, R. V. and Folk, J. E. (1964). Circadian metabolic patterns in cultured
hamster adrenal glands. Comp. Biochem. Physiol. 11:393–409.

Andrezsyuk, H. I. (1968). Quoted in Schaeffer, K. E., Kerr, C. M., Buss, D., and
Haus, E. (1979). Effect of 18-h watch schedules on circadian cycles of
physiological functions during submarine patrols. Undersea Biomed.
Res.:S81–S90. (Submarine Supplement)

Anonymous (1978). Sinking spells. U.S. Army Aviation Digest 24:12–13.

Applewhite, P. B., Satter, R. L., and Galston, A. W. (1973). Protein synthesis
during endogenous rhythmic leaflet movement in Albizzia. J. Gen. Physiol.
62:707–713.

Aschoff, J. (1951). Die 24-stunden periodik der maus unter konstanten umge-
bungs bedingungen. Naturwissenschaften 38:506–507.

Aschoff, J. (1954). Zeitgeber der tierischen tagesperiodik. Naturwissenschaften
41:49–56.

Aschoff, J. (1955). Der tagesgang der korpertemperatur beim menschen. Klin.
Wochenschr. 33:545–551.

Aschoff, J. (1960). Exogenous and endogenous components in circadian
rhythms. Cold Spring Harbor Symp. Quant. Biol. 25:11–26.

Aschoff, J. (1965a). Response curves in circadian periodicity. In *Circadian Clocks* (Aschoff, J., ed.). Amsterdam: North-Holland. 95–111.

Aschoff, J. (1965b). Circadian rhythms in man. *Science* 148:1427–1432.

Aschoff, J. (1970). Circadian rhythm of activity and body temperature. In *Physiological and Behavioral Temperature Regulation* (Hardy, J. D., Gagge, A. P., and Stolwijk, J. A. J., eds.). Springfield, Ill.: Thomas. 905–919.

Aschoff, J. (1978a). Circadian rhythms within and outside their ranges of entrainment. In *Environmental Endocrinology* (Assenmacher, I. and Farner, D. S., eds.). New York: Springer-Verlag. 172–181.

Aschoff, J. (1978b). Problems of re-entrainment of circadian rhythms: asymmetry effect, dissociation and partition. In *Environmental Endocrinology* (Assenmacher, I. and Farner, D. S., eds.). New York: Springer-Verlag. 185–195.

Aschoff, J. (1978c). Features of circadian rhythms relevant for the design of shift work schedules. *Ergonomics* 21:739–754.

Aschoff, J. (1979a). Circadian rhythms: general features and endocrinological aspects. In *Endocrine Rhythms* (Krieger, D. T., ed.). New York: Raven Press. 1–61.

Aschoff, J. (1979b). Influences of internal and external factors on the period measured in constant conditions. *Z. Tierpsychol.* 49:225–249.

Aschoff, J., Biebach, H., Heise, A., and Schmidt, T. (1974). Day–night variation in heat balance. In *Heat Loss From Animals and Man* (Monteith, J. L. and Mount, L. E., eds.). London: Butterworths. 147–172.

Aschoff, J., Gerecke, U., Kureck, A., Pohl, H., Rieger, P., von Saint-Paul, U., and Wever, R. (1971). Interdependent parameters of circadian activity rhythms in birds and man. In *Biochronometry*, vol. 3 (Menaker, M., ed.). Washington, D.C.: National Academy of Sciences. 3–29.

Aschoff, J., Gerecke, U., and Wever, R. (1967). Desynchronization of human circadian rhythms. *Jpn. J. Physiol.* 17:450–457.

Aschoff, J. and Heise, A. (1972). Thermal conductance in man: its dependence on time of day and ambient temperature. In *Advances in Climatic Physiology* (Itoh, S., Ogata, K., and Yoshimura, H., eds.). New York: Springer-Verlag. 334–348.

Aschoff, J., Klotter, K., and Wever, R. (1965). Circadian vocabulary. In *Circadian Clocks* (Aschoff, J., ed.). Amsterdam: North-Holland. (pp x–xix)

Aschoff, J. and Pohl, H. (1970). Rhythmic variations in energy metabolism. *Fed. Proc.* 29:1541–1552.

Aschoff, J. and Pohl, H. (1978). Phase relations between a circadian rhythm and its zeitgeber within the range of entrainment. *Naturwissenschaften* 65:80–84.

Aschoff, J., Poppel, E., and Wever, R. (1969). Circadiane periodik des menschen unter dem einfluss von licht-dunkel-wechseln unterschiedlicher periode. *Pfluegers Arch.* 306:58–70.

Aschoff, J., Saint-Paul, U., and Wever, R. (1971). Die lebensdauer von fliegen unter dem einfluss von zeitverschiebungen. *Naturwissenschaften* 58:574.

Aschoff, J. and Wever, R. (1962). Spontanperiodik des menschen bei ausschluss aller zeitgeber. *Naturwissenschaften* 49:337-342.

Aserinsky, E. and Kleitman, N. (1953). Regularly occurring periods of eye motility and concomitant phenomena during sleep. *Science* 118:273-274.

Ashkenazi, I. E., Hartman, H., Strulovitz, B., and Dar, O. (1975). Activity rhythms of enzymes in human red blood cell suspensions. *J. Interdiscipl. Cycle Res.* 6:291-301.

Atkinson, M., Kripke, D. F., and Wolf, S. R. (1975). Autorhythmometry in manic depressives. *Chronobiologia* 2:325-335.

Axelrod, J. (1974). The pineal gland: a neurochemical transducer. *Science* 184:1341-1348.

Axelson, J. F., Gerall, A. A., and Albers, H. E. (1981). Effect of progesterone on the estrous activity cycle of the rat. *Physiol. Behav.* 26:631-635.

Baker, J. R. and Ranson, R. M. (1932). Factors affecting the breeding of the field mouse. (*Microtus agrestis*). I. Light. *Proc. R. Soc. Lond. (Biol.)* 110:313-322.

Baker, R. R. (1980). Goal orientation by blindfolded humans after a long-distance displacement: possible involvement of a magnetic sense. *Science* 210:554-557.

Balagura, S. and Davenport, L. D. (1970). Feeding patterns of normal and ventromedial hypothalamic lesioned male and female rats. *J. Comp. Physiol. Psychol.* 71:357-364.

Banks, J. A., Mick, C., and Freeman, M. E. (1980). A possible cause for the differing responses of the luteinizing hormone surge mechanism of ovariectomized rats to short term exposure to estradiol. *Endocrinology* 106:1677-1681.

Barger, A. C., Berlin, R. D., and Tulenko, J. F. (1958). Infusion of aldosterone, 9-fluorohydrocortisone and antidiuretic hormone into the renal artery of normal and adrenalectomized unanesthetized dogs: effect on electrolyte and water excretion. *Endocrinology* 62:804-815.

Barnes, C. A., McNaughton, B. L., Goddard, G. V., Douglas, R. M., and Adamec, R. (1977). Circadian rhythm of synaptic excitability in rat and monkey central nervous system. *Science* 197:91-92.

Barry, J. and Dubois, M. P. (1976). Immunoreactive LRF neurosecretory pathways in mammals. *Acta Anat.* 94:497-503.

Bartter, F. C., Chan, J. C. M., and Simpson, H. W. (1979). Chronobiological aspects of plasma renin activity, plasma aldosterone, and urinary electrolytes. In *Endocrine Rhythms* (Krieger, D. T., ed.). New York: Raven Press. 225-245.

Bartter, F. C., Delea, C. S., and Halberg, F. (1962). A map of blood and urinary changes related to circadian variations in adrenal cortical function in normal subjects. *Ann. N.Y. Acad. Sci.* 98:969-983.

Bayliss, R. I. S. (1955). General discussion. *Ciba Found. Colloquia Endocrinol.* 8:649-650.

Beckett, A. H. and Rowland, M. (1964). Rhythmic urinary excretion of amphetamine in man. *Nature* 204:1203-1204.

Behl, P. N. and Garg, B. R. (1967). Diurnal rhythm of corticosteroid therapy. *J. Indian Med. Assoc.* 48:375-378.

Beling, I. (1929). Uber das zeitgedachtnis der bienen. *Z. Vergl. Physiol.* 9:259-338.

Beljan, J. R., Rosenblatt, L. S., Hetherington, N. W., Layman, J., Flaim, S. E. T., Dale, G. T., and Holley, D. C. (1972). *Human Performance in the Aviation Environment.* NASA Contract no. 2-6657, Pt. Ia, 253-259.

Bellonici, J. (1888). Uber die zentrale endigung des Nervus opticus bei den vertebraten. *Zeitschr. fur Wissensch. Zoologie* 47:1-46.

Bennett, C. M., Brenner, B. M., and Berliner, R. W. (1968). Micropuncture study of nephron function in the rhesus monkey. *J. Clin. Invest.* 47:203-216.

Berger, H. (1930). Ueber das elektroenkephalogramm des menschen. *J. Psychol. Neurol.* 40:160-179.

Berliner, R. W., Kennedy, T. S., and Orloff, J. (1951). Relationship between acidification of the urine and potassium metabolism. *Am. J. Med.* 11:274-282.

Berlucchi, G., Maffei, E., Moruzzi, G., and Strata, P. (1964). EEG and behavioral effects elicited by cooling effects of medulla and pons. *Arch. Ital. Biol.* 102:372-392.

Bernard, C. (1878). *Les Phenomenes de la Vie.* Paris: J. B. Bailliere.

Bernstein, I. L. (1975). Relationship between activity, rest, and free feeding in rats. *J. Comp. Physiol. Psychol.* 89:253-257.

Binkley, S., Riebman, J. B., and Reilly, K. B. (1978). The pineal gland: a biological clock in vitro. *Science* 202:1198-1201.

Bissonette, T. H. (1932). Modification of mammalian sexual cycles; reactions of ferrets (*Putorius vulgaris*) of both sexes to electric light added after dark in November and December. *Proc. R. Soc. Lond. (Biol.)* 110:322-336.

Bissonette, T. H. (1935). Modification of mammalian sexual cycles. III. Reversal of the cycle in male ferrets (*Putorius vulgaris*) by increasing periods of exposure to light between October second and March thirtieth. *J. Exp. Zool.* 71:341-373.

Bissonette, T. H. (1941). Experimental modification of breeding cycles in goats. *Physiol. Zool.* 14:379-383.

Bixler, E. O., Kales, A., Soldatos, C. R., Kales, J. D., and Healey, S. (1979). Prevalence of sleep disorders in the Los Angeles metropolitan area. *Am. J. Psychiatry* 136:1257-1262.

Bjerner, B., Holm, A., and Swensson, A. (1955). Diurnal variation in mental performance—a study of three-shift workers. *Br. J. Ind. Med.* 12:103-110.

Black, I. B. and Reis, D. J. (1971). Central neural regulation by adrenergic nerves of the daily rhythm in hepatic tyrosine transaminase activity. *J. Physiol. (London)* 219:267-280.

Blass, E. M. and Epstein, A. N. (1971). A lateral preoptic osmosensitive zone for thirst in the rat. *J. Comp. Physiol. Psychol.* 76:378-394.

Bloch, M. (1964). Rhythmic diurnal variation in limb blood flow in man. *Nature* 202:398-399.

Bobillier, P. and Mouret, J. R. (1971). The alterations of the diurnal variations of brain tryptophan, biogenic amines and 5-hydroxyindole acetic acid in the rat under limited time feeding. *Int. J. Neurosci.* 2:271–282.

Bolles, R. C. and Stokes, L. W. (1965). Rat's anticipation of diurnal and a-diurnal feeding. *J. Comp. Physiol. Psychol.* 60:290–294.

Borbély, A. and Huston, J. P. (1974). Effects of two-hour light-dark cycles on feeding, drinking and motor activity of the rat. *Physiol. Behav.* 13:795–802.

Borst, J. G. G. and DeVries, L. A. (1950). The three types of "natural" diuresis. *Lancet* 2:1–6.

Boswell, J. (1785). Entry for Sept. 14, 1773 in *The Journal of the Tour of the Hebrides with Samuel Johnson.* New ed. 1936. New York: Viking Press.

Boulos, Z., Rosenwasser, A. M., and Terman, M. (1980). Feeding schedules and the circadian organization of behavior in the rat. *Behav. Brain Res.* 1:39–65.

Bovet, J. and Oertli, E. F. (1974). Free-running circadian activity rhythms in free-living beaver (*Castor canadensis*). *J. Comp. Physiol.* 92:1–10.

Bremer, F. (1935). Cerveau "isole" et physiologie du sommeil. *C. R. Soc. Biol.* (*Paris*) 122:460–463.

Brenner, B. M. and Berliner, R. W. (1973). The transport of potassium. In *Handbook of Physiology,* Section 8: *Renal Physiology* (Orloff, J. and Berliner, R. W., eds.). Washington, D.C.: American Physiological Society. 497–519.

Bretzl, H. (1903). *Botanische Forschungen des Alexanderzuges.* Leipzig: B. G. Teubner.

Breuer, H., Kaulhausen, H., Muhlbauer, W., Fritzsche, G., and Vetter, H. (1974). Circadian rhythm of the renin-angiotensin-aldosterone system. In *Chronobiological Aspects of Endocrinology* (Aschoff, J., Ceresa, F., and Halberg, F., eds.). Stuttgart: Schattauer-Verlag. 101–109.

Bridges, R. S. and Goldman, B. D. (1975). Diurnal rhythms in gonadotropins and progesterone in lactating and photoperiod induced acyclic hamsters. *Biol. Reprod.* 13:617–622.

Brockhaus, H. (1942). Beitrag zur normalen anatomie des hypothalamus und der zona incerta beim menschen. *J. f. Psychol. u. Neurol.* 51:96–196.

Brown, F. A. (1957). Biological chronometry. *Am. Natural.* 91:129–133.

Brown, F. A. (1972). The clocks timing biological rhythms. *Am. Sci.* 60:756–766.

Brown, H., Englert, E., Wallach, S., and Simmons, E. L. (1957). Metabolism of free and conjugated 17-hydroxycorticosteroids in normal subjects. *J. Clin. Endocrinol. Metab.* 17:1191–1201.

Browne, R. C. (1949). The day and night performance of teleprinter switchboard operators. *Occup. Psychol.* 23:1–6.

Bruce, V. G. (1960). Environmental entrainment of circadian rhythms. *Cold Spring Harbor Symp. Quant. Biol.* 25:29–48.

Bruce, V. G. (1972). Mutants of the biological clock in *Chlamydomonas reinhardi. Genetics* 70:537–548.

Buhl, A. E., Norman, R. L., and Resko, J. A. (1978). Sex differences in estrogen induced gonadotropin release in hamsters. *Biol. Reprod.* 18:592-597.

Buijs, D. F., Swaab, D. F., Dogterom, J., and Van Leeuwen, F. W. (1978). Intra- and extrahypothalamic vasopressin and oxytocin pathways in the rat. *Cell Tissue Res.* 186:423-433.

Bünning, E. (1932). Uber die erblichkeit der tagesperiodizitat bei den *Phaseolus* blattern. *Jb. wiss. Bot.* 77:283-320.

Bünning, E. (1935a). Zur kenntis der erblichen tagesperiodiztat bei den primarblattern von *Phaseolus multiflorus*. *Jb. wiss. Bot.* 81:411-418.

Bünning, E. (1935b). Zur kenntnis der endogenen tagesrhythmik bei insekten und pflanzen. *Berl. Dtsch. Bot. Ges.* 53:594-623.

Bünning, E. (1936). Die endogene tagesrhythmik als grundlage der photoperiodischen reaktion. *Berl. Dtsch. Bot. Ges.* 54:590-607.

Bünning, E. (1958). Das weiterlaufen der "physiologischen uhr" im saugerdarm ohne zentrale steuerung. *Naturwissenschaften* 45:68.

Bünning, E. (1960). Circadian rhythms and the time measurement in photoperiodism. *Cold Spring Harbor Symp. Quant. Biol.* 25:249-256.

Bünning, E. (1973). *The Physiological Clock*. London: English University Press.

Bünning, E. (1977). Fifty years of research in the wake of Wilhelm Pfeffer. *Ann. Rev. Plant Physiol.* 28:1-22.

Bünning, E. and Chandrashekaran, M. K. (1975). Pfeffer's views on rhythms. *Chronobiologia* 2:160-167.

Bünning, E. and Moser, I. (1972). Influence of valinomycin on circadian leaf movements of *Phaseolus*. *Proc. Natl. Acad. Sci. USA* 69:2732-2733.

Bünning, E. and Moser, I. (1973). Light-induced phase shifts of circadian leaf movements of *Phaseolus*: comparison with the effects of potassium and of ethyl alcohol. *Proc. Natl. Acad. Sci. USA* 70:3387-3389.

Burchard, J. E. (1958). Re-setting a biological clock. Ph.D. thesis, Princeton University.

Cabanac, M., Hildebrandt, G., Massonnet, B., and Strempel, H. (1976). A study of the hypothermal cycle of behavioral temperature regulation in man. *J. Physiol. (London)* 257:275-291.

Cahn, A. A., Folk, G. E., and Huston, P. E. (1968). Age comparisons of human day-night physiological differences. *Aviat. Space Environ. Med.* 39:608-610.

Campbell, C. B. G. and Ramally, J. A. (1974). Retinohypothalamic projections: correlations with onset of the adrenal rhythms in infant rats. *Endocrinology* 94:1201-1204.

Cannon, W. B. (1929). Organization for physiological homeostasis. *Physiol. Rev.* 9:399-431.

Cannon, W. H. and Jensen, O. G. (1975). Terrestrial timekeeping and general relativity—a discovery. *Science* 188:317-328.

Card, J. P., Riley, J. N., and Moore, R. Y. (1980). The suprachiasmatic hypothalamic nucleus: ultrastructure of relations to optic chiasm. *Neurosci. Abstr.* 6:758.

Cardoso, S. S., Scheving, L. E., and Halberg, F. (1970). Mortality of mice as influenced by the hour of the day of the drug (ara-C) administration. *Pharmacologist* 12:302.

Carlsson, A. and Serin, F. (1950a). Time of day as a factor influencing the toxicity of nikethamide. *Acta Pharmacol. Toxicol.* 6:181–186.

Carlsson, A. and Serin, F. (1950b). The toxicity of nikethamide at different times of day. *Acta Pharmacol. Toxicol.* 6:187–193.

Cheifetz, P., Gaffud, N., and Dingan, J. F. (1968). Effects of bilateral adrenalectomy and continuous light on the circadian rhythm of corticotropin in female rats. *Endocrinology* 82:1117–1124.

Chouvet, G., Mouret, J., Coindet, J., Siffre, M., and Jouet, M. (1974). Periodicite bicircadienne du cycle vielle-sommeil dans des condition hors du temps. *Electroencephalogr. Clin. Neurophysiol.* 37:367–380.

Chu, H. N. (1932). The cell masses of the diencephalon of the opossum, *Didelphis virg. Monographs of the National Res. Inst. of Psychol.* 2:1–37.

Cicerone, C. M. (1976). Cones survive rods in the light-damaged eye of the albino rat. *Science* 194:1183–1185.

Clark, W. E. (1938). Morphological aspects of the hypothalamus. In *The Hypothalamus: Morphological, Functional, Clinical and Surgical Aspects* (Clark, W. E., Beattie, J., Riddoch, G., and Dott, N. M., eds.). Edinburgh: Oliver and Boyd. 1–68.

Clarke, J. R. and Kennedy, J. P. (1967). Effect of light and temperature upon gonad activity in the vole (*Microtus agrestis*). *Gen. Comp. Endocrinol.* 8:474–488.

Coale, A. J. (1974). The history of the human population. *Sci. Am.* 231:40–51.

Cohen, M. I. and Feldman, J. L. (1977). Models of respiratory phase-switching. *Fed. Proc.* 36:2367–2374.

Cohn, C., Webb, L., and Joseph, D. (1970). Diurnal rhythm in urinary electrolyte excretion by the rat: influence of feeding habits. *Life Sci.* 9:803–809.

Coindet, J., Chouvet, G., and Mouret, J. (1975). Effects of lesions of the suprachiasmatic nuclei on paradoxical sleep and slow wave sleep circadian rhythms in the rat. *Neurosci. Lett.* 1:243–247.

Colin, J., Timbal, J., Boutelier, C., Houdas, Y., and Siffre, M. (1968). Rhythm of the rectal temperature during a 6-month free-running experiment. *J. Appl. Physiol.* 25:170–176.

Cone, A. L. (1974). Feeding time entrainment of activity and self-produced illumination change in a squirrel monkey. *Bull. Psychonom. Soc.* 2:389–391.

Conrad, L. C. and Pfaff, D. W. (1976). Efferents from the medial basal forebrain and hypothalamus in the rat. II. An autoradiographic study of the anterior hypothalamus. *J. Comp. Neurol.* 169:221–261.

Conroy, R. T. W. L. and Mills, J. N. (1970). *Human Circadian Rhythms.* London: J. A. Churchill.

Cormia, F. E. (1952). Experimental histamine pruritus. I. Influence of physical and psychological factors on threshold reactivity. *J. Invest. Dermatol.* 19:21–33.

Cornelius, G. (1980). On the existence of circadian rhythms in human erythrocyte suspensions. *J. Interdiscipl. Cycle Res.* 11:55–68.

Cranston, W. I. and Brown, W. (1963). Diurnal variation in plasma volume in normal and hypertensive subjects. *Clin. Sci. Mol. Med.* 25:107–114.

Crowley, T. J., Halberg, F., Kripke, D. F., and Pegram, G. V. (1971). Individual variation in circadian rhythms of sleep, EEG, temperature and activity among monkeys: implications for regulatory mechanisms. In *Biochronometry* (Menaker, M., ed.). Washington, D. C.: National Academy of Sciences. 30–54.

Crowley, T. J., Kripke, D. F., Halberg, F., Pegram, G. V., and Schildkraut, J. J. (1972). Circadian rhythms of *Macaca mulatta*: sleep, EEG, body and eye movement, and temperature. *Primates* 13: 149–168.

Czeisler, C. A. (1978). Internal organization of temperature, sleep–wake, and neuroendocrine rhythms monitored in an environment free of time cues. Ph.D. thesis, Stanford University.

Czeisler, C. A. and Guilleminault, C. (1979). 250 years ago: tribute to a new discipline (1729–1979). *Sleep.* 2:155–160.

Czeisler, C. A. and Guilleminault, C., eds. (1980). *REM Sleep: Its Temporal Distribution.* New York: Raven Press.

Czeisler, C. A., Moore-Ede, M. C., Regestein, Q. R., Kisch, E. S., Fang, V. S., and Ehrlich, E. N. (1976). Episodic 24-hour cortisol secretory patterns in patients awaiting elective cardiac surgery. *J. Clin. Endocrinol. Metab.* 42:273–283.

Czeisler, C. A., Richardson, G. S., Coleman, R. M., Zimmerman, J. C., Moore-Ede, M. C., Dement, W. C., and Weitzman, E. D. (1981). Chronotherapy: resetting the circadian clocks of patients with delayed sleep phase insomnia. *Sleep* 4:1–21.

Czeisler, C. A., Richardson, G. S., Zimmerman, J. C., Moore-Ede, M. C., and Weitzman, E. D. (1981). Entrainment of human circadian rhythms by light–dark cycles: a reassessment. *Photochem. Photobiol.* 34:239–247.

Czeisler, C. A., Weitzman, E. D., Moore-Ede, M. C., and Krause, A. L. (1977). Relationship of the circadian rhythms of skin and core body temperature under entrained and free-running conditions in man. *Fed. Proc.* 36:423.

Czeisler, C. A., Weitzman, E. D., Moore-Ede, M. C., Zimmerman, J. C., and Knauer, R. S. (1980). Human sleep: its duration and organization depend on its circadian phase. *Science* 210:1264–1267.

Czeisler, C. A., Zimmerman, J. C., Ronda, J., Moore-Ede, M. C., and Weitzman, E. D. (1980). Timing of REM sleep is coupled to the circadian rhythm of body temperature in man. *Sleep* 2.329–340.

Daan, S. and Berde, C. (1978). Two coupled oscillators: simulations of the circadian pacemaker in mammalian activity rhythms. *J. Theor. Biol.* 70:297–313.

Daan, S., Damassa, D., Pittendrigh, C. S., and Smith, E. (1975). An effect of castration and testosterone replacement on a circadian pacemaker in mice (*Mus musculus*). *Proc. Natl. Acad. Sci. USA* 72:3744–3747.

Daan, S. and Pittendrigh, C. S. (1976a). A functional analysis of circadian pacemakers in nocturnal rodents. II. The variability of phase response curves. *J. Comp. Physiol.* 106:253–266.

Daan, S. and Pittendrigh, C. S. (1976b). A functional analysis of circadian pacemakers in nocturnal rodents. III. Heavy water and constant light: homeostasis of frequency? *J. Comp. Physiol.* 106:267–290.

Dallman, M. F., Engeland, W. C., and McBride, M. H. (1977). The neural regulation of compensatory adrenal growth. *Ann. N.Y. Acad. Sci.* 297:373–392.

Dallman, M. F., Engeland, W. C., Rose, J. C., Wilkinson, C. W., Shinsako, J., and Siedenburg, F. (1978). Nycthemeral rhythm in adrenal responsiveness to ACTH. *Am. J. Physiol.* 235:210–218.

Dark, J. G. and Asdourian, D. (1975). Entrainment of the rat's activity rhythm by cyclic light following lateral geniculate nucleus lesions. *Physiol. Behav.* 15:295–301.

Darwin, C. (1880). *On the Power of Movement in Plants.* London: John Murray.

David-Nelson, G. A. and Brodish, A. (1969). Evidence of diurnal rhythm of corticotropin releasing factor in the hypothalamus. *Endocrinology* 85:861–866.

Davis, F. C. (1981). Ontogeny of circadian rhythms. In *Handbook of Behavioral Neurobiology: Biological Rhythms,* vol. 4 (Aschoff, J., ed.). New York: Plenum. 257–274.

Davis, F. C. and Menaker, M. (1980). Hamsters through time's window: temporal structure of hamster locomotor rhythmicity. *Am. J. Physiol.* 239: R149–R155.

Davis, G. J. and Meyer, R. K. (1972). The effect of daylength on pituitary FSH and LH and gonadal development of snowshoe hares. *Biol. Reprod.* 6:264–269.

Davis, W. M. (1962). Day–night periodicity in pentobarbital response of mice and influence of socio-psychological conditions. *Experientia* 18:235–237.

Davis, W. M. and Webb, O. L. (1963). A circadian rhythm of chemoconvulsive response thresholds in mice. *Med. Exp.* 9:263–267.

Dawber, T. R. (1980). *The Framingham Study: The Epidemiology of Atherosclerotic Disease.* Cambridge, Mass.: Harvard University Press.

De Candolle, A. P. (1832). *Physiologie Vegetale,* vol. 2. Paris: Bechet Jeune.

De Castro, J. M. (1978). Diurnal rhythms of behavioral effects on core temperature. *Physiol. Behav.* 21:883–886.

DeCoursey, P. J. (1959). Daily activity rhythms in the flying squirrel. Ph.D. thesis, University of Wisconsin.

DeCoursey, P. J. (1960a). Daily light sensitivity rhythm in a rodent. *Science* 131:33–35.

DeCoursey, P. J. (1960b). Phase control of activity in a rodent. *Cold Spring Harbor Symp. Quant. Biol.* 25:49–55.

Defendini, R. and Zimmerman, E. A. (1978). The magnocellular neurosecretory system of the mammalian hypothalamus. In *The Hypothalamus* (Reichlin,

S., Baldessarini, R. J., and Martin, J. B., eds.). New York: Raven Press. 137–152.

Deguchi, T. (1975). Ontogenesis of a biological clock for serotonin: acetyl coenzyme A N-acetyltransferase in pineal gland of rat. *Proc. Natl. Acad. Sci. USA* 72:2814–2818.

Deguchi, T. (1978). Ontogenesis of circadian rhythm of melatonin synthesis in pineal gland of rat. *J. Neural. Transm.* 13:115–128.

Deguchi, T. (1979). Ontogenesis and the phylogenesis of circadian rhythm of the serotonin N-acetyltransferase activity in the pineal gland. In *Biological Rhythms and Their Central Mechanism* (Suda, M., Hayaishi, O., and Nakagawa, H., eds.). New York: Elsevier/North-Holland. 159–168.

De Lacerda, L., Kowarski, A., and Migeon, C. J. (1973). Diurnal variation of the metabolic clearance rate of cortisol. Effect on measurement of the cortisol production rate. *J. Clin. Endocrinol. Metab.* 36:1043–1049.

DeMairan, J. (1729). Observation botanique. *Histoire de L'Academie Royale des Sciences.* 35–36.

Dement, W. C. (1972). *Some Must Watch While Some Must Sleep.* Stanford, Calif.: Stanford Alumni Association.

Dement, W. C., Ferguson, J., Cohen, H., and Barchas, J. (1969). Nonchemical methods and data using a biochemical model: the REM quanta. In *Methods and Theory in Psychochemical Research in Man* (Mandell, A., ed.). New York: Academic Press. 275–325.

Dement, W. C., Guilleminault, C., and Zarcone, V. (1975). The pathologies of sleep: a case series approach. In *The Nervous System: The Clinical Neurosciences,* vol. 2 (Tower, D. B., ed.). New York: Raven Press. 501–518.

Dement, W. C. and Kleitman, N. (1957). Cyclic variations in EEG during sleep and their relation to eye movements, body motility, and dreaming. *Electroencephalogr. Clin. Neurophysiol.* 9:673–690.

Dement, W. C., Mitler, M., and Henriksen, S. (1972). Sleep changes during chronic administration of parachlorophenylalanine. *Rev. Can. Biol.* 31:239–246.

De Moor, P., Heirwegh, K., Heremans, J. F., and Declerck-Raskin, M. (1962). Protein binding of corticoids studied by gel filtration. *J. Clin. Invest.* 41:816–827.

Dettli, L. and Spring, P. (1967). Diurnal variations in the elimination rate of a sulphonamide in man. *Helv. Med. Acta* 33:291–306.

Dicker, J. E. and Nunn, J. (1957). The role of anti-diuretic hormone during water deprivation in rats. *J. Physiol. (London)* 136:235–248.

Dierickx, K. and Vandesande, F. (1977). Immunocytochemical localization of the vasopressin-neurophysin and oxytocin-neurophysin neurons in the human hypothalamus. *Cell Tissue Res.* 184:15–27.

Dierickx, K. and Vandesande, F. (1979). Immunocytochemical demonstration of separate vasopressin-neurophysin and oxytocin-neurophysin neurons in the human hypothalamus. *Cell Tissue Res.* 196:203–212.

DiRaimondo, V. C. and Forsham, P. H. (1956). Some clinical implications of the spontaneous diurnal variation in adrenal cortical secretory activity. *Am. J. Med.* 21:321–323.

DiRaimondo, V. C. and Forsham, P.H. (1958). Pharmacophysiologic principles in the use of corticoids and adrenocorticotrophin. *Metabolism* 7:5–24.

Doe, R. P. (1965). A study of the circadian variation in adrenal function and related rhythms. Ph.D. thesis, University of Minnesota.

Doe, R. P., Vennes, J. A., and Flink, E. B. (1960). Diurnal variations of 17-hydroxycorticosteroids, sodium, potassium, magnesium and creatinine in normal subjects and in cases of treated adrenal insufficiency and Cushing's syndrome. *J. Clin. Endocrinol. Metab.* 20:263–265.

Dowse, H. B. and Palmer, J. D. (1969). Entrainment of circadian activity rhythms in mice by electrostatic fields. *Nature* 222:564–566.

Drachman, R. H. and Tepperman, J. (1954). Aurothioglucose obesity in the mouse. *Yale J. Biol. Med.* 26:394–409.

Dubois, E. L. and Adler, D. C. (1963). Single daily dose oral administration of corticosteroids in rheumatic disorders: an analysis of its advantages, efficacy, and side effects. *Curr. Ther. Res.* 5:43–56.

Duesberg, R. and Weiss, W. (1939). Quoted in Rutenfranz, J., Knauth, P., and Colquhoun, W. P. (1976). Hours of work and shiftwork. *Ergonomics* 19:331–340.

Duhamel DuMonceau, H. L. (1759). *La Physique des Arbres.* Paris: HL Guerin and LF Delatour.

Dunlap, J. C., Hastings, J. W., Sulzman, F. M., and Moore-Ede, M. C. (1976). Oscillations in activities of enzymes from red blood cells maintained in vitro. Paper presented at La Jolla Conference on Circadian Rhythms, La Jolla, Calif.

Dunn, J., Dyer, R., and Bennett, M. (1971). Diurnal variation in plasma corticosterone following long-term exposure to continuous illumination. *Endocrinology* 90:1660–1663.

Dunn, J., Scheving, L., and Millet, P. (1972). Circadian variation in stress evoked increases in plasma corticosterone. *Am. J. Physiol.* 223:402–406.

Dunn, J. D., Castro, A. J., and McNulty, J. A. (1977). Effect of suprachiasmatic ablation on the daily temperature rhythm. *Neurosci. Lett.* 6:345–348.

Dushkov, B. A. and Komolinskii, F. P. (1968). Rational establishment of cosmonaut work schedules. In *The Psychophysiology of the Labor of Astronauts* (Gworski, M. N., ed.). Foreign Division Clearinghouse, Department of Commerce, AD684–690.

Eastman, C. I. (1980). Circadian rhythms of temperature, waking, and activity in the rat: dissociations, desynchronizations, and disintegrations. Ph.D. thesis, University of Chicago.

Ebihara, S., Tsuji, K., and Kondo, K. (1978). Strain differences of the mouse's free running circadian rhythm in continuous darkness. *Physiol. Behav.* 20:795–799.

Edmonds, S. C. and Adler, N. T. (1977). Food and light as entrainers of circadian running activity in the rat. *Physiol. Behav.* 18:915–919.

Ehret, C. F. (1981). New approaches to chronohygiene for the shift worker in the nuclear power industry. In *Night- and Shift-Work. Biological and Social Aspects* (Reinberg, A., Vieux, N., and Andlauer, T., eds.). Oxford: Pergamon Press. 263–270.

Ehret, C. F., Groh, K. R., and Meinert, J. C. (1978). Circadian dysynchronism and chronotypic ecophilia as factors in aging and longevity. In *Aging and Biological Rhythms* (Samis, H. V. and Capobianco, S., eds.). New York: Plenum Press. 185–213.

Ehret, C. F., Meinert, J. C., and Groh, K. R. (1979). The chronopharmacology of l-dopa: implications for orthochronal therapy in the prevention of circadian dysynchronism. In *Proceedings of the International Conference on Chronopharmacology* (Walker, C. A., Soliman, K. F. A., and Winget, C. M., eds.). Tallahassee: University Presses of Florida.

Ehret, C. F., Potter, V. R., and Dobra, K. W. (1975). Chronotypic action of theophylline and of pentobarbital as circadian zeitgebers in the rat. *Science* 188: 1212–1215.

Eik-Nes, K. and Clark, L. D. (1958). Diurnal variation of plasma 17-hydroxycorticosteroids in subjects suffering from severe brain damage. *J. Clin. Endocrinol. Metab.* 18:764–768.

Elliott, A. L., Mills, J. N., Minors, D. S., and Waterhouse, J. M. (1972). The effect of real and simulated timezone shifts upon circadian rhythms of body temperature, plasma 11-hydroxycorticosteroids and renal excretion in human subjects. *J. Physiol. (London)* 221:227–257.

Elliott, A. L., Mills, J. N., and Waterhouse, J. M. (1971). A man with too long a day. *J. Physiol. (London)* 212:30–33.

Elliott, J. (1979). Finally: some details on in vitro fertilization. *J.A.M.A.* 241:868–869.

Elliott, J. A. (1974). Photoperiodic regulation of testis function in the golden hamster: relation to circadian system. Univ. of Texas Ph.D. thesis, Austin.

Elliott, J. A. (1976). Circadian rhythms and photoperiodic time measurement in mammals. *Fed. Proc.* 35:2339–2346.

Elliott, J. A., Stetson, M. H., and Menaker, M. (1972). Regulation of testis function in golden hamsters: a circadian clock measures photoperiodic time. *Science* 178:771–773.

Ellis, G. B., Losee, S. H., and Turek, F. W. (1979). Prolonged exposure of castrated male hamsters to a nonstimulatory photoperiod: spontaneous change in sensitivity of the hypothalamic-pituitary axis to testosterone feedback. *Endocrinology* 104:631–635.

Ellis, G. B., McKlveen, R. E., and Turek, F. (1982). Dark pulses affect the circadian rhythm of activity in hamsters kept in constant light. *Am. J. Physiol.*

Ellis, G. B. and Turek, F. W. (1979). Time course of the photoperiod-induced change in sensitivity of the hypothalamic-pituitary axis to testosterone feedback in castrated male hamsters. *Endocrinology* 104:625–630.

Emlen, S. T. and Kem, W. (1963). Activity rhythm in Peromyscus: its influence on rates of recovery from nembutal. *Science* 142:1682–1683.

Engelmann, W. (1973). A slowing down of circadian rhythms by lithium ions. *Z. Naturforsch.* 28:733–736.

Enright, J. T. (1965). Synchronization and ranges of entrainment. In *Circadian Clocks* (Aschoff, J., ed.). Amsterdam: North-Holland. 112–124.

Enright, J. T. (1971). The internal clock of drunken isopods. *Z. Vergl. Physiol.* 75:332–346.

Enright, J. T. (1980). *The Timing of Sleep and Wakefulness*. New York: Springer-Verlag.

Epstein, A. N. (1971). The lateral hypothalamic syndrome: its implications for the physiological psychology of hunger and thirst. In *Progress in Physiological Psychology* (Stellar, E. and Sprague, J. M., eds.). New York: Academic Press. 263–317.

Epstein, A. N. (1973). Retrospect and prognosis. In *The Neuropsychology of Thirst* (Epstein, A. N., Kissileff, H. R., and Stellar, E., eds.). Washington, D.C.: Winston and Sons. 315–332.

Epstein, A. N. (1978). Consensus, controversies and curiosities. *Fed. Proc.* 37:2711–2716.

Epstein, A. N., Spector, D., Samman, A., and Goldblum, C. (1964). Exaggerated prandial drinking in the rat without salivary glands. *Nature* 201:1342–1343.

Epstein, M. (1978). Renal effects of head-out water immersion in man: implications for an understanding of volume homeostasis. *Physiol. Rev.* 58:529–581.

Erkert, H. G. (1970). Der einfluss der schwingungsbreite von licht-dunkel-cyclen auf phasenlage und resynchronisation der circadianen aktivitatsperiodik dunkelaktiver tiere. *J. Interdiscipl. Cycle Res.* 7:71–91.

Erkert, H. G. (1981). Effects of the zeitgeber pattern on the resynchronization behavior on dark-active mammals (*Roesettus aegyptiacus*). *Int. J. Chronobiol.*, in press.

Erlich, S., Weitzman, E. D., and McGregor, P. (1974). A comparison of the twenty-four hour cortisol secretory pattern during ambulatory functional activity and minimal activity at bed rest. *Sleep Res.* 3:168.

Eskin, A. (1972). Phase-shifting a circadian rhythm in the eye of *Aplysia* by high potassium pulses. *J. Comp. Physiol.* 80:353–376.

Everett, J. W. (1948). Progesterone and estrogen in the experimental control of ovulation time and other features of the estrous cycle in the rat. *Endocrinology* 43:389–405.

Everett, J. W. and Sawyer, C. H. (1950). A 24-hour periodicity in the "LH-release apparatus" of female rats, disclosed by barbiturate sedation. *Endocrinology* 47:198–218.

Farrer, D. N. and Ternes, J. W. (1969). Illumination intensity and behavioral circadian rhythms. In *Circadian Rhythms in Nonhuman Primates*, vol. 9 (Rohles, F. H., ed.). Basel: S. Karger. 1–7.

Feder, H. H., Siegel, H., and Wade, G. N. (1974). Uptake of (6, 7 ^{3}H) estradiol-17 in ovariectomized rats, guinea pigs and hamsters: correlations with species differences in behavioral responsiveness to estradiol. *Brain Res.* 71:93–103.

Feldman, J. F. (1967). Lengthening the period of the biological clock of *Euglena* by cycloheximide, an inhibitor of protein synthesis. *Proc. Natl. Acad. Sci. USA* 57:1080–1087.

Feldman, J. F. (1971). New mutations affecting circadian rhythmicity in *Neurospora*. In *Biochronometry* (Menaker, M., ed.). Washington, D.C.: National Academy of Sciences. 652–656.

Feldman, J. F. and Hoyle, M. N. (1973). Isolation of circadian clock mutants of *Neurospora crassa*. *Genetics* 75:605–613.

Fimognari, G. M., Fanestil, D. D., and Edelman, I. S. (1967). Induction of RNA and protein synthesis in the action of aldosterone in the rat. *Am. J. Physiol.* 213:954–962.

Finkelstein, J. W., Roffwarg, R. P., Boyar, R. M., Kream, J., and Hellman, L. (1972). Age related changes in the 24-hour spontaneous secretion of growth hormone. *J. Clin. Endocrinol. Metab.* 35:665–670.

Finkenstaedt, J. T., Dingman, J. F., Jenkins, D., Laidlow, J. C., and Merrill, J. P. (1954). The effect of intravenous hydrocortisone and corticosterone in the diurnal rhythm in renal function and electrolyte equilibria in normal and Addisonian subjects. *J. Clin. Invest.* 33:933.

Finn, M. G. C., Berrizbeitia, L. D., Coploys, M., Moore-Ede, M. C., and Moore, F. D. (1982). Circadian timing of intravenous feeding: day-night comparisons of nutrient effectiveness in the squirrel monkey. Unpublished ms.

Fitzgerald, K. M. and Zucker, I. (1976). Circadian organization of the estrous cycle of the golden hamster. *Proc. Natl. Acad. Sci. USA* 73:2923–2927.

Fitzsimons, J. T. (1960). The control of drinking in the rat. Ph.D. thesis, University of Cambridge.

Fitzsimons, J. T. (1961). Drinking by rats depleted of body fluid without increase in osmotic pressure. *J. Physiol.* (London) 159:297–309.

Fitzsimons, J. T. (1964). Drinking caused by constriction of the inferior vena cava in the rat. *Nature* 204:479–480.

Fitzsimons, J. T. (1969). The role of a renal thirst factor in drinking induced by extra-cellular stimuli. *J. Physiol.* (London) 201:349–368.

Fitzsimons, J. T. (1971). The physiology of thirst: a review of the extraneural aspects of the mechanisms of drinking. In *Progress in Physiological Psychology*, vol. 4 (Stellar, E. and Sprague, J. M., eds.). New York: Academic Press. 119–201.

Fitzsimons, J. T. and LeMagnen, J. (1969). Eating as a regulatory control of drinking in the rat. *J. Comp. Physiol. Psychol.* 67:273–283.

Flink, E. B. and Doe, R. P. (1959). Effect of sudden time displacement by air travel on synchronization of adrenal function. *Proc. Soc. Exp. Biol. Med.* 100:498–501.

Foix, C. and Nicolesco, L. (1925). *Anatomie Cerebrale. Les Noyaux Gris Centraux et la Region Mesencephalo-sous-optique*. Paris: Masson.

Folkard, S., Monk, T. H., Knauth, P., and Rutenfranz, J. (1976). The effect of memory load on circadian variation in performance efficiency under a rapidly rotating shift system. *Ergonomics* 19:479–488.

Forel, A. H. (1910). *Das Sinnesleben der Insekten* (Semon, M., trans.). Munchen: Ernst Reinhardt Verlag.

Foulks, E. F. (1972). *The Arctic Hysterias of the North Alaskan Eskimo.* Washington, D.C.: American Anthropological Association.

Fox, R. H., Woodward, P. M., Exton-Smith, A. N., Green, M. F., Donnison, D. V., and Wicks, M. H. (1973). Body temperatures in the elderly: a national study of physiological, social, and environmental conditions. *Br. Med. J.* 1:200–206.

Franks, R. (1967). Diurnal variation of plasma 17-OHCS in children. *J. Clin. Endocrinol. Metab.* 27:75–78.

Freeman, L. W. (1949). Metabolism of calcium in patients with spinal cord injuries. *Ann. Surg.* 129:177–184.

Frelinger, J. G., Matulsky, H., and Woodward, D. O. (1976). Effects of chloramphenicol on the circadian rhythm of *Neurospora crassa. Plant Physiol.* 58:592–594.

Frey, S. (1929). Der tod des menschen in seinen beziehungen zu den tages und jahreszeiten. *Dtsch. Z. Chir.* 218:336–369.

Friedemann, M. (1912). Die cytoarchitektonik des zwischen hirns des cercopitheken mit besonderer zerucksichtigung des thalamus opticus. *J. f. Psychol. u. Neurol.* 18:309–378.

Fuchs, J. L. and Moore, R. Y. (1980). Development of circadian rhythmicity and light responsiveness in the rat suprachiasmatic nucleus: a study using the 2-deoxy (1–^{14}C) glucose method. *Proc. Natl. Acad. Sci. USA* 77:1204–1208.

Fuller, C. A., Horowitz, J. M., and Horwitz, B. A. (1977). Spinal cord thermosensitivity and sorting of neural signals in cold-exposed rats. *Am. J. Physiol.* 42:154–158.

Fuller, C. A., Lydic, R., Sulzman, F. M., Albers, H. E., Tepper, B., and Moore-Ede, M. C. (1981). Circadian rhythm of body temperature persists after suprachiasmatic lesions in the squirrel monkey. *Am. J. Physiol.,* in press.

Fuller, C. A., Sulzman, F. M., and Moore-Ede, M. C. (1978a). Active and passive responses of circadian rhythms in body temperature to light-dark cycles. *Fed. Proc.* 37:832. (Abstract)

Fuller, C. A., Sulzman, F. M ., and Moore-Ede, M. C. (1978b). Thermoregulation is impaired in an environment without circadian time cues. *Science* 199:794–796.

Fuller, C. A., Sulzman, F. M., and Moore-Ede, M. C. (1979a). Circadian control of thermoregulation in the squirrel monkey (*Saimiri sciureus*). *Am. J. Physiol.* 236:R153–R161.

Fuller, C. A., Sulzman, F. M., and Moore-Ede, M. C. (1979b). Effective thermoregulation in primates depends upon internal circadian synchronization. *Comp. Biochem. Physiol.* (A) 63:207–212.

Fuller, C. A., Sulzman, F. M., and Moore-Ede, M. C. (1979c). Sound cycles exert circadian phase control after partial suprachiasmatic nucleus lesions in the squirrel monkey. *Physiologist* 22:41.

Fuller, C. A., Sulzman, F. M., and Moore-Ede, M. C. (1980). Circadian body temperature rhythms of primates in warm and cool environments. *Fed. Proc.* 39:989.

Fuller, C. A., Sulzman, F. M., and Moore-Ede, M. C. (1981). Shift-work and the jet-lag syndrome: conflicts between environmental and body time. In *The Twenty-Four Hour Workday: Proceedings of a Symposium on Variations in Work-Sleep Schedules* (Johnson, L. C., Tepas, D. I., Colquhoun, W. P., and Colligan, M. J., eds.). Cincinnati: U.S. Department of Health and Human Services (National Institute for Occupational Safety and Health) publication no. 81-127. 305–320.

Fuller, R. W. and Snoddy, H. D. (1968). Feeding schedule alteration of daily rhythm in tyrosine alpha-ketoglutamate transaminase of rat liver. *Science* 159:738.

Fulton, J. F. and Bailey, P. (1929). Tumors in the region of the third ventricle: their diagnosis and relation to pathological sleep. *J. Nerv. Ment. Dis.* 69:1–25, 145–164, 261–277.

Fuxe, K. (1965). The distribution of monoamine terminals in the central nervous system. *Acta Physiol. Scand.* 64:37–85.

Gallagher, T. F., Yoshida, K., Roffwarg, H. D., Fukushima, D. K., Weitzman, E. D., and Hellman, L. (1973). ACTH and cortisol secretory patterns in man. *J. Clin. Endocrinol. Metab.* 36:1058–1073.

Gander, P. H. (1979). The ciradian locomotor activity rhythm of *Hemideina thoracica* (Orthoptera): the effects of temperature perturbations. *Chronobiologia* 6:243–262.

Gander, P. H. (1980). Circadian organization in the regulation of locomotor activity and reproduction in *Rattus exulans*. Ph.D. thesis, University of Auckland, N.Z.

Garrod, O. and Burston, R. A. (1952). The diuretic response to ingested water in Addison's disease and panhypopituitarism and the effect of cortisone thereon. *Clin. Sci. Mol. Med.* 11:113–128.

Gauer, O., Henry, J., Siedker, H., and Eendt, W. (1970). The regulation of extracelluar fluid volume. *Ann. Rev. Physiol.* 32:547–595.

Gauthrie, M. (1973). Circadian rhythm in the vasomotor oscillations of skin temperature in man. *Int. J. Chronobiol.* 1:103–139.

Gerritzen, F. (1962). The diurnal rhythm in water, chloride, sodium and potassium excretion during a rapid displacement from East to West and vice versa. *Aviat. Space Environ. Med.* 33:697–701.

Gibbs, F. P. (1980). Temperature dependence of rat circadian pacemaker. *Am. J. Physiol.* 241:R17–R20.

Gibbs, F. P. and Van Brunt, P. (1975). Correlation of plasma corticosterone ("B") levels with running activity in blinded rats. *Fed. Proc.* 34:301.

Giebisch, G. (1971). Renal potassium excretion. In *The Kidney*, vol. 3 (Rouillier, C. and Muller, A. F., eds.). New York: Academic Press. 329–378.

Gierse, A. (1842). Quoniam sit ratio caloris organici. Doctoral thesis, Halle.

Gillespie, R. D. (1930). *Sleep and the Treatment of its Disorders.* New York: William Wood and Company. 266 pages.

Glotzbach, S. F. and Heller, H. C. (1976). Central nervous regulation of body temperature during sleep. *Science* 194:537–539.

Goetz, F., Bishop, J., Halberg, F., Sothern, R., Brunning, R., Senske, B., Greenberg, B., Minors, D., Stoney, P., Smith, I., Rosen, G., Cressey, D., Haus, E., and Apfelbaum, M. (1976). Timing of single daily meal influences relations among human circadian rhythms in urinary cyclic AMP and hemic glucagon, insulin and iron. *Experientia* 32:1081–1084.

Gologoski, R. (1979). Fatigue: the unseen enemy feared by airline crews. *San José Mercury News,* Nov. 4.

Goodman, L. S. and Gilman, A., eds. (1975). *The Pharmacological Basis of Therapeutics.* New York: Macmillan. 1497–1498.

Gordon, R. D., Spinks, J., Dulmanis, A., Hudson, B., Halberg, F., and Bartter, F. C. (1968). Amplitude and phase relations of several circadian rhythms in human plasma and urine: demonstration of rhythm for tetrahydrocortisol and tetrahydrocorticosterone. *Clin. Sci. Mol. Med.* 35:307–324.

Graef, V. and Golf. S. W. (1979). Circadian rhythm of the hepatic steroid metabolizing enzyme activities in the rat. *J. Steroid Biochem.* 11:1299–1302.

Greer, M. A., Panton, P., and Allen, C. F. (1972). Relationship of nycthemeral cycles of running activity and plasma corticosterone concentration following basal hypothalamic isolation. *Horm. Behav.* 3:289–295.

Grodins, F. S. (1963). *Control Theory in Biological Systems.* New York: Columbia University Press.

Groos, G. A. and Hendriks, J. (1979). Regularly firing neurons in the rat suprachiasmatic nucleus. *Experientia* 35:1597–1598.

Groos, G. A. and Mason, R. (1978). Maintained discharge of rat suprachiasmatic neurones at different adaptation levels. *Neurosci. Lett.* 8:59–64.

Groos, G. A. and Mason, R. (1980). The visual properties of rat and cat suprachiasmatic neurons. *J. Comp. Physiol.* 135:349–356.

Guldner, F. H. (1976). Synaptology of the rat suprachiasmatic nucleus. *Cell Tissue Res.* 165:509–544.

Guldner, F. H. and Wolff, J. R. (1974). Dendro-dendritic synapses in the suprachiasmatic nuclei of the rat hypothalamus. *J. Neurocytol.* 3:245–250.

Guldner, F. H. and Wolff, J. R. (1978a). Self-innervation of dendrites in the rat suprachiasmatic nucleus. *Exp. Brain Res.* 32:77–82.

Guldner, F. H. and Wolff, J. R. (1978b). Retinal afferents form Gray-type-I and type-II synapses in the suprachiasmatic nucleus (rat). *Exp. Brain Res.* 32:83–89.

Gulevich, G., Dement, W. C., and Johnson, I. (1966). Psychiatric and EEG observations on a case of prolonged (264 hours) wakefulness. *Archiv. Gen. Psychiatry* 15:29–35.

Gwinner, E. (1974). Testosterone induces 'splitting' of circadian locomotor activity rhythms in birds. *Science* 185:72–74.

Gwinner, E. (1978). Effects of pinealectomy on circadian locomotor activity rhythms in European starlings, *Sturnus vulgaris*. *J. Comp. Physiol.* 126:123–129.

Halberg, F. (1959). Physiologic 24-hour periodicity in human beings and mice, the lighting regimen and daily routine. In *Photoperiodism and Related Phenomena in Plants and Animals* (Withrow, R. B., ed.). Washington: American Association for the Advancement of Science, 803–878.

Halberg, F. (1960). Temporal coordination of physiologic function. *Cold Spring Harbor Symp. Quant. Biol.* 25:289–310.

Halberg, F. (1969). Chronobiology. *Ann. Rev. Physiol.* 31:675–725.

Halberg, F., Frank, G., Harner, R., Matthews, J., Aaker, H., Gravem, H., and Melby, J. (1961). The adrenal cycle in men on different schedules of motor and mental activity. *Experientia* 17:282–284.

Halberg, F., Galicich, J. H., Ungar, F., and French, L. A. (1965). Circadian rhythmic pituitary adrenocorticotropic activity, rectal temperature and pinnal mitosis of starving, dehydrated C mice. *Proc. Soc. Exp. Biol. Med.* 118:414–419.

Halberg, F., Johnson, E., Brown, W., and Bittner, J. J. (1960). Susceptibility rhythm to *E. coli* endotoxin and bioassay. *Proc. Soc. Exp. Biol. Med.* 103:142–144.

Halberg, F., Nelson, W. L., and Cadotte, L. (1977). Living routine shifts simulated on mice by weekly manipulation of light-dark cycle. In *XII International Conference Proceedings, International Society for Chronobiology* (Halberg, F., ed.). Milan: Il Ponte. 133–138.

Halberg, F., Visscher, M. B., and Bittner, J. J. (1954). Relation of visual factors to eosinophil rhythm in mice. *Am. J. Physiol.* 179:229–235.

Halberg, F., Zander, H. A., Houglum, M. W., and Muhlemann, H. R. (1954). Daily variations in tissue mitoses, blood eosinophils, and rectal temperatures in rats. *Am. J. Physiol.* 177:361–366.

Hammel, H. T., Jackson, D. C., Stolwijk, J. A. J., Hardy, J. D., and Stromme, S. B. (1963). Temperature regulation by hypothalamic proportional control with an adjustable set point. *J. Appl. Physiol.* 18:1146–1154.

Hammond, J. (1954). Light regulation of hormone secretion. *Vitam. Horm.* (Leipzig) 12:157–206.

Hamner, K. C., Finn, J. C., Sirohi, G. S., Hoshizaki, T., and Carpenter, B. H. (1962). The biological clock at the South Pole. *Nature* 195:476–480.

Hardeland, R. (1973a). Circadian rhythmicity in cultured liver cells. I. Rhythms in tyrosine aminotransferase activity and inducibility and in [^{3}H] leucine incorporation. *Int. J. Biochem.* 4:581–590.

Hardeland, R. (1973b). Circadian rhythmicity in cultured liver cells. II. Reinduction of rhythmicity in tyrosine aminotransferase activity. *Int. J. Biochem.* 4:591–595.

Hardy, J. D. (1973). Posterior hypothalamus and the regulation of body temperature. *Fed. Proc.* 32:1564-1571.

Harris, W. (1977). Fatigue, circadian rhythm, and truck accidents. In *Vigilance Theory, Operational Performance, and Physiological Correlates* (Mackie, R., ed.). New York: Plenum Press. 133-146.

Hartenbower, D. L. and Coburn, J. W. (1973). Natural daytime variation in renal clearances of phosphorus, sodium, calcium, magnesium and potassium in normal and parathyroidectomized dogs. *Fed. Proc.* 32:326.

Harter, J. G., Reddy, W. J., and Thorn, G. W. (1963). Studies on an intermittent corticosteroid dosage regime. *N. Engl. J. Med.* 269:591-596.

Hartman, H., Ashkenazi, I., and Epel, B. L. (1976). Circadian changes in membrane properties of human red blood cells in vitro, as measured by a membrane probe. *FEBS Lett.* 67:161-163.

Hastings, J. W. (1960). Biochemical aspects of rhythms: phase shifting of chemicals. *Cold Spring Harbor Symp. Quant. Biol.* 25:131-143.

Hastings, J. W. (1971). Cellular-biochemical clock hypothesis. In *Biological Clock: Two Views* (Brown, F., Hastings, J. W., and Palmer, J. D., eds.). New York: Academic Press. 61-91.

Hastings, J. W. and Sweeney, B. M. (1957). On the mechanisms of temperature independence in a biological clock. *Proc. Natl. Acad. Sci. USA* 43: 804-811.

Hastings, J. W. and Sweeney, B. M. (1958). A persistent diurnal rhythm of luminescence in *Gonyaulax polyedra*. *Biol. Bull.* 115:440-458.

Haus, E. (1976). Pharmacological and toxicological correlates of circadian synchronization and desynchronization. In *Shift Work and Health* (Rentos, P. G. and Shepard, R. D., eds.). Washington, D.C.: U.S. Department of Health, Education and Welfare. 87-117.

Haus, E., Halberg, F., Kuhl, J. F. W., and Lakatua, D. J. (1974). Chronopharmacology in animals. In *Chronobiological Aspects of Endocrinology* (Aschoff, J., Ceresa, F., and Halberg, F., eds.). Stuttgart: Schattauer-Verlag. 269-304.

Hawking, F. and Lobban, M. C. (1970). Circadian rhythms in *Macaca* monkeys (physical activity, temperature, urine and microfilarial levels). *J. Interdiscipl. Cycle Res.* 1:267-290.

Hawking, F., Lobban, M. C., Gamage, K., and Worms, M. J. (1971). Circadian rhythms (activity, temperature, urine and microfilariae) in dog, cat, hen, duck, *Thamnomys* and *Gerbillus*. *J. Interdiscipl. Cycle Res.* 2:455-473.

Hayden, P. and Lindberg, R. G. (1969). Circadian rhythm in mammalian body temperature entrained by cyclic pressure changes. *Science* 164:1288-1289.

Haymaker, W., Anderson, E., and Nauta, W. J. H., eds. (1969). *The Hypothalamus*. Springfield, Ill.: Charles C. Thomas.

Hellbrugge, T. (1960). The development of circadian rhythms in infants. *Cold Spring Harbor Symp. Quant. Biol.* 25:311-323.

Hellman, L., Nakada, F., Curti, J., Weitzman, E. D., Kream, J., Roffwarg, H., Ellman, S., Fukushima, D. K., and Gallagher, T. F. (1970). Cortisol is secreted episodically by normal man. *Electroencephalogr. Clin. Neurophysiol.* 30:411–422.

Hemmingsen, A. M. and Krarup, N. B. (1937). Rhythmic diurnal variations in the oestrous phenomena of the rat and their susceptibility to light and dark. *Klg. Dansk. Vidensk. Selskab. Biol. Meddel.* 13:1–61.

Hendricks, J. C., Bowker, R. M., and Morrison, A. R. (1976). Functional characteristics of cats with pontine lesions during sleep and wakefulness and their usefulness for sleep research. In *Sleep: 3rd European Congress of Sleep Research* (Koella, W. P. and Levin, P., eds.). Basel: S. Karger. 207–210.

Hendrickson, A. E., Wagoner, N., and Cowan, W. M. (1972). An autoradiographic and electron microscope study of retino-hypothalamic connections. *Z. Zellforsch.* 135:1–26.

Hery, F., Chouvet, G., Kan, J. P., Pujol, J. F., and Glowinski, J. (1977). Daily variations of various parameters of serotonin metabolism in the rat brain. II. Circadian variations in serum and cerebral tryptophase levels: lack of correlation with 5-HT turnover. *Brain Res.* 123:137–145.

Hess, W. R. (1957). *The Functional Organization of the Diencephalon* (Hughes, J. R., ed.). New York: Grune and Stratton.

Hetherington, A. W. and Ranson, S. W. (1939). Experimental hypothalamico-hypophyseal obesity in the rat. *Proc. Soc. Exp. Biol. Med.* 41:465–466.

Hetzel, M. R. and Clark, T. J. H. (1979). The clinical importance of circadian factors in severe asthma. In *Chronopharmacology* (Reinberg, A. and Halberg, F., eds.). New York: Pergamon Press. 213–221.

Heusner, A. (1956). Mise en evidence d'une variation nycthemerale de la calorification independant du cycle de l'activitie chez le rat. *C.R. Soc. Biol.* (Paris) 150:1246–1248.

Hicks, R. A., Lindseth, K., and Hawkins, J. (1980). Change to and from daylight savings time is associated with increases in traffic accidents. Abstracts of papers presented at the 146th AAAS annual meeting, San Francisco. 168.

Hildebrandt, G., Rohmert, W., and Rutenfranz, J. (1974). 12 and 24 h rhythms in error frequency of locomotive drivers and the influence of tiredness. *Int. J. Chronobiol.* 2:175–180.

Hilfenhaus, M. and Hertig, T. (1979). Effect of inverting the light–dark cycle on circadian rhythm of urinary excretion of aldosterone, corticosterone and electrolytes in the rat. In *Chronopharmacology* (Reinberg, A. and Halberg, F., eds.). New York: Pergamon Press. 49–55.

Hillman, W. S. (1956). Injury of tomato plants by continuous light and unfavorable photoperiodic cycles. *Am. J. Bot.* 43.09–90.

Hiroshige, T., Sakakura, M., and Itoh, S. (1969). Diurnal variation of corticotropin releasing activity in the rat hypothalamus. *Endocrinol. Jpn.* 16:465–467.

Hiroshige, T. and Sato, T. (1970). Postnatal development of circadian rhythm of corticotropin-releasing activity in the rat hypothalamus. *Neuroendocrinology* 17:1.

Hiroshige, T. and Wada, S. (1974). Modulation of the circadian rhythm of CRF activity in the rat hypothalamus. In *Chronobiological Aspects of Endocrinology* (Aschoff, J., Ceresa, F., and Halberg, F., eds.). Stuttgart: Schattauer-Verlag. 51–63.

Hobson, J. A. (1975). The sleep dream cycle: a neurobiological rhythm. In *Pathobiology Annual 1975* (Joachim, H. L., ed.). New York: Appleton-Century-Crofts. 369–403.

Hobson, J. A., McCarley, R. W., and Wyzinski, P. W. (1975). Sleep cycle oscillation: reciprocal discharge by two brainstem neuronal groups. *Science* 189:55–58.

Hobson, J. A., McCarley, R. W., Wyzinski, P. W., and Pivik, R. T. (1973). Reciprocal tonic firing by FTG and LC neurons during the sleep/waking cycle. *Sleep Res.* 2:29.

Hoffman, J. C. (1968). Effect of photoperiod on estrous cycle length in the rat. *Endocrinology* 83:1355–1357.

Hoffmann, K. (1965). Overt circadian frequencies and the circadian rule. In *Circadian Clocks* (Aschoff, J., ed.). Amsterdam: North-Holland. 87–94.

Hoffmann, K. (1969). Zum einfluss der zeitgeberstarke auf die phasenlage der synchronisierten circadianen periodik. *Z. Vergl. Physiol.* 62:93–110.

Hoffmann, K. (1971). Splitting of the circadian rhythm as a function of light intensity. In *Biochronometry* (Menaker, M., ed.). Washington, D.C.: National Academy of Sciences. 134–150.

Hoffmann, K., Gunderoth-Palmowski, M., Wiedenmann, G., and Engelmann, W. (1978). Further evidence for period lengthening effect of Li^+ ion on circadian rhythms. *Z. Naturforsch.* 33:231–234.

Honma, K. and Hiroshige, T. (1978). Endogenous ultradian rhythms in rats exposed to prolonged continuous light. *Am. J. Physiol.* 235:R250–R256.

Honma, K. and Hiroshige, T. (1979). Participation of brain catecholaminergic neurons in a self-sustained circadian oscillation of plasma corticosterone in the rat. *Brain Res.* 169:519–529.

Honma, K., Watanabe, K., and Hiroshige, T. (1979). Effects of parachlorophenylalanine and 5,6-dihydroxytryptamine on the free-running rhythms of locomotor activity and plasma corticosterone in the rat exposed to continuous light. *Brain Res.* 169:531–544.

Horne, J. A. and Ostberg, O. (1977). Individual differences in human circadian rhythms. *Biol. Psych.* 5:179–190.

Horowitz, J. M., Fuller, C. A., and Horwitz, B. A. (1976). Central neural pathways and the control of nonshivering thermogenesis. In *Regulation of Depressed Metabolism and Thermogenesis* (Jansky, L. and Musacchia, X. J., eds.). Springfield, Ill.: C. C. Thomas. 3–25.

Horseman, N. D., Meinert, J. C., and Ehret, C. F. (1979). Corticosteroid injections phase-shift the circadian thermoregulatory rhythm of rat. *Am. Zool.* 19:896.

Hunter, J. (1778). Of the heat of animals and vegetables. *Phil. Trans. R. Soc. Lond. (Biol.)* 68:7–49.

Ibuka, N., Inouye, S. T., and Kawamura, H. (1977). Analysis of sleep-wakefulness rhythms in male rats after suprachiasmatic nucleus lesions and ocular enucleation. *Brain Res.* 122:33–47.

Ibuka, N. and Kawamura, H. (1975). Loss of circadian rhythm in sleep-wakefulness cycle in the rat by suprachiasmatic nucleus lesions. *Brain Res.* 96:76–81.

Ingle, D. J. (1952). The role of the adrenal cortex in homeostasis. *J. Endocrinol.* 8:23–37.

Ingram, D. L., Walters, D. E., and Legge, K. F. (1975). Variations in behavioral thermoregulation in the young pig over 24 hour periods. *Physiol. Behav.* 14:689–695.

Ingram, W. R. (1940). Nuclear organization and chief connections of the primate hypothalamus. In *The Hypothalamus and Central Levels of Autonomous Function* (Fulton, J. F., Ranson, S. W., and Frantz, A. M., eds.). Baltimore: Williams and Wilkins. 195–244.

Inouye, S. T. and Kawamura, H. (1979). Persistence of circadian rhythmicity in mammalian hypothalamic "island" containing the suprachiasmatic nucleus. *Proc. Natl. Acad. Sci. USA* 76:5961–5966.

Institute of Medicine, National Academy of Sciences (1979). *Sleeping Pills, Insomnia and Medical Practice.* Washington, D.C.: NAS Office of Publications.

Issekutz, B., Jr., Blizzard, B. B., Birkhead, N. C., and Rodahl, K. (1966). Effect of prolonged bedrest on urinary calcium output. *J. Appl. Physiol.* 21:1013–1020.

Ixart, G., Szafarczyk, A., Belugou, J., and Assenmacher, I. (1977). Temporal relationships between the diurnal rhythm of hypothalamic corticotrophin releasing factor, pituitary corticotrophin and plasma corticosterone in the rat. *J. Endocrinol.* 72:113–120.

Janowitz, H. D. and Grossman, M. I. (1949). Some factors affecting food intake of normal dogs and dogs with esophagostomy and gastric fistula. *Am. J. Physiol.* 159:143–148.

Jarrett, R. J. (1979). Rhythms in insulin and glucose. In *Endocrine Rhythms* (Krieger, D. T., ed.). New York: Raven Press. 247–258.

Jenny, E. (1933). Tagesperiodische einflusse auf geburt und tod. *Schweiz. Med. Wochenschr.* 63:15–17.

Johnson, L. C. (1979). Sleep disturbances in humans. In *Sleep, Wakefulness and Circadian Rhythms,* vol. 105 (Nicholson, A. N., ed.). Neuilly sur Seine, France: NATO Advisory Group for Aerospace Research and Development. 4.1–4.16.

Johnson, L. C., Tepas, D. I., Colquhoun, W. P., and Colligan, M. J., eds. (1981). *The Twenty-Four Hour Workday: Proceedings of a Symposium on Variations in Work–Sleep Schedules.* Cincinnati: U.S. Department of Health and Human Services (National Institute for Occupational Safety and Health) publication no. 81-127.

Johnson, M. S. (1939). Effect of continuous light on periodic spontaneous activity of white-footed mice (*Peromyscus*). *J. Exp. Zool.* 82:315–318.

Johnsson, A. and Engelmann, W. (1980). Influence of lithium ions on human circadian rhythms. Z. Naturforsch. 35:503-507.

Johnsson, A., Pflug, B., Engelmann, W., and Klemke, W. (1979). Effect of lithium carbonate on circadian periodicity in humans. Pharmakopsychiatr. Neuropsychopharmakol. 12:423-425.

Johnston, R. L. and Washeim, H. (1924). Studies in gastric secretion. II. Gastric secretion in sleep. Am. J. Physiol. 70:247-253.

Jones, B. E., Bobillier, P., and Jouvet, M. (1969). Effets de la destruction des neurones contenant des catecholamines du mesencephale sur le cycle veille-sommeils du chat. C.R. Soc. Biol. (Paris) 163:176-180.

Jori, A., DiSalle, E., and Santini, V. (1971). Daily rhythmic variation and liver drug metabolism in rats. Biochem. Pharmacol. 20:2965-2969.

Jouvet, M. (1962). Recherches sur les structures nerveuses et les mecanismes responsables des differentes phases du sommeil physiologique. Arch. Ital. Biol. 100:125-207.

Jouvet, M. (1974). The role of monoaminergic neurons in the regulation and function of sleep. In Basic Sleep Mechanisms (Petre-Quadeno, O. and Schlag, J. D., eds.). New York: Academic Press. 207-236.

Jouvet, M., Mouret, J., Chouvet, G., and Siffre, M. (1974). Toward a 48-hour day: experimental bicircadian rhythm in man. In Circadian Oscillations and Organization in Nervous Systems (Pittendrigh, C. S., ed.). Cambridge, Mass.: The MIT Press. 491-497.

Jouvet, M. and Renault, J. (1966). Insomnic persistante apres lesions des noyaux du raphe chez le chat. C.R. Soc. Biol. (Paris) 160:1461-1465.

Juergensen, T. (1873). Die korperwarme des gesunden menschen. Ph.D. thesis, Leipzig.

Jusatz, J. H. and Eckhardt, E. (1934). Die haufigste todesstude. Munch. Med. Wochenschr. 81:709-710.

Kahn, E. and Fisher, C. (1969). The sleep characteristics of the normal aged male. J. Nerv. Ment. Dis. 148:477-494.

Kaiser, I. H. and Halberg, F. (1962). Circadian periodic aspects of birth. Ann. N.Y. Acad. Sci. 98:1056-1058.

Kaneko, M., Zechman, F. W., and Smith, R. E. (1968). Circadian variation in human peripheral blood flow levels and exercise responses. J. Appl. Physiol. 25:109-114.

Kapen, S., Boyar, R., Hellman, L., and Weitzman, E. (1975). Twenty-four-hour patterns of luteinizing hormone secretion in humans: ontogenetic and sexual considerations. Prog. Brain Res. 42:103-113.

Kappagoda, C., Linden, R., and Snow, H. (1972). A reflex increase in heart rate from distension of the junction between the superior vena cava and the right atrium. J. Physiol. (London) 220:177-197.

Karakashian, M. W. and Schweiger, H. G. (1976). Evidence for a cycloheximide-sensitive component in the biological clock of Acetabularia. Exp. Cell Res. 98:303-312.

Karsch, F. J. (1980). Seasonal reproduction: a saga of reversible fertility. *Physiologist* 23:29–38.

Kasal, C., Menaker, M., and Perez-Polo, R. (1979). Circadian clock in culture: N-acetyltransferase activity of chick pineal glands oscillates in vitro. *Science* 203:656–658.

Kass, D. A. (1980). Renal responses to lower body positive pressure in the conscious primate: implications for acute and chronic volume homeostasis. M.D. thesis, Yale University School of Medicine.

Kass, D. A., Sulzman, F. M., Fuller, C. A., and Moore-Ede, M. C. (1980a). Renal responses to central vascular expansion are suppressed at night in conscious primates. *Am. J. Physiol.* 239:F343–F351.

Kass, D. A., Sulzman, F. M., Fuller, C. A., and Moore-Ede, M. C. (1980b). Are ultradian and circadian rhythms in renal potassium excretion related? *Chronobiologia* 7:343–355.

Katz, F. H. (1964). Adrenal function during bedrest. *Aviat. Space Environ. Med.* 35:849–851.

Kavanau, J. L. (1962). Twilight transitions and biological rhythmicity. *Nature* 194:1293–1295.

Kavanau, J. L. (1971). Locomotion and activity phasing of some medium sized mammals. *J. Mammal.* 52:386–403.

Kawamura, H. and Ibuka, W. (1978). The search for circadian rhythm pacemakers in the light of lesion experiments. *Chronobiologia* 5:69–88.

Kellogg, R. (1953). Studies on isotonic saline diuresis. Ph.D. thesis, Harvard University.

Kenagy, G. J. (1978). Seasonality of endogenous circadian rhythms in a diurnal rodent *Ammospermophilus leucurus* and a nocturnal rodent *Dipodomys merriami*. *J. Comp. Physiol.* 128:21–36.

Kennedy, G. C. (1953). The role of depot fat in the hypothalamic control of food intake in the rat. *Proc. R. Soc. Lond. (Biol.)* 140:578–592.

King, J. C., Tobet, S. A., Snavely, F. L., and Arimura, A. A. (1981). The LHRH system: cells and pathways related to the hypothalamus in the rat. *J. Comp. Neurol.* in press.

Kissileff, H. R. (1970). Free feeding in normal and "recovered lateral" rats monitored by a pellet-detecting eatometer. *Physiol. Behav.* 5:163–173.

Kleiber, M. (1975). *The Fire of Life: An Introduction to Animal Energetics.* New York: Krieger.

Klein, D. C. (1979). Circadian rhythms in the pineal gland. In *Endocrine Rhythms* (Krieger, D. T., ed.). New York: Raven Press. 203–223.

Klein, D. C. and Moore, R. Y. (1979). Pineal N acetyltransferase and hydroxy-indole-O-methyltransferase: control by the retinohypothalamic tract and the suprachiasmatic nucleus. *Brain Res.* 174:245–262.

Klein, K. E. and Wegmann, H. M. (1974). The resynchronization of human circadian rhythms after transmeridian flights as a result of flight direction and mode of activity. In *Chronobiology* (Scheving, L. E., Halberg, F., and Pauly, J. E., eds.). Tokyo: Igaku. 564–570.

Klein, K. E. and Wegmann, H. M. (1979). Circadian rhythms in air operations. In *Sleep, Wakefulness and Circadian Rhythms*, vol. 105 (Nicholson, A. N., ed.). Neuilly sur Seine, France: NATO Advisory Group for Aerospace Research and Development. 10. 1–10.25.

Klein, K. E., Wegmann, H. M., and Bruner, H. (1968). Circadian rhythm in indices of human performance, physical fitness and stress resistance. *Aviat. Space Environ. Med.* 39:512–518.

Klein, K. E., Wegmann, H. M., and Hunt, B. I. (1972). Desynchronization of body temperature and performance circadian rhythm as a result of outgoing and homegoing transmeridian flights. *Aviat. Space Environ. Med.* 43:119–132.

Kleinhoonte, A. (1932). Untersuchungen uber die autonomen bewegungen der primarblatter von canavalia enfiformis. *Jb. wiss. Bot.* 75:679–725.

Kleitman, N. (1963). *Sleep and Wakefulness*. Chicago: The University of Chicago Press.

Kleitman, N. and Engelmann, T. G. (1953). Sleep characteristics of infants. *J. Appl. Physiol.* 6:269–282.

Kobberling, J. and von zur Muhlen, A. (1974). The circadian rhythm of free cortisol determined by urine sampling at two hour intervals in normal subjects and in patients with severe obesity or Cushing's syndrome. *J. Clin. Endocrinol. Metab.* 38:313–319..

Koikegami, H. (1938). Beitrage zur kenntnis der kerne des hypothalamus bei saugetieren. *Arch. Psychiatr. Nervenkr.* 107:742–774.

Koizumi, K. and Nishino, H. (1976). Circadian and other rhythmic activity of neurones in the ventromedial nuclei and lateral hypothalamic area. *J. Physiol.* (London) 263:331–356.

Kokkoris, C. P., Weitzman, E. D., Pollak, C. P., Spielman, A. J., Czeisler, C. A., and Bradlow, H. (1978). Long-term ambulatory temperature monitoring in a subject with a hypernychthemeral sleep-wake cycle disturbance. *Sleep* 1:177–190.

Konopka, R. J. (1979). Genetic dissection of the *Drosophila* circadian system. *Fed. Proc.* 38:2602–2605.

Konopka, R. J. and Benzer, S. (1971). Clock mutants of *Drosophila melanogaster*. *Proc. Natl. Acad. Sci. USA* 68:2112–2116.

Konopka, R. J. and Wells, S. (1980). *Drosophila* clock mutations affect the morphology of a brain neurosecretory cell group. *J. Neurobiol.* 11:411–415.

Kramer, G. (1952). Experiments on bird orientation. *Naturwissenschaften* 94:265–285.

Kramm, K. R. (1971). Circadian activity in the antelope ground squirrel, *Ammospermophilus leucurus*. Ph.D. thesis, University of California, Irvine.

Kraus, R. F. and Buffler, P. A. (1979). Sociocultural stress and the American native in Alaska: an analysis of changing patterns of psychiatric illness and alcohol abuse among Alaska natives. *Culture, Medicine and Psychiatry* 3:111–151.

Kreider, M. B., Buskirk, E. R., and Bass, D. E. (1958). Oxygen consumption and body temperatures during the night. *J. Appl. Physiol.* 12:361–366.

Kreider, M. B. and Iampietro, P. F. (1959). Oxygen consumption and body temperature during sleep in cold environments. *J. Appl. Physiol.* 14:765–767.

Krey, L. C. and Everett, J. W. (1973). Multiple ovarian responses to single estrogen injections early in rat estrous cycles: impaired growth, luteotropic stimulation and advanced ovulation. *Endocrinology* 93:377–384.

Krieg, W. J. S. (1932). The hypothalamus of the albino rat. *J. Comp. Neurol.* 55:19–89.

Krieger, D. T. (1974). Food and water restriction shifts corticosterone, temperature, activity, and brain amine periodicity. *Endocrinology* 95:1195–1201.

Krieger, D. T., ed. (1979a). *Endocrine Rhythms.* New York: Raven Press.

Krieger, D. T. (1979b). Rhythms in CRF, ACTH, and corticosteroids. In *Endocrine Rhythms* (Krieger, D. T., ed.). New York: Raven Press. 123–142.

Krieger, D. T. (1980). Ventromedial hypothalamic lesions abolish food-shifted circadian adrenal and temperature rhythmicity. *Endocrinology* 106:649–654.

Krieger, D. T., Hauser, H., and Krey, L. C. (1977). Suprachiasmatic nuclear lesions do not abolish food-shifted circadian adrenal and temperature rhythmicity. *Science* 197:398–399.

Krieger, D. T. and Krieger, H. P. (1967). Circadian patterns of urinary electrolyte excretion in central nervous system disease. *Metabolism* 16:815–823.

Kripke, D. F., Mullaney, D. J., Atkinson, M., and Wolf, S. (1978). Circadian rhythm disorders in manic-depressives. *Biol. Psychiatry* 13:335–351.

Kripke, D. F., Wyborney, V. G., and McEachron, D. (1979). Lithium slows rat activity rhythms. *Chronobiologia* 6:122.

Kronauer, R. E., Czeisler, C. A., Pilato, S. F., Moore-Ede, M. C., and Weitzman, E. D. (1982). Mathematical model of the human circadian system with two interacting oscillators. *Am. J. Physiol.,* in press.

Kronauer, R. E., Moore-Ede, M. C., and Menser, M. S. (1978). Ultradian cortisol rhythms in monkeys: synchronized or not synchronized? *Science* 202:1001–1002.

Kukorelli, T. and Juhasz, G. (1977). Sleep induced by intestinal stimulation in cats. *Physiol. Behav.* 19:355–358.

Kupfer, D. J., Foster, F. G., Coble, P., McPartland, R. J., and Ulrich, R. F. (1978). The application of EEG sleep for the differential diagnosis of affective disorders. *Am. J. Psychiatry* 135:64–74.

Langner, R. and Rensing, L. (1972). Circadian rhythms of oxygen consumption in rat liver suspension culture: changes of pattern. *Z. Naturforsch.* 27:1117–1118.

Lees, A. D. (1966). Photoperiodic timing mechanisms in insects. *Nature* 210:986–989.

Legan, S. J., Coon, G. A., and Karsch, F. J. (1975). Role of estrogen as initiator of daily LH surges in the ovariectomized rat. *Endocrinology* 96:50–56.

Le Magnon, J. and Tallon, S. (1966). La periodicite spontanee de la prise d'aliments *ad libitum* du rat blanc. *J. Physiol. (Paris)* 58:323-349.

Lengvari, I. and Liposits, Z. S. (1977). Return of diurnal plasma corticosterone rhythm long after frontal isolation of the medial basal hypothalamus in the rat. *Neuroendocrinology* 23:279-284.

Levy, M. S., Power, M. H., and Kepler, E. J. (1946). The specificity of the "water test" as a diagnostic procedure in Addison's disease. *J. Clin. Endocrinol. Metab.* 6:607-632.

Lewis, A. A. G. (1953). The control of the renal excretion of water. *Ann. R. Coll. Surg. Engl.* 13:36-54.

Lewis, H. E. and Masterton, J. P. (1955). British North Greenland Expedition. Medical and physiological aspects. *Lancet* 2:494-500, 549-556.

Lewis, P. R. and Lobban, M. C. (1957). The effects of prolonged periods of life on abnormal time routines upon excretory rhythms in human subjects. *J. Exp. Physiol.* 42:356-371.

Liddle, G. W. (1966). Analysis of circadian rhythms in human adrenocortical secretory activity. *Arch. Intern. Med.* 117:739-743.

Linazosoro, J. M., Jimenez Diaz, C., and Castro Mendoza, H. (1954). The kidney and thirst regulation. *Bull. Inst. Med. Res. Univ. Madrid* 7:53-61.

Lincoln, D. W., Church, J., and Mason, C. A. (1975). Electrophysiological activation of suprachiasmatic neurones by changes in retinal illumination. *Acta Endocrinol. (suppl., Kbh.)* 199:184.

Lincoln, G. A., Rowe, P. H., and Racey, R. A. (1974). The circadian rhythm in plasma testosterone concentration in man. In *Chronobiological Aspects of Endocrinology* (Aschoff, J., Ceresa, F., and Halberg, F., eds.). Stuttgart: Schattauer-Verlag. 137-152.

Lindberg, R. G. and Hayden, P. (1974). Thermoperiodic entrainment of arousal from torpor in the little pocket mouse, *Perognathus longimembris*. *Chronobiologia* 1:356-361.

Lindsley, D., Bowden, J., and Magoun, H. (1949). Effect upon EEG of acute injury to the brainstem activating system. *Electroencephalogr. Clin. Neurophysiol.* 1:475-498.

Lindsley, D., Schreiner, L., Krowles, W., and Magoun, H. (1950). Behavioral and EEG changes following chronic brainstem lesions in the cat. *Electroencephalogr. Clin. Neurophysiol.* 2:483-498.

Linnaeus, C. (1751). *Philosophia Botanica.* Stockholm: Godofr. Kiesewetter.

Lobban, M. C. (1960). The entrainment of circadian rhythms in man. *Cold Spring Harbor Symp. Quant. Biol.* 25:325-332.

Lobban, M. C. (1976). Seasonal variations in daily patterns of urinary excretion by Eskimo subjects. In *Circumpolar Health* (Shephard, R. J. and Itoh, S., eds.). Toronto: University of Toronto Press. 17-23.

Lobban, M. C. and Tredre, B. E. (1967). Diurnal rhythms of renal excretion and of body temperature in aged subjects. *J. Physiol. (London)* 188:48-49.

Lodge, J. R. and Salisbury, G. W. (1970). Seasonal variation and male reproductive efficiency. In *The Testis,* vol. 3 (Johnson, A. D., Gomes, W. R., and Van Demark, N. L., eds.). New York: Academic Press. 139–167.

Longson, D. and Mills, J. N. (1953). The failure of the kidney to respond to respiratory acidosis. *J. Physiol.* (London) 122:81–92.

Loo, Y. T. (1931). The forebrain of the opossum, *Didelphys virginiana.* 3. Histology. *J. Comp. Neurol.* 52:1–14.

Loomis, A. L., Harvey, E. N., and Hobart, G. A. (1935). Electrical potentials of the human brain. *J. Exp. Psychol.* 19:249–279.

Lucas, E. A. (1978). A study of the daily sleep and waking patterns of the laboratory cat. *Sleep Res.* 7:142.

Lund, R. (1974). Personality factors and desynchronization of circadian rhythms. *Psychosom. Med.* 36:224–228.

Lunell, N. O., Cunningham, A. W., and Rylander, B. J. (1961). Long-term cyclic variations in the electrical behavior of heart tissue in culture. *Acta Pathol. Microbiol. Scand.* 53:129–138.

Lutsch, E. F. and Morris, R. W. (1967). Circadian periodicity in susceptibility to lidocaine hydrochloride. *Science* 156:100–102.

Lydic, R., Albers, H. E., Tepper, B., and Moore-Ede, M. C. (1980). Comparative three dimensional morphology of the mammalian suprachiasmatic nuclei. *Neurosci. Abstr.* 6:832.

Lydic, R., Albers, H. E., Tepper, B., and Moore-Ede, M. C. (1981). Three-dimensional structure of mammalian suprachiasmatic nuclei: a comparative study of five species. *J. Comp. Neurol.* in press.

Lydic, R. and Moore-Ede, M. C. (1980). Three dimensional structure of the suprachiasmatic nuclei in the diurnal squirrel monkey (*Saimiri sciureus*). *Neurosci. Lett.* 17:295–299.

Lydic, R., Schoene, W. C., Czeisler, C. A., and Moore-Ede, M. C. (1980). Suprachiasmatic region of the human hypothalamus: homolog to the primate circadian pacemaker? *Sleep* 2:355–362.

Mack, P. B. and Lachance, P. L. (1967). Effect of recumbency and space flight on bone density. *Am. J. Clin. Nutr.* 20:1194–1205.

Macnish, R. (1836). *The Philosophy of Sleep.* Glasgow: W. R. M'Phun.

Maddison, S., Wood, R. J., Rolls, E. T., Rolls, B. J., and Gibbs, J. (1980). Drinking in the rhesus monkey: peripheral factors. *J. Comp. Physiol. Psychol.* 94:365–374.

Magnes, J., Moruzzi, G., and Pompeiano, O. (1961). Synchronization of the EEG produced by low frequency stimulation of the region of the solitary tract. *Arch. Ital. Biol.* 99:113–123.

Maini, B., Blackburn, G. L., Bistrian, B. R., Page, J. G., Benotti, P., and Reinhoff, H. Y. (1976). Cyclic hyperalimentation: an optimal technique for preservation of visceral protein. *J. Surg. Res.* 20:515–525.

Malnic, G., Klose, R. M., and Giebisch, G. (1964). Micropuncture study of renal potassium excretion in the rat. *Am. J. Physiol.* 206:674–686.

Manchester, R. C. (1933). The diurnal rhythm in water and mineral exchange. *J. Clin. Invest.* 12:995–1008.

Manshardt, J. and Wurtman, R. J. (1968). Daily rhythm in the noradrenaline content of rat hypothalamus. *Nature* 217:574–575.

Marimuthu, G., Subbaraj, R., and Chandrashekaran, M. K. (1978). Social synchronization of the activity rhythm in a cave-dwelling insectivorous bat. *Naturwissenschaften* 65:600.

Marotta, S. F., Hiles, L. G., Lanuza, D. M., and Boonayathap, U. (1975). The relation of hepatic in vitro inactivation of corticosteroids to the circadian rhythm of plasma corticosterone. *Horm. Metab. Res.* 7:334–337.

Marrone, B. L., Gentry, R. T., and Wade, G. N. (1976). Gonadal hormones and body temperature in rats: effects of estrous cycles, castration and steroid replacement. *Physiol. Behav.* 17:419–425.

Marte, E. (1961). Circadian susceptibility in relation to administration of pharmacologic agents. In *Circadian Systems, 39th Ross Conference on Pediatric Research* (Forman, S. J., ed.). Columbus, Ohio: Ross Laboratories. 52–55.

Marte, E. and Halberg, F. (1961). Circadian susceptibility rhythm to librium. *Fed. Proc.* 20:305.

Martel, P. J., Sharp, G. W. G., Slorach, S. A., and Vipond, H. J. (1962). A study of the roles of adrenocortical steroids and glomerular filtration rate in the mechanism of the diurnal rhythm of water and electrolyte excretion. *J. Endocrinol.* 24:159–169.

Martin, J. B. (1979). Brain mechanisms for integration of growth hormone secretion. *Physiologist* 22:23–29.

Martin, M. M., Mintz, D. H., and Tamagaki, H. (1963). Effects of altered thyroid function upon steroid circadian rhythm in man. *J. Clin. Endocrinol. Metab.* 23:242–247.

Martinez, J. L. (1972). Effects of selected illumination levels on circadian periodicity in the rhesus monkey (*Macaca mulatta*). *J. Interdiscipl. Cycle Res.* 3:47–59.

Matthews, J. H., Marte, E., and Halberg, F. (1964). A circadian susceptibility-resistance cycle to fluothane in male B1 mice. *Can. Anaesth. Soc. J.* 11:280–290.

Mauri, P. (1977). *The Sleep Disorders*. Michigan: Upjohn Co.

Mayer, J. (1953). Genetic, traumatic and environmental factors in the etiology of obesity. *Physiol. Rev.* 33:472–508.

Mayer, W., Gruner, R., and Strubel, H. (1975). Period-lengthening and phase-shifting of the circadian rhythm of *Phaseolus coccineus* L by theophylline. *Planta* 125:141–148.

Mayer, W. and Scherer, I. (1975). Phase shifting effect of caffeine in the circadian rhythm of *Phaseolus coccineus* L. *Z. Naturforsch.* 30:855–856.

McCarley, R. W. and Hobson, J. A. (1975). Neuronal excitability modulation over the sleep cycle: a structural and mathematical model. *Science* 189:58–60.

McCarley, R. W., Hobson, J. A., and Pivik, R. T. (1972). Activation of individual brain stem neurons during repeated sleep-wake cycles. *Sleep Res.* 1:26.

McCormack, C. E. and Sridaran, R. (1978). Timing of ovulation in rats during exposure to continuous light: evidence for a circadian rhythm of luteinizing hormone. *J. Endocrinol.* 76:135–144.

McGinty, D. J. and Harper, R. M. (1976). Dorsal raphe neurons: depression of firing during sleep in cats. *Brain Res.* 101:569–575.

McGuire, R. A., Rand, W. M., and Wurtman, R. J. (1973). Entrainment of the body temperature rhythm in rats: effect of color and intensity of environmental light. *Science* 181:956–957.

McNew, J. J., Burson, R. C., Hoshizaki, T., and Adey, W. R. (1972). Sleep-wake cycle of an unrestrained isolated chimpanzee under entrained and free running conditions. *Aviat. Space Environ. Med.* 43:155–161.

Meier, A. H. (1977). Daily variations in plasma corticosteroid concentrations in hypophysectomized fish and rats. In *XII International Conference Proceedings, International Society for Chronobiology* (Halberg, F., ed.). Milan: Il Ponte. 235–238.

Meier-Koll, A., Hall, U., Hellwig, U., Kott, G., and Meier-Koll, V. (1978). A biological oscillator system and the development of sleep-waking behavior during early infancy. *Chronobiologia* 5:425–440.

Mellinkoff, S. M., Frankland, D., Boyle, D., and Griepel, M. (1956). Relationship between serum amino acid concentration and fluctuations in appetite. *J. Appl. Physiol.* 8:535–538.

Menaker, M. (1959). Endogenous rhythms of body temperature in hibernating bats. *Nature* 184:1251–1252.

Menaker, M. (1974). Aspects of the physiology of circadian rhythmicity in the vertebrate nervous system. In *The Neurosciences: Third Study Program* (Schmitt, F. O. and Worden, F. G., eds.). Cambridge, Mass.: MIT Press. 479–489.

Menaker, M. and Eskin, A. (1966). Entrainment of circadian rhythms by sound in *Passer domesticus*. *Science* 154:1579–1581.

Menaker, M., Takahashi, J. S., and Eskin, A. (1978). The physiology of circadian pacemakers. *Ann. Rev. Physiol.* 40:501–526.

Mendels, J. and Cochrane, C. (1968). The nosology of depression: the endogenous-reactive concept. *Am. J. Psychiatry* 124:1–11.

Meyer, A. (1968). Einfluss von schall auf die tagesperiodische aktivitat des Goldhamsters. *Naturwissenschaften* 55:234–235.

Meyer, D. C. and Quay, W. B. (1976a). Effects of continuous light and darkness, and of pinealectomy, adrenalectomy and gonadectomy on uptake of ^{3}H-serotonin by the suprachiasmatic nuclei region of male rats. *Neuroendocrinology* 22:231–239.

Meyer, D. C. and Quay, W. B. (1976b). Hypothalamic and suprachiasmatic uptake of serotonin in vitro: twenty-four-hour changes in male and proestrous female rats. *Endocrinology* 98:1160–1165.

Migeon, C. J., Tyler, F. H., Mahoney, J. P., Florentin, A. A., Castle, H., Bliss, E. L., and Samuels, L. T. (1956). The diurnal variation of plasma levels and urinary excretion of 17-hydroxycorticosteroids in normal subjects, night workers, and blind subjects. *J. Clin. Endocrinol. Metab.* 16:622-633.

Miles, G. H. (1962). Telemetering techniques for periodicity studies. *Ann. N.Y. Acad. Sci.* 98:858-865.

Miles, L. E. M., Raynal, D. M., and Wilson, M. A. (1977). Blind man living in normal society has circadian rhythms of 24.9 hours. *Science* 198:421-423.

Miller, N. E. (1957). Experiments on motivation. *Science* 126:1271-1278.

Mills, J. N. (1951). Diurnal rhythms in urine flow. *J. Physiol.* (London) 112:53.

Mills, J. N. (1966). Human circadian rhythms. *Physiol. Rev.* 46:128-171.

Mills, J. N. (1973). Transmission processes between clock and manifestations. In *Biological Aspects of Circadian Rhythms* (Mills, J. N., ed.). New York: Plenum Press. 27-84.

Mills, J. N., Minors, D. S., and Waterhouse, J. M. (1974). The circadian rhythms of human subjects without timepieces or indication of the alternation of day and night. *J. Physiol.* (London) 240:567-594.

Mills, J. N., Minors, D. S., and Waterhouse, J. M. (1977). The physiological rhythms of subjects living on a day of abnormal length. *J. Physiol.* (London) 268:803-826.

Mills, J. N., Minors, D. S., and Waterhouse, J. M. (1978a). Adaptation to abrupt time shifts of the oscillator(s) controlling human circadian rhythms. *J. Physiol.* (London) 285:455-470.

Mills, J. N., Minors, D. S., and Waterhouse, J. M. (1978b). Exogenous and endogenous influences on rhythms after sudden time shift. *Ergonomics* 21:755-762.

Mills, J. N., Morgan, R., Minors, D. S., and Waterhouse, J. M. (1977). The free-running circadian rhythms of two schizophrenics. *Chronobiologia* 4:353-360.

Mills, J. N. and Stanbury, S. W. (1952). Persistent 24 hour renal excretory rhythm on a 12 hour cycle of activity. *J. Physiol.* (London) 117:22-37.

Mills, J. N. and Stanbury, S. W. (1954). A reciprocal relationship between K^+ and H^+ excretion in the diurnal excretory rhythm in man. *Clin. Sci. Mol. Med.* 13:177-186.

Mills, J. N. and Stanbury, S. W. (1955). Rhythmic diurnal variations in the behaviour of the human renal tubule. *Acta Med. Scand.* (suppl.) 307:95-96.

Mills, J. N., Thomas, S., and Williamson, K. S. (1960). The acute effect of hydrocortisone, deoxycorticosterone, and aldosterone upon the excretion of sodium, potassium and acid by the human kidney. *J. Physiol.* (London) 151:312-331.

Mitler, M. M., Lund, R., Sokolove, P. G., Pittendrigh, C. S., and Dement, W. C. (1977). Sleep and activity rhythms in mice: a description of circadian patterns and unexpected disruptions in sleep. *Brain Res.* 131:129-145.

Moline, M. L. (1981). Luteinizing hormone rhythms in female golden hamsters: circadian, photoperiodic and endocrine interactions. Ph.D. thesis, Harvard University.

Moline, M. L., Albers, H. E., Todd, R. B., and Moore-Ede, M. C. (1981). Light-dark entrainment of proestrous LH surges and circadian locomotor activity in female hamsters. *Horm. Behav.* 15 (4):451–458.

Monk, T. H. (1980). Traffic accident increases as a possible indicant of desynchronosis. *Chronobiologia* 7:527–529.

Monk, T. H. and Aplin, L. C. (1980). Spring and autumn Daylight Saving Time changes: studies of adjustment in sleep timings, mood, and efficiency. *Ergonomics* 23:167–178.

Monk, T. H. and Folkard, S. (1976). Adjusting to the changes to and from Daylight Saving Time. *Nature* 261:688–689.

Monnier, M., Hatt, A. M., Cueni, L. B., and Schoenenberger, G. A. (1972). Purification and assessment of a hypnogenic fraction of "sleep dialysate" (factor delta). *Arch. Ges. Physiol.* 331:257–265.

Moore, R. Y. (1973). Retinohypothalamic projections in mammals: a comparative study. *Brain Res.* 49:403–409.

Moore, R. Y. (1974). Visual pathways and the central neural control of diurnal rhythms. In *The Neurosciences: Third Study Program* (Schmitt, F. O. and Worden, F. G., eds.). Cambridge, Mass.: MIT Press. 537–542.

Moore, R. Y. (1978). Central neural control of circadian rhythms. In *Frontiers in Neuroendocrinology, vol. 5* (Ganong, W. F. and Martini, L., eds.). New York: Raven Press. 185–206.

Moore, R. Y. (1979). The anatomy of central neural mechanisms regulating endocrine rhythms. In *Endocrine Rhythms* (Krieger, D., ed.). New York: Raven Press. 63–87.

Moore, R. Y. (1980). Suprachiasmatic nucleus, secondary synchronizing stimuli and the central neural control of circadian rhythms. *Brain Res.* 183:13–28.

Moore, R. Y., Card, J. P., and Riley, J. N. (1980). The suprachiasmatic hypothalamic nucleus: neuronal ultrastructure. *Neurosci. Abstr.* 6:758.

Moore, R. Y. and Eichler, V. B. (1972). Loss of a circadian adrenal corticosterone rhythm following suprachiasmatic lesions in the rat. *Brain Res.* 42:201–206.

Moore, R. Y. and Eichler, V. B. (1976). Central neural mechanisms in diurnal rhythm regulation and neuroendocrine responses to light. *Psychoneuroendocrinology* 1:265–279.

Moore, R. Y. and Klein, D. C. (1974). Visual pathways and the central neural control of a circadian rhythm in pineal serotonin N-acetyltransferase activity. *Brain Res.* 71:17–33.

Moore, R. Y. and Lenn, N. J. (1972). A retinohypothalamic projection in the rat. *J. Comp. Neurol.* 146:1–14.

Moore, R. Y., Marchand, E. R. and Riley, J. N. (1979). Suprachiasmatic nucleus afferents in rat: An HRP retrograde transport study. *Soc. Neurosci. Symposia* 5:232.

Moore-Ede, M. C. (1973). Circadian rhythms of drug effectiveness and toxicity. *Clin. Pharmacol. Ther.* 14:925–935.

Moore-Ede, M. C. (1976a). Circadian rhythms in drug effectiveness and toxicity in shiftworkers. In *Shift Work and Health* (Rentos, P. G. and Shepard,

R. D., eds.). National Institute of Occupational Safety and Health Report. Cincinnati: Department of Health, Education and Welfare. 140–144.

Moore-Ede, M. C. (1976b). Perspectives on shift work and future research plans: Medico toxicological group report. In *Shift Work and Health* (Rentos, P. G. and Shepard, R. D., eds.). National Institute of Occupational Safety and Health Report. Cincinnati: Department of Health, Education and Welfare. 254–257.

Moore-Ede, M. C. (1981). Hypothermia: a timing disorder of circadian thermoregulatory rhythms? In *The Nature and Treatment of Hypothermia* (Pozos, R. S., ed.). Minneapolis: University of Minnesota Press. In press.

Moore-Ede, M. C., Brennan, M. F., and Ball, M. R. (1975). Circadian variation of intercompartmental potassium fluxes in man. *J. Appl. Physiol.* 38:163–170.

Moore-Ede, M. C. and Burr, R. G. (1973a). Circadian rhythm of urinary calcium excretion during immobilization. *Aviat. Space Environ. Med.* 44:495–498.

Moore-Ede, M. C. and Burr, R. G. (1973b). Circadian rhythm of therapeutic effectiveness of oxymetholone in paraplegic patients. *Clin. Pharmacol. Ther.* 14:448–454.

Moore-Ede, M. C. and Czeisler, C. A., eds. (1982). *Mathematical Modelling of Circadian Systems.* New York: Raven Press. In press.

Moore-Ede, M. C., Czeisler, C. A., Schmelzer, W. S., and Kass, D. A. (1977). Circadian internal desynchronization induced by circadian arrhythmias in synchronizing mediators: an etiological hypothesis. In *XII International Conference Proceedings, International Society for Chronobiology* (Halberg, F., ed.). Milan: Il Ponte. 477–482.

Moore-Ede, M. C., Faulkner, M. M., and Tredre, B. E. (1972). An intrinsic rhythm of urinary calcium excretion and the specific effect of bedrest on the excretory pattern. *Clin. Sci. Mol. Med.* 42:433–445.

Moore-Ede, M. C. and Fuller, C. A. (1980). Evaluation of aircraft accident reports. *Forum Int. Soc. Air Safety Invest.* 13:13–15.

Moore-Ede, M. C., Gander, P., and Czeisler, C. A. (1980). Importance of circadian time of day effects in scheduling aircrew: a proposed modification to the FAA rules. Comments on Federal Aviation Administration Docket #17669.

Moore-Ede, M. C., Gander, P. H., Eagan, S. M., and Martin, P. (1981). Evidence for weak circadian organization in the cat sleep-wake cycle. *Sleep Res.* In press.

Moore-Ede, M. C. and Herd, J. A. (1977). Renal electrolyte circadian rhythms: independence from feeding and activity patterns. *Am. J. Physiol.* 232:F128–F135.

Moore-Ede, M. C., Kass, D. A., and Herd, J. A. (1977). Transient circadian internal desynchronization after light-dark phase shift in monkeys. *Am. J. Physiol.* 232:R31–R37.

Moore-Ede, M. C., Lydic, R., Czeisler, C. A., Fuller, C. A., and Albers, H. E. (1980). Structure and function of suprachiasmatic nuclei (SCN) in human and non-human primates. *Neurosci. Abstr.* 6:708.

Moore-Ede, M. C., Meguid, M. M., Fitzpatrick, G. F., Boyden, C. M., and Ball, M. R. (1978). Circadian variation in response to potassium infusion. *Clin. Pharmacol. Ther.* 23:218–227.

Moore-Ede, M. C., Schmelzer, W. S., Kass, D. A., and Herd, J. A. (1976). Internal organization of the circadian timing system in multicellular animals. *Fed. Proc.* 35:2333–2338.

Moore-Ede, M. C., Schmelzer, W. S., Kass, D. A., and Herd, J. A. (1977). Cortisol mediated synchronization of circadian rhythm in urinary potassium excretion. *Am. J. Physiol.* 233:R230–R238.

Moore-Ede, M. C. and Sulzman, F. M. (1977). The physiological basis of circadian timekeeping in primates. *Physiologist* 20:17–25.

Moore-Ede, M. C. and Sulzman, F. M. (1981). Internal temporal order. In *Handbook of Behavioral Neurobiology, Biological Rhythms*, vol. 4 (Aschoff, J., ed.). New York: Plenum. 215–241.

Moore-Ede, M. C., Sulzman, F. M., and Fuller, C. A. (1979a). Uncoupling of circadian oscillators at the limits of light-dark cycle entrainment in the squirrel monkey. *Fed. Proc.* 38:1318.

Moore-Ede, M. C., Sulzman, F. M., and Fuller, C. A. (1979b). Circadian organization in the squirrel monkey: the internal coupling between oscillators. In *Biological Rhythms and Their Central Mechanism* (Suda, M., Hayaishi, O., and Nakagawa, H., eds.). New York: Elsevier North-Holland, 405–419, 435–438.

Morgan, R. and Cheadle, A. J. (1976). Circadian body temperature in chronic schizophrenia. *Br. J. Psychiatry* 129:350–354.

Morgan, R. and Drew, C. D. A. (1970). Early to bed . . . ? *Social Psychiat.* 5:99–101.

Morimoto, T. and Shiraki, K. (1970). Circadian variation in circulating blood volume. *Jpn. J. Physiol.* 20:550–559.

Morimoto, Y., Oishi, T., Ansue, K., and Yamamura, Y. (1979). Effect of food restriction and its withdrawal on the circadian adrenocortical rhythm in rats under constant dark or constant lighting conditions. *Neuroendocrinology* 29:77–83.

Morin, L. P., Fitzgerald, K. M., and Zucker, I. (1977). Estradiol shortens the period of hamster circadian rhythms. *Science* 196:305–307.

Morris, R. W. and Lutsch, E. F. (1967). Susceptibility to morphine-induced analgesia in mice. *Nature* 216:494–495.

Morrison, A. A. and Pompeiano, O. (1965). An analysis of the supraspinal influences acting on motor neurons during sleep in the unrestrained cat: responses of the alpha motorneurons to direct electrical stimulation during sleep. *Arch. Ital. Biol.* 103:497–516.

Morrison, S. D. (1968). Regulation of water intake by rats deprived of food. *Physiol. Behav.* 3:75–81.

Moruzzi, G. and Magoun, H. (1949). Brainstem reticular formation and activation of the EEG. *Electroencephalogr. Clin. Neurophysiol.* 1:455–473.

Mosko, S. S., Erikson, G. F., and Moore, R. Y. (1980). Dampened circadian rhythms in reproductive senescent female rats. *Behav. and Neural Biol.* 28:144.

Mosko, S. S. and Moore, R. Y. (1978a). Neonatal suprachiasmatic nucleus lesions: effects on the development of circadian rhythms in the rat. *Brain Res.* 164:17–38.

Mosko, S. S. and Moore, R. Y. (1978b). Neonatal SCN ablation: effects on the development of the pituitary-gonadal axis in the female rat. *Neurosci. Abstr.* 4:350.

Mouret, J., Coindet, J., Debilly, G., and Chouret, G. (1978). Suprachiasmatic nuclei lesions in the rat: alterations in sleep circadian rhythms. *Electroencephalogr. Clin. Neurophysiol.* 45:402–408.

Muller, A. F., Manning, E. L., and Riondel, A. M. (1958). Influence of position and activity on the secretion of aldosterone. *Lancet* 1:711–713.

Mullin, C. S., Jr. (1960). Some psychological aspects of isolated Antarctic living. *Am. J. Psychiatry* 117:323–325.

Munson, E. S., Martucci, R. W., and Smith, R. E. (1970). Circadian variations in anesthetic requirement and toxicity in rats. *Anesthesiology* 32:507–514.

Murray, E. J., Williams, H. L., and Lubin, A. (1958). Body temperature and psychological ratings during sleep deprivation. *J. Exp. Psychol.* 56:271–273.

Nabarro, J. D. N. (1956). The adrenal cortex and renal function. In *Modern Views on the Secretion of Urine* (Winston, F. R., ed.). Boston: Little, Brown. 148–185.

Nagai, K., Nishio, T., Nakagawa, H., Nakamura, S., and Fukuda, Y. (1978). Effect of bilateral lesions of the suprachiasmatic nuclei on the circadian rhythm of food intake. *Brain Res.* 142:384–389.

Nair, V. and Casper, R. (1969). The influence of light on daily rhythm in hepatic drug metabolizing enzymes in rat. *Life Sci.* 8:1291–1298.

Nakayama, T., Arai, S., and Yamamoto, K. (1979). Body temperature rhythm and its central mechanism. In *Biological Rhythms and Their Central Mechanism* (Suda, M., Hayaishi, O., and Nakagawa, H., eds.). New York: Elsevier North-Holland. 395–403.

Naquet, R., Denavit, M., and Albe-Fessard, D. (1966). Comparison entre le role du subthalamus et celui des differentes structures bulbomesencephaliques dans le maintien de la vigilance. *Electroencephalogr. Clin. Neurophysiol.* 20:149–164.

Nauta, W. J. H. (1946). Hypothalamic regulation of sleep in rats. An experimental study. *J. Neurophysiol.* 9:285–315.

Negus, N. C. and Berger, P. J. (1972). *Biology of Reproduction: Basic and Clinical Studies.* Mexico City: Bay Publishers.

Nelson, W., Nichols, G., Halberg, F., and Kattke, G. (1973). Interacting effects of lighting (LD 12:12) and restricted feeding (4H/24H) on circadian temperature rhythm of mice. *Int. J. Chronobiol.* 1:347.

Nichols, C. T., Nugent, C. A., and Tyler, F. (1965). Diurnal variation in suppression of adrenal function by glucocorticoids. *J. Clin. Endocrinol. Metab.* 25:343–349.

Nichols, C. T. and Tyler, F. H. (1967). Diurnal variation in adrenal cortical function. *Ann. Rev. Med.* 18:313–324.

Nishino, H. and Koizumi, K. (1976). Circadian and other rhythmic activity of neurones in the ventromedial nuclei and lateral hypothalamic area. *J. Physiol.* (London) 263:331–356.

Nishino, H. and Koizumi, K. (1977). Responses of neurons in the suprachiasmatic nuclei of the hypothalamus to putative transmitters. *Brain Res.* 120:167–172.

Nishino, H., Koizumi, K., and Brooks. C. M. (1976). The role of suprachiasmatic nuclei of the hypothalamus in the production of circadian rhythm. *Brain Res.* 112:45–59.

Njus, D., McMurry, L., and Hastings, J. W. (1977). Conditionality of circadian rhythmicity: synergistic action of light and temperature. *J. Comp. Physiol.* 117:335–344.

Njus, D., Sulzman, F. M., and Hastings, J. W. (1974). Membrane model for the circadian clock. *Nature* 248:116–119.

Noble, D. (1975). *The Initiation of the Heartbeat.* Oxford: Oxford University Press.

Norman, R. L., Blake, C. A., and Sawyer, C. H. (1973). Effects of hypothalamic deafferentation on LH secretion and the estrous cycle in the hamster. *Endocrinology* 91:95–100.

Norn, M. (1929). Untersuchungen uber das verhalten des kalium im organismus. II. Uber schwankungen der kalium, natrium-, und chloridausscheidung durch die niere im laufe des tages. *Skand. Arch. Physiol.* 55:184–210.

Nunez, A. A. and Stephan, F. K. (1977). The effects of hypothalamic knife cuts on drinking rhythms and the estrous cycle of the rat. *Behav. Biol.* 20:224–234.

Oatley, K. (1971). Dissociation of the circadian drinking pattern from eating. *Nature* 229:494–496.

Ogle, J. W. (1866). On the diurnal variations in the temperature of the human body in health. *St. George's Hosp. Rep.* 1:220–245.

Oliverio, A. and Malorni, M. (1979). Wheel running and sleep in two strains of mice: plasticity and rigidity in the expression of circadian rhythmicity. *Brain Res.* 163:121–133.

Oomura, Y., Ono, T., Nishino, H., Kita, H., Shimizu, N., Ishizuka, S., and Sasaki, K. (1979). Hypothalamic control of feeding behavior: modulation by the suprachiasmatic nucleus. In *Biological Rhythms and Their Central Mechanism* (Suda, M., Hayaishi, O., and Nakagawa, H., eds.). New York: Elsevier North-Holland. 295–000.

Ortavant, R., Mauleon, P., and Thibault, C. (1964). Photoperiodic control of gonadal and hypophyseal activity in domestic mammals. *Ann. N.Y. Acad. Sci.* 177:157–193.

Orth, D. N., Island, D. P., and Liddle, G. W. (1967). Experimental alteration of the circadian rhythm in plasma cortisol concentration in man. *J. Clin. Endocrinol. Metab.* 27:549–555.

Page, R. B., Galicich, J. H., and Grunt, J. A. (1973). Alteration of circadian temperature rhythm with third ventricular obstruction. *J. Neurosurg.* 38:309-319.

Paintal, A. S. (1954). A study of gastric stretch receptors. Their role in the peripheral mechanism of satiation of hunger and thirst. *J. Physiol.* (London) 126:255-270.

Pappenheimer, J. R. (1976). The sleep factor. *Sci. Am.* 235:24-29.

Parmeggiani, P. L. and Rabini, C. (1970). Sleep and environmental temperature. *Arch. Ital. Biol.* 108:369-387.

Parmeggiani, P. L., Rabini, C., and Cattalani, M. (1969). Sleep phases at low environmental temperature. *Arch. Sci. Biol.* 53:277-290.

Pate, J. R. (1937). Trans-neural atrophy of the nucleus ovoideus following eye removal in cats. *Anat. Rec.* 67:39 (suppl. no. 3).

Patrick, G. T. W. and Gilbert, J. A. (1896). On the effects of loss of sleep. *Psychol. Rev.* 3:469-483.

Pauly, J. E. and Scheving, L. E. (1964). Temporal variations in the susceptibility of white rats to pentobarbital sodium and tremorine. *Int. J. Neuropharmacol.* 3:651-658.

Pauly, J. E., Scheving, L. E., Burns, E. R., Landon, J., and Stone, J. E. (1977). Diurnal variations of serum testosterone levels in intact and gonadectomized male and female rhesus monkeys. *Steroids* 29:21-32.

Pavlidis, T. (1973). *Biological Oscillators: Their Mathematical Analysis.* New York: Academic Press.

Peck, J. W. and Novin, D. (1971). Evidence that osmoreceptors mediating drinking in rabbits are in the lateral preoptic area. *J. Comp. Physiol. Psychol.* 74:134-147.

Perkins, M. N. and Whitehead, S. A. (1978). Responses and pharmacological properties of preoptic/anterior hypothalamic neurons following medial forebrain bundle stimulation. *J. Physiol.* (London) 279:347-360.

Perkoff, G. T., Eik-Nes, K., Nugent, C. A., Fred, H. L., Nimer, R. A., Rush, L., Samuels, L. T., and Tyler, F. H. (1959). Studies of the diurnal variation of plasma 17-hydroxycorticosteroids in man. *J. Clin. Endocrinol. Metab.* 19:432-443.

Peterson, R. E. (1957). Plasma corticosterone and hydroxycortisone levels in man. *J. Clin. Endocrinol. Metab.* 17:1150-1157.

Pfeffer, W. (1873). *Phisiologische Untersuchungen.* Leipzig: Wilhelm Engelmann.

Pfeffer, W. (1915). Beitrage zur kenntnis der enstenhung der schlafbewegungen der blattorgane. *Abh. Math. Phys. Kl. Koenig. Saech. Akad. Wissensch.* 34:1-154.

Phelps, C. P., Lengvari, I., Carrillo, A. J., and Sawyer, C. H. (1977). Changes in diurnal thyroid stimulating hormone and corticosterone following anterior hypothalamic deafferentation in the rat. *Brain Res.* 141:283-292.

Phillips, J. L. M. and Mikulka, P. J. (1979). The effects of restricted food access upon locomotor activity in rats with suprachiasmatic nucleus lesions. *Physiol. Behav.* 23:257–262.

Pincus, G. (1943). A diurnal rhythm in the excretion of urinary ketosteroids by young men. *J. Clin. Endocrinol. Metab.* 3:195–199.

Pittendrigh, C. S. (1954). On temperature independence in the clock-system controlling emergence in *Drosophila. Proc. Natl. Acad. Sci. USA* 40:1018–1029.

Pittendrigh, C. S. (1958). Perspectives in the study of biological clocks. In *Perspectives in Marine Biology* (Buzzati-Traverso, A. A., ed.). California: Scripps Institution of Oceanography. 239–268.

Pittendrigh, C. S. (1960). Circadian rhythms and the circadian organization of living systems. *Cold Spring Harbor Symp. Quant. Biol.* 25:159–182.

Pittendrigh, C. S. (1965). On the mechanism of entrainment of a circadian rhythm by light cycles. In *Circadian Clocks* (Aschoff, J., ed.). Amsterdam: North-Holland Publishing Co. 277–297.

Pittendrigh, C. S. (1967a). Circadian rhythms, space research and manned space flight. *Life Sci. Space Res.* 5:122–134.

Pittendrigh, C. S. (1967b). Circadian systems. 1. The driving oscillation and its assay in *Drosophila pseudoobscura. Proc. Natl. Acad. Sci. USA* 58:1762–1767.

Pittendrigh, C. S. (1974). Circadian oscillations in cells and the circadian organization of multicellular systems. In *The Neurosciences Third Study Program* (Schmitt, F. O. and Worden, F. G., eds.). Cambridge, Mass.: MIT Press. 437–458.

Pittendrigh, C. S., ed. (1981). *Atlas of Phase Response Curves.* In press.

Pittendrigh, C. S. and Bruce, V. G. (1957). An oscillator model for biological clocks. In *Rhythmic and Synthetic Processes in Growth* (Rudnick, D., ed.). Princeton: Princeton University Press. 75–109.

Pittendrigh, C. S. and Caldarola, P. C. (1973). General homeostasis of the frequency of circadian oscillations. *Proc. Natl. Acad. Sci. USA* 70:2697–2701.

Pittendrigh, C. S. and Daan, S. (1974). Circadian oscillations in rodents: a systematic increase of their frequency with age. *Science* 186:548–550.

Pittendrigh, C. S. and Daan, S. (1976a). A functional analysis of circadian pacemakers in nocturnal rodents. I. The stability and lability of spontaneous frequency. *J. Comp. Physiol.* 106:223–252.

Pittendrigh, C. S. and Daan, S. (1976b). A functional analysis of circadian pacemakers in nocturnal rodents. IV. Entrainment: pacemaker as clock. *J. Comp. Physiol.* 106:291–331.

Pittendrigh, C. S. and Daan, S. (1976c). A functional analysis of circadian pacemakers in nocturnal rodents. V. Pacemaker structure: a clock for all seasons. *J. Comp. Physiol.* 106:333–355.

Pittendrigh, C. S. and Minis, D. H. (1972). Circadian systems: longevity as a function of circadian resonance in *Drosophila melanogaster. Proc. Natl. Acad. Sci. USA* 69:1537–1539.

Pohl, C. R. and Gibbs, F. P. (1978). Circadian rhythms in blinded rats: correlation between pineal and activity cycles. *Am. J. Physiol.* 234:R110–R114.

Pohl, H. (1972). Die aktivitatsperiodik von zwei tagaktiven nagern, *Funambulus palmarum* und *Eutamias sibiricus* unter dauerlichtbedingungen. *J. Comp. Physiol.* 78:60–74.

Post, W. (1931). *Around the World in Eight Days.* New York: Rand McNally.

Powell, E. W., Pasley, R. N., Brockway, B., Scheving, L. E., Lubanovic, W., and Halberg, F. (1977). Suprachiasmatic dinuclear lesion alters circadian temperature rhythm's amplitude and timing in light-dark synchronized rats. *Chronobiologia* 4:270.

Puizillout, J. and Ternaux, J. (1974). Variations d'activites toniques phasiques et respiratoires, au niveau bulbaire pendant l'endormement de la preparation encephale isole. *Brain Res.* 66:67–83.

Pujol, J. E., Stein, D., Blondaux, C., Petitjean, F., Froment, J. L., and Jouvet, M. (1973). Biochemical evidence for interaction phenomena between noradrenergic and serotonergic systems in the cat brain. In *Frontiers in Catecholamine Research* (Usdin, E. and Snyder, S. H., eds.). New York: Pergamon. 771–772.

Quay, W. B. (1970). Precocious entrainment and associated characteristics of activity patterns following pinealectomy and reversal of photoperiod. *Physiol. Behav.* 5:1281–1290.

Quay, W. B. (1972). Pineal homeostatic regulation of shifts in the circadian activity rhythms during maturation and aging. *Trans. N.Y. Acad. Sci.* 34:239–254.

Radzialowski, F. M. and Bousquet, W. F. (1967). Circadian rhythm in hepatic drug metabolizing activity in the rat. *Life Sci.* 6:2545–2548.

Radzialowski, F. M. and Bousquet, W. F. (1968). Daily rhythmic variation in hepatic drug metabolism in the rat and mouse. *J. Pharmacol. Exp. Ther.* 163:229–238.

Raisman, G. and Brown-Grant, K. (1977). The 'suprachiasmatic syndrome': endocrine and behavioural abnormalities following lesions of the suprachiasmatic nuclei in the female rat. *Proc. R. Soc. Lond. (Biol.)* 198:297–314.

Ramon y Cajal, S. (1911). *Histologie du Systeme Nerveux de l'Homme et des Vertebres.* Paris: A. Maloine.

Ratte, J. M., Halberg, F., Haus, E., and Najarian, J. S. (1974). Circadian urinary rhythms in rats with renal grafts. *Chronobiologia* 1:62–73.

Rawson, K. S. (1956). Homing behavior and endogenous activity rhythms. Ph.D. thesis, Harvard University.

Rawson, K. S. (1960). Effects of tissue temperature on mammalian activity rhythms. *Cold Spring Harbor Symp. Quant. Biol.* 25:105–113.

Rebar, R. W. and Yen, S. S. C. (1979). Endocrine rhythms in gonadotrophins and ovarian steroids with reference to reproductive processes. In *Endocrine Rhythms* (Krieger, D. T., ed.). New York: Raven Press. 259–298.

Rechtschaffen, A., Wolpert, E. A., Dement, W. C., Mitchell, S. A., Fisher, C.

(1963). Nocturnal sleep of narcoleptics. *Electroencephalogr. Clin. Neurophysiol.* 15:599–609.

Reinberg, A. (1967). The hours of changing responsiveness or susceptibility. *Perspect. Biol. Med.* 11:111–126.

Reinberg, A., Ghata, J., Halberg, F., Apfelbaum, M., Gervais, P., Boudon, P., Abulker, C., and Dupont, J. (1971). Distribution temporelle du traitement de l'insuffisance corticosurrenalienne. Essai de chemotherapeutique. *Ann. Endocrinol. (Paris)* 32:566–573.

Reinberg, A. and Halberg, F. (1971). Circadian chronopharmacology. *Ann. Rev. Pharmacol. Toxicol.* 11:455–492.

Reinberg, A. and Sidi, E. (1966). Circadian changes in the inhibitory effects of an antihistamine drug in man. *J. Invest. Dermatol.* 46:415–419.

Reinberg, A., Sidi, E., and Ghata, J. (1965). Circadian reactivity rhythms of human skin to histamine or allergen and the adrenal cycle. *J. Allergy Clin. Immunol.* 36:273–283.

Reinberg, A., Vieux, N., Ghata, J., Chaumont, A. J., and LaPorte, A. (1978a). Is the rhythm amplitude related to the ability to phase-shift circadian rhythms of shift-workers? *J. Physiol.* (Paris) 274:405–409.

Reinberg, A., Vieux, N., Ghata, J., Chaumont, A. J., and LaPorte, A. (1978b). Circadian rhythm amplitude and individual ability to adjust to shift work. *Ergonomics* 21:763–766.

Reinberg, A., Zagula-Mally, Z. W., Ghata, J., and Halberg, F. (1967). Circadian rhythm in duration of salicylate excretion referred to phase of excretory rhythms and routine. *Proc. Soc. Exp. Biol. Med.* 124:826–832.

Reinberg, A., Zagula-Mally, Z. W., Ghata, J., and Halberg, F. (1969). Circadian reactivity rhythm of human skin to house dust, penicillin, and histamine. *J. Allergy Clin. Immunol.* 44:292–306.

Reindl, K., Falliers, C., Halberg, F., Chai, H., Hillman, D., and Nelson, W. (1969). Circadian acrophases in peak expiratory flow rate and urinary electrolyte excretion of asthmatic children: phase-shifting of rhythms by prednisone given in different circadian system phases. *Rass. Neuro. Veg.* 23:5–26.

Reis, D. J., Weinbren, M., and Corvelli, A. (1968). A circadian rhythm of norepinephrine regionally in cat brain: its relationship to environmental lighting and to regional diurnal variations in brain serotonin. *J. Pharmacol. Exp. Ther.* 164:135–145.

Reiter, R. J. (1974). Pineal-mediated regression of the reproductive organs of female hamsters exposed to natural photoperiods during the winter months. *Am. J. Obstet. Gynecol.* 118:878–880.

Ronoo, B. and Lobor, W. (1975) Arrhythmically singing crickets: thermoperiodic reentrainment after bilobectomy. *Science* 190:385–387.

Renner, M. (1955). Ein transozeanversuch zum zeitsinn der honigbiene. *Naturwissenschaften* 42:540–541.

Renner, M. (1960). The contribution of the honey bee to the study of time sense and astronomical orientation. *Cold Spring Harbor Symp. Quant. Biol.* 25:361–367.

Rensing, L., Goedeke, K., Wassmann, G., and Broich, G. (1974). Presence and absence of daily rhythms of nuclear size and DNA synthesis of different normal and transformed cells in culture. *J. Interdiscipl. Cycle Res.* 5:267–276.

Reppert, S. M., Artman, H. G., Swaminathan, S., and Fisher, D. A. (1981). Vasopressin exhibits a rhythmic daily pattern in cerebrospinal fluid but not in blood. *Science* 1256–1257.

Reppert, S. M., Chez, R. A., Anderson, A., and Klein, D. C. (1979). Maternal-fetal transfer of melatonin in the non-human primate. *Pediatr. Res.* 13:788–791.

Reppert, S. M., Perlow, M. J., Ungerleider, L. G., Mishkin, M., Tamarkin, L., Orloff, D. G., Hoffman, H. J., and Klein, D. C. (1981). Effects of damage to the suprachiasmatic area of the anterior hypothalamus on the daily melatonin and cortisol rhythms in the Rhesus monkey. *J. Neuroscience.* In press.

Retiene, K., Frohns, T., and Schulz, F. (1967). Experimentelle ergebnisse zur chronopharmakologie von corticosteroiden. *Ven. Med.* 73:990–993.

Ribak, C. E. and Peters, A. (1975). An autoradiographic study of the projections from the lateral geniculate body of the rat. *Brain Res.* 92:341–368.

Rice, R. W., Abe, K., and Critchlow, V. (1978). Abolition of plasma growth hormone response to stress and the circadian rhythm in pituitary-adrenal function in female rats with preoptic-anterior hypothalamic lesions. *Brain Res.* 148:129–141.

Richter, C. P. (1922). A behavioristic study of the activity of the rat. *Comp. Psych. Monographs* 1:1–55.

Richter, C. P. (1927). Animal behavior and internal drives. *Q. Rev. Biol.* 2:307–342.

Richter, C. P. (1965). *Biological Clocks in Medicine and Psychiatry.* Springfield, Illinois: C. C. Thomas.

Richter, C. P. (1967). Sleep and activity: their relation to the 24-hour clock. *Proc. Assoc. Res. Nerv. Ment. Dis.* 45:8–27.

Richter, C. P. (1968). Inherent twenty-four hour and lunar clocks of a primate—the squirrel monkey. *Comm. Behav. Biol.* 1:305–332.

Richter, C. P. (1975). Deep hypothermia and its effect on the 24-hour clock and hamsters. *Johns Hopkins Med. J.* 136:1–10.

Richter, C. P. (1977). Heavy water as a tool for study of the forces that control length of period of the 24-hour clock of the hamster. *Proc. Natl. Acad. Sci. USA* 74:1295–1299.

Righetti, R. (1903). Contributo clinico e anatomo-patologico allo studio dei gliomi cerebrali e all'anatomia delle vie ottiche centrale. *Riv. Patol. Nerv. Ment.* 8:241–267; pt. 2, 289–312.

Riley, J. N. and Moore, R. Y. (1977). Organization of the suprachiasmatic nucleus in the albino rat. *Neurosci. Abstr.* 3:355.

Rioch, D. M. (1929). Studies of the diencephalon of carnivora. I. The nuclear configuration of the thalamus, epithalamus and hypothalamus of the dog and cat. *J. Comp. Neurol.* 49:1–153.

Roberts, P., Turnbull, M. J., and Winterburn, A. (1970). Diurnal variation in sensitivity to and metabolism of barbiturates in the rat—lack of correlation between in vivo and in vitro findings. *Eur. J. Pharmacol.* 12:375–377.

Roberts, S. K. (1962). Circadian activity rhythm in cockroaches: entrainment and phase-shifting. *J. Cell. Comp. Physiol.* 59:175–186.

Roberts, W. (1860). Observations on some of the daily changes of the urine. *Edin. B. Med. J.* 5:817–825; 906–923.

Rockwell, D. A., Hodgson, M. G., Beljan, J. R., and Winget, C. (1976). Psychologic and psychophysiologic response to 105 days of social isolation. *Aviat. Space Environ. Med.* 47:1087–1093.

Roffwarg, H. P., Muzio, J. N., and Dement, W. C. (1966). Ontogenetic development of the human sleep-dream cycle. *Science* 152:604–619.

Rose, M. (1935). Das zwischenhirn des kaninchens. Extrait des *Mémoires de l'Académie Polonaise des Sciences et des Lettres Classe des Sciences Mathématiques et Naturelles*. Série B, Sciences Naturelles. 1–108.

Rosenbaum, J. D., Papper, S., and Ashley, M. M. (1955). Variations in renal excretion of sodium independent of change in adrenocortical hormone dosage in patients with Addison's disease. *J. Clin. Endocrinol. Metab.* 15:1459–1474.

Rosenberg, G. D. and Runcorn, S. K. (1975). Conclusions. In *Growth Rhythms and the History of the Earth's Rotation* (Rosenberg, G. D. and Runcorn, S. K., eds.). London: John Wiley and Sons. 535–538.

Rothig, P. (1911). Beitrage zum studium des zentralnervensystems der wirbeltiere. 3. Zur phylogenese des hypothalamus. *Folia neurobiol.* 5:913–927.

Rothman, B. S. and Strumwasser, F. (1976). Phaseshifting the circadian rhythm of neuronal activity in the isolated *Aplysia* eye with puromycin and cycloheximide. *J. Gen. Physiol.* 68:359–384.

Rowland, N. (1976a). Endogenous circadian rhythms in rats recovered from lateral hypothalamic lesions. *Physiol. Behav.* 16:257–266.

Rowland, N. (1976b). Circadian rhythms and partial recovery of regulatory drinking in rats after lateral hypothalamic lesions. *J. Comp. Physiol. Psychol.* 90:382–393.

Rubinstein, E. H. and Sonnenschein, R. R. (1971). Sleep cycles and feeding behavior in the cat: role of gastrointestinal hormones. *Acta Cient. Venez.* 22:125–128.

Rusak, B. (1977). The role of the suprachiasmatic nuclei in the generation of circadian rhythms in the golden hamster, *Mesocricetus auratus. J. Comp. Physiol.* 118:145–164.

Rusak, B. (1979). Neural mechanisms for entrainment and generation of mammalian circadian rhythms. *Fed. Proc.* 38:2589–2595.

Rusak, B. and Morin, L. P. (1976). Testicular responses to photoperiod are blocked by lesions of the suprachiasmatic nuclei in golden hamsters. *Biol. Reprod.* 15:366–374.

Rusak, B. and Zucker, I. (1979). Neural regulation of circadian rhythms. *Physiol. Rev.* 59:449–526.

Rutenfranz, J., Colquhoun, W. P., Knauth, P., and Ghata, J. N. (1977). Biomedical and psychological aspects of shift work. *Scand. J. Work Environ. Health* 3:165–182.

Rutenfranz, J., Knauth, P., and Colquhoun, W. P. (1976). Hours of work and shiftwork. *Ergonomics* 19:331–340.

Rutledge, J. T. and Angle, M. J. (1977). Persistence of circadian activity rhythms in pinealectomized European starlings (*Sturnus vulgaris*). *J. Exp. Zool.* 202:333–338.

Sacher, G. A. and Duffy, P. H. (1978). Age changes in rhythms of energy metabolism, activity, and body temperature in *Mus* and *Peromyscus*. In *Aging and Biological Rhythms* (Samis, H. V. and Capobianco, S., eds.). New York: Plenum Press. 105–124.

Saint-Paul, U. and Aschoff, J. (1978). Longevity among blowflies *Phormia terraenovae* R.D. kept in non-24-hour light–dark cycles. *J. Comp. Physiol.* 127:191–195.

Saito, M. and Noma, H. (1980). Food intake and growth of rats fed with adiurnal periodicity. *Physiol. Behav.* 24:87–91.

Saleh, M. A., Hard, P. J., and Winget, C. M. (1977). Loss of circadian rhythmicity and locomotor activity following suprachiasmatic lesions in the rat. *J. Interdiscipl. Cycle Res.* 8:341–346.

Saper, C. B., Swanson, L. W., and Cowan, W. M. (1976). The efferent connections of the ventromedial nucleus of the hypothalamus of the rat. *J. Comp. Neurol.* 169:409–442.

Sasaki, T. (1972). Circadian rhythm in body temperature. In *Advances in Climatic Physiology* (Itoh, S., Ogata, K., and Yoshimura, H., eds.). New York: Springer-Verlag. 319–333.

Sassin, J. F., Frantz, A. G., Weitzman, E. D., and Kapen, S. (1972). Human prolactin: 24-hour pattern with increased release during sleep. *Science* 177:1205–1207.

Satinoff, E. (1978). Neural organization and evolution of thermal regulation in mammals. *Science* 201:16–22.

Satter, R. L. and Galston, A. W. (1971). Potassium flux: a common feature of *Albizzia* leaf movement controlled by phytochrome or endogenous rhythm. *Science* 174:518–520.

Satter, R. L., Marinoff, P., and Galston, A. W. (1970). Phytochrome controlled nyctinasty in *Albizzia julibrissin*. 2. Potassium flux as a basis for leaflet movement. *Am. J. Bot.* 57:916–926.

Saunders, D. S. (1977). *An Introduction to Biological Rhythms*. London: Blackie.

Schaeffer, K. E., Kerr, C. M., Buss, D., and Haus, E. (1979). Effect of 18-h watch schedules on circadian cycles of physiological functions during submarine patrols: circadian cycles in submarine patrols. *Undersea Biomed. Res.* S81–S90 (Submarine Supplement).

Scheving, L. E., Burns, E. R., Pauly, J. E., Halberg, F., and Haus, E. (1977). Survival and cure of leukemic mice after circadian optimization with treatment of cyclophosphamide and 1-B-D arabinofuranosylcytosine. *Cancer Res.* 37:3648–3655.

Scheving, L. E., Halberg, F., and Kanabrocki, E. L. (1977). Circadian rhythmometry on 42 variables of thirteen presumably healthy young men. In *XII International Conference Proceedings, International Society for Chronobiology* (Halberg, F., ed.). Milan: Il Ponte. 47–71.

Scheving, L. E., Vedral, D. F., and Pauly, J. E. (1968). Daily circadian rhythms in rats to D-amphetamine sulphate: effect of blinding and continuous illumination on the rhythm. *Nature* 219:621–622.

Schmidek, W. R., Hoshino, K., Schmidek, M., and Timo-Ionia, C. (1972). Influence of environmental temperature on the sleep–wakefulness cycle in the rat. *Physiol. Behav.* 8:363–371.

Schmidt-Nielson, K. (1975). *Animal Physiology: Adaptation and Environment.* Cambridge: Cambridge University Press.

Schmitt, M. (1973). Circadian rhythmicity in responses of cells in the lateral hypothalamus. *Am. J. Physiol.* 225:1096–1101.

Schulz, F. and Retiene, K. (1971). Tierexperimentelle und klinische befunde bei therapeutischer anwendung von corticosteroiden zu verschiedenen tageszeiten. *Klin. Wochenschr.* 49:100–105.

Schwartz, W. J., Davidsen, L. C., and Smith, C. B. (1980). In vivo metabolic activity of a putative circadian oscillator, the rat suprachiasmatic nucleus. *J. Comp. Neurol.* 189:157–167.

Schwartz, W. J. and Gainer, H. (1977). Suprachiasmatic nucleus: use of ^{14}C-labeled deoxyglucose uptake as a functional marker. *Science* 197:1089–1091.

Schwartz, W. J., Smith, C. B., and Davidsen, L. C. (1979). In vivo glucose utilization of the suprachiasmatic nucleus. In *Biological Rhythms and Their Central Mechanism* (Suda, M., Hayaishi, O., and Nakagawa, H., eds.). New York: Elsevier North-Holland, 355–367.

Schweig (1843). Quoted in Speck (1882). Untersuchungen uber die beziehungen der geistigen thatigkeit zum stoffwechsel. *Arch. Exp. Pathol. Pharmakol.* 15:81–145.

Schweiger, E., Wallraff, H. G., and Schweiger, H. G. (1964). Endogenous circadian rhythm in cytoplasm of *Acetabularia*: influence of the nucleus. *Science* 146:658–659.

Segre, E. J. and Klaiber, E. L. (1966). Therapeutic utilization of the diurnal variation in pituitary-adrenocortical activity. *Calif. Med.* 101:363 365.

Setalo, G. S., Vigh, A. V., Schally, A., Arimura, and Flerko, B. (1976). Immunohistological study of the origin of LH-RH-containing nerve fibers of the rat hypothalamus. *Brain Res.* 103:597–602.

Sharma, K. N., Anand, B. K., Dua, S., and Singh, B. (1961). Role of stomach in regulation of activities of hypothalamic feeding centers. *Am. J. Physiol.* 201:593–598.

Sharp, G. W. G. (1960). Reversal of diurnal rhythms of water and electrolyte excretion in man. *J. Endocrinol.* 21:97–106.

Sharp, G. W. G., Slorach, S. A., and Vipond, H. J. (1961). Diurnal rhythms of keto- and ketogenic steroid excretion and the adaptation to changes of the activity-sleep routine. *J. Endocrinol.* 22:377–385.

Sherrington, C. S. (1900). Cutaneous sensations. In *Textbook of Physiology*, vol. 2 (Schafer, E. A., ed.). Edinburgh: Young J. Penland. 920–1001.

Shift Work Committee, Japan Association of Industrial Health (1979). Opinion on night work and shift work. *J. Sci. Labour* 55:1–55.

Shiotsuka, R., Jovonovich, J., and Jovonovich, J. A. (1974). In vitro data on drug sensitivity: circadian and ultradian corticosterone rhythms in adrenal organ cultures. In *Chronobiological Aspects of Endocrinology* (Aschoff, J., Ceresa, F., and Halberg, F., eds.). Stuttgart: Schattauer-Verlag. 255–267.

Shivers, B., Harlan, R., and Moss, R. L. (1979). Effects on ovulation of horizontal knife cuts above the suprachiasmatic nuclei in the rat. *Fed. Proc.* 38:1108.

Sieber, W. (1976). Synchronisierte und autonome circadiane periodik physiologischer funktionen bei blinden unter besonderer berucksichtigung des freien urin-cortisols. Doctoral thesis, University of Munich.

Siegel, J. M. and McGinty, D. J. (1977). Pontine reticular formation neurons: relationship of discharge to motor activity. *Science* 196:678–680.

Siffre, M. (1964). *Beyond Time* (Briffault, H., ed. and trans.). New York: McGraw-Hill.

Silver, J. and Brand, S. (1979). A route for direct retinal input to the preoptic hypothalamus: dendritic projections into the optic chiasm. *Am. J. Anat.* 155:391–402.

Silverstone, T., ed. (1976). *Appetite and Food Intake: Dahlem Workshop on Appetite and Food Intake.* Berlin: Abakon Verlagsgesellschaft.

Simpson, G. E. (1924). Diurnal variations in the rate of urine excretion for two-hour intervals: some associated factors. *J. Biol. Chem.* 59:107–122.

Simpson, G. E. (1926). The effect of sleep on urinary chloride and pH. *J. Biol. Chem.* 67:505–516.

Simpson, H. W., Lobban, M. C., and Halberg, F. (1970). Urinary near 24 hour rhythms in subjects living on a 21 hour routine in the Arctic. *Arctic Anthropol.* 7:144–164.

Simpson, S. and Galbraith, J. J. (1906). Observations on the normal temperature of the monkey and its diurnal variation, and on the effect of changes in the daily routine on this variation. *Trans. Roy. Soc. Edinburgh* 45:65–106.

Sjostrand, T. (1952). The regulation of the blood distribution in man. *Acta Physiol. Scand.* 26:312–327.

Slonaker, J. R. (1908). Description of an apparatus for recording the activity of small mammals. *Anat. Rec.* 2:116–122.

Slonaker, J. R. (1924). The effect of pubescence, oestruation and menopause on the voluntary activity in the albino rat. *Am. J. Physiol.* 68:294–315.

Smith, H. W. (1951). *The Kidney, Structure and Function in Health and Disease.* New York: Oxford University Press.

Smith, R. E. (1969). Circadian variations in human thermoregulatory responses. *J. Appl. Physiol.* 26:554–559.

Smith, R. E. and Wekstein, D. R. (1969). Circadian variations of physiological variables in isolated and non-isolated *Macaca nemestrina.* In *Circadian Rhythms in Nonhuman Primates* (Rohles, F. H., ed.). Basel: S. Karger. 75–90.

Snyder, F. (1969). Sleep and REM as biological enigmas. In *Sleep Physiology and Pathology* (Kales, A., ed.). Philadelphia: J. B. Lippincott Co. 266–280.

So, K. F., Schneider, G. E., and Frost, D. O. (1971). Postnatal development of retinal projections to the lateral geniculate body in Syrian hamsters. *Brain Res.* 142:343–353.

Spiegel, E. A. and Zweig, H. (1917). Zur cytoarchitektonik des tuber cinereum. *Arb. Neurol. Inst. Wiener Univ.* 22:278.

Stamm, D. (1967). Tagesschwankungen der normalbereiche diagnostisch wichtiger bluthestandteile. *Ven. Med.* 73:982–989.

Stanbury, S. W. and Thomson, A. E. (1951). Diurnal variations in electrolyte excretion. *Clin. Sci. Mol. Med.* 10:267–293.

Stephan, F. K. and Nunez, A. A. (1977). Elimination of circadian rhythms in drinking, activity, sleep and temperature by isolation of the suprachiasmatic nuclei. *Behav. Biol.* 20:1–16.

Stephan, F. K., Swann, J. M., and Sisk, C. L. (1979a). Anticipation of 24-hour feeding schedules in rats with lesions of the suprachiasmatic nucleus. *Behav. and Neural Biol.* 25:346–363.

Stephan, F. K., Swann, J. M., and Sisk, C. L. (1979b). Entrainment of circadian rhythms by feeding schedules in rats with suprachiasmatic lesions. *Behav. and Neural Biol.* 25:545–554.

Stephan, F. K. and Zucker, I. (1972a). Rat drinking rhythms: central visual pathways and endocrine factors mediating responsiveness to environmental illumination. *Physiol. Behav.* 8:315–326.

Stephan, F. K. and Zucker, I. (1972b). Circadian rhythms in drinking behavior and locomotor activity of rats are eliminated by hypothalamic lesions. *Proc. Natl. Acad. Sci. USA* 69:1583–1586.

Sterman, M. B. and Clemente, C. D. (1962). Forebrain inhibitory mechanisms: sleep patterns induced by basal forebrain stimulation. *Exp. Neurol.* 6:91–102.

Sterman, M. B., Knauss, T., Lehmann, D., and Clemente, C. D. (1965). Circadian sleep and waking patterns in the laboratory cat. *Electroencephalogr. Clin. Neurophysiol.* 19:509–517.

Stetson, M. H. and Watson-Whitmyre, M. (1976). Nucleus suprachiasmaticus: the biological clock in the hamster? *Science* 191:197–199.

Stetson, M. H. and Watson-Whitmyre, M. (1977). The neural clock regulating estrous cyclicity in hamsters: gonadotropin release following barbiturate blockade. *Biol. Reprod.* 16:536–542.

Stewart, C. C. (1898). Variations in daily activity produced by alcohol and by changes in barometric pressure and diet, with a description of recording methods. *Am. J. Physiol.* 1:40–56.

Stewart, M. C. and Reeder, W. C. (1968). Temperature and light synchronization experiments with circadian activity rhythms in two color forms of the rock pocket mouse. *Physiol. Zool.* 41:149–156.

Subbaraj, R. and Chandrashekaran, M. (1978). Pulses of darkness shift the phase of a circadian rhythm in an insectivorous bat. *J. Comp. Physiol.* 127:239–243.

Suda, M. and Saito, M. (1979). Coordinative regulation of feeding behavior and metabolism by a circadian timing system. In *Biological Rhythms and Their Central Mechanism* (Suda, M., Hayaishi, O., and Nakagawa, H., eds.). New York: Elsevier North-Holland. 263–271.

Sulzman, F. M., Fuller, C. A., Hiles, L. G., and Moore-Ede, M. C. (1978). Circadian rhythm dissociation in an environment with conflicting temporal information. *Am. J. Physiol.* 235:R175–R180.

Sulzman, F. M., Fuller, C. A., and Moore-Ede, M. C. (1977a). Feeding time synchronizes primate circadian rhythms. *Physiol. Behav.* 18:775–779.

Sulzman, F. M., Fuller, C. A., and Moore-Ede, M. C. (1977b). Environmental synchronizers of squirrel monkey circadian rhythms. *J. Appl. Physiol.* 43:795–800.

Sulzman, F. M., Fuller, C. A., and Moore-Ede, M. C. (1977c). Spontaneous internal desynchronization of circadian rhythms in the squirrel monkey. *Comp. Biochem. Physiol. (A)* 58:63–67.

Sulzman, F. M., Fuller, C. A., and Moore-Ede, M. C. (1978a). Preliminary characterization of persisting circadian rhythms during spaceflight: *Neurospora* as a model system. In *Spacelab Mission 1: Experimental Descriptions* (Craven, P. D., ed.). NASA Report TM-78137.

Sulzman, F. M., Fuller, C. A., and Moore-Ede, M. C. (1978b). Extent of circadian synchronization by cortisol in the squirrel monkey. *Comp. Biochem. Physiol. (A)* 59:279–283.

Sulzman, F. M., Fuller, C. A., and Moore-Ede, M. C. (1978c). Comparison of synchronization of primate circadian rhythms by light and food. *Am. J. Physiol.* 234:R130–R135.

Sulzman, F. M., Fuller, C. A., and Moore-Ede, M. C. (1979a). Tonic effects of light on the circadian system of the squirrel monkey. *J. Comp. Physiol.* 129:43–50.

Sulzman, F. M., Fuller, C. A., and Moore-Ede, M. C. (1979b). One second light pulses entrain circadian rhythms in a diurnal primate. *Physiologist* 224:66.

Sulzman, F. M., Fuller, C. A., and Moore-Ede, M. C. (1981). Effects of phasic and tonic light inputs on the circadian organization of the squirrel monkey. *Photochem. Photobiol.* 34:249–256.

Swade, R. H. and Pittendrigh, C. S. (1967). Circadian locomotor rhythms of the rodent in the Arctic. *Am. Natural.* 101:431–464.

Swanson, L. W. (1977). Immunohistochemical evidence for a neurophysin-containing autonomic pathway arising in the paraventricular nucleus of the hypothalamus. *Brain Res.* 128:346–353.

Swanson, L. W. and Cowan, W. M. (1975). The efferent connections of the suprachiasmatic nucleus of the hypothalamus. *J. Comp. Neurol.* 160:1–12.

Swanson, L. W., Cowan, W. M., and Jones, E. G. (1974). An autoradiographic study of the efferent connections of the ventral geniculate nucleus in the albino rat and the cat. *J. Comp. Neurol.* 156:143–164.

Sweeney, B. M. (1974). The potassium content of *Gonyaulax polyhedra* and phase changes in the circadian rhythm of stimulated bioluminescence by short exposure to ethanol and valinomycin. *Plant Physiol.* 53:337–342.

Sweeney, B. M. and Hastings, J. W. (1960). Effect of temperature upon diurnal rhythms. *Cold Spring Harbor Symp. Quant. Biol.* 25:87–104.

Sweeney, B. M. and Haxo, F. T. (1961). Persistence of a photosynthetic rhythm in enucleated *Acetabularia*. *Science* 134:1361–1363.

Szabo, I., Kovats, T. G., and Halberg, F. (1978). Circadian rhythm in murine reticuloendothelial function. *Chronobiologia* 5:137–143.

Szafarczyk, A., Ixart, G., Malaval, F., Nouguier-Soule, J., and Assenmacher, I. (1979). Effects of lesions of the suprachiasmatic nuclei and of p-chlorophenylalanine on the circadian rhythms of adrenocorticotrophic hormone and corticosterone in the plasma, and on locomotor activity of rats. *J. Endocrinol.* 83:1–16.

Takahashi, J. S. and Menaker, M. (1980). Interaction of estradiol and progesterone: effects on circadian locomotor rhythm of female golden hamsters. *Am. J. Physiol.* 239:R497–R504.

Takahashi, K., Hanada, K., and Takahashi, Y. (1979). Factors setting the phase of the circadian, adrenocortical rhythm in rats. In *Biological Rhythms and Their Central Mechanism* (Suda, M., Hayaishi, O., and Nakagawa, H., eds.). New York: Elsevier North-Holland. 189–198.

Takebe, K., Sakakura, M., and Mashimo, K. (1972). Continuance of diurnal rhythmicity of CRF activity in hypophysectomized rats. *Endocrinology* 90:1515–1520.

Takebe, K., Setaishi, C., and Hirami, M. (1966). Effects of a bacterial pyrogen on the pituitary-adrenal axis at various times in the 24 hours. *J. Clin. Endocrinol. Metab.* 26:437–442.

Tarquini, B., Fantini, F., and Cagnoni, M. (1967). Atrofia corticosurrenalica da trattamento steroideo: variazione degli effetti in rapporto all'ora di somministrazione. *Sperimentale* 116:373–380.

Tarttelin, M. F. and Gorski, R. A. (1971). Variations in food and water intake in the normal and acyclic female rat. *Physiol. Behav.* 7:847–852.

Taub, J. M. and Berger, R. J. (1974). Acute shifts in sleep-wakefulness cycle: effects on performance and mood. *Psychosom. Med.* 36:164–173.

Teitelbaum, P. (1971). The encephalization of hunger. In *Progress in Physiological Psychology*, vol. 4 (Stellar, E. and Sprague, J. M., eds.). New York: Academic Press. 319–350.

Tharp, G. D. and Folk, G. E., Jr. (1965). Rhythmic changes in rate of the mammalian heart and heart cells during prolonged isolation. *Comp. Biochem. Physiol.* 14:255–273.

Thorpe, P. A. (1975). The presence of a retinohypothalamic projection in the ferret. *Brain Res.* 85:343–346.

Tigges, J., Bos, J., and Tigges, M. (1977). An autoradiographic investigation of the subcortical visual system in the chimpanzee. *J. Comp. Neurol.* 172:367–380.

Tokura, H. and Aschoff, J. 1980. Quoted in: Aschoff, J. (1979). Circadian rhythms: general features and endocrinological aspects. In *Endocrine Rhythms* (Krieger, D. T., ed.). New York: Raven Press. 1–61.

Tribukait, B. (1956). Die aktivitatsperiodik der wissen maus im kunnsttag von 16 bis 29 stunden lange. *Z. Vergl. Physiol.* 38:479–490.

Tsang, Y. C. (1938). Hunger motivation in gastrectomized rats. *J. Comp. Physiol. Psychol.* 26:1–17.

Turek, F. W. and Campbell, C. S. (1979). Photoperiodic regulation of neuroendocrine-gonadal activity. *Biol. Reprod.* 20:32–50.

Turek, F. W., Jacobson, C. D., and Gorski, R. A. (1980). Lesions of the suprachiasmatic nuclei affect photoperiod-induced changes in the sensitivity of the hypothalamic-pituitary axis to testosterone feedback. *Endocrinology* 107:942–947.

Ungar, F. (1967). "In vitro" studies of circadian rhythms in hypothalamo-pituitary-adrenal systems. *Rass. Neuro. Veg.* 21:57–70.

Vagnucci, A., Shapiro, A. P., and McDonald, R. H., Jr. (1969). Effects of upright posture on renal electrolyte cycles. *J. Appl. Physiol.* 26:720–731.

Vance, W. B. (1965). Observations on the role of salivary secretions in the regulation of food and fluid intake in the white rat. *Psych. Monographs* 79:1–22.

van den Hoed, J. and Boukamp, B. A. (1980). Response to travel across time zones in an infant at 3.5 and 16.5 months of age. *Sleep Res.* 9:278.

Vandesande, F., Dierickx, K., and Demey, J. (1975). Identification of the vasopressin-neurophysin producing neurons of the rat suprachiasmatic nuclei. *Cell Tissue Res.* 156:377–380.

Van Itallie, T. B. and Hashim, S. A. (1960). Biochemical concomitants of hunger and satiety in man. *Am. J. Clin. Nutr.* 8:587–594.

Van Tienhoven, A. (1968). *Reproductive Physiology of Vertebrates*. Philadelphia: Saunders.

Vernikos-Danellis, J., Leach, C. S., Winget, C. M., Rambaut, P.C., and Mack, P. B. (1972). Thyroid and adrenal cortical rhythmicity during bedrest. *J. Appl. Physiol.* 33:644–648.

Vernikos-Danellis, J. and Winget, C. M. (1979). The importance of light, postural and social cues in the regulation of the plasma cortisol rhythms in man. In *Chronopharmacology* (Reinberg, A. and Halbert, F., eds.). New York: Pergamon Press. 101–106.

Vernon, H. M. (1921). *Industrial Fatigue and Efficiency*. New York: E. P. Dutton.

Vesell, E. S. (1968). Genetic and environmental factors affecting hexobarbital metabolism in mice. *Ann. N.Y. Acad. Sci.* 151:900-912.

Vogel, J. (1863). Quoted in Speck, S. (1882). Klinische untersuchungen uber den stoffwechsel bei gesunden und kranken menschen uberhaupt und den durch den urin in besondere. *Arch. Exp. Pathol. Pharmakol.* 15:81-145.

Volker, H. (1927). Uber die tagesperiodischen schwankungen einiger lebens-vorgange des menschen. *Arch. Ges. Physiol.* 215:43-77.

von Economo, C. (1929). Schlaftheorie. *Ergeb. Physiol.* 28:312-339.

von Holst, E. (1939). Relative coordination as a phenomenon and as a method for analysis of central nervous functions. *Ergeb. Physiol.* 42:228-306.

von Marilaun, A. K. (1895). *The Natural History of Plants,* vol. 2. New York: Henry Holt.

von Mayersbach, H. (1974). Discussion. In *Chronobiological Aspects of Endocrinology* (Aschoff, J., Ceresa, F., and Halberg, F., eds.). New York: Schat-tauer-Verlag. 111-116.

Wade, G. N. and Zucker, I. (1970). Modulation of food intake and locomotor activity in female rats by diencephalic hormone implants. *J. Comp. Physiol. Psychol.* 72:328-336.

Wahlstrom, G. (1965). Experimental modifications of the internal clock in the canary, studied by self-selection of light and darkness. In *Circadian Clocks (Aschoff, J., ed.). Amsterdam: North-Holland. 324-328.*

Walcott, C., Gould, J. L., and Kirschvink, J. L. (1979). Pigeons have magnets. Science 205:1027-1028.

Wang, G. H. (1923). The relation between "spontaneous" activity and oestrous cycle in the white rat. *Comp. Psych. Monographs* 2:1-27.

Warden, A. W. and Sachs, B. D. (1974). Circadian rhythms of self-selected lighting in hamsters. *J. Comp. Physiol.* 91:127-134.

Watkins, W. B. (1976). Localization of neurosecretory pathways in the hypothalamus. In *Progress in Neuropathology, vol. 3* (Zimmerman, H. M., ed.). New York: Grune and Stratton. 384-446.

Webb, O. L. and Russell, R. L. (1966). Diurnal chemo-convulsive responses and central inhibition. *Arch. Int. Pharmacodyn. Ther.* 159:471-476.

Webb, W. B. and Agnew, H. W. (1969). Measurements and characteristics of nocturnal sleep. *Prog. Clin. Psych.* 8:2-27.

Webb, W. B. and Agnew, H. W. (1974). Sleep and waking in a time free environment. *Aviat. Space Environ. Med.* 45:617-622.

Weber, A. L., Cary, M. S., Connor, N., and Keyes, P. (1980). Human non-24-hour cloop walkc cyclcs in an everyday environment. *Sleep* 2:347-354.

Wegmann, H. M., Klein, K. E., and Kuklinski, P. (1973). Quoted in Klein, K. E. and Wegmann, H. M. (1979). Circadian rhythms in air operations. In *Sleep, Wakefulness and Circadian Rhythms,* vol. 105 (Nicholson, A. N., ed.). Neuilly sur Seine, France: NATO Advisory Group for Aerospace Research and Development. 10.1-10.25.

Wehr, T. A., Wirz-Justice, A., Goodwin, F. K., Duncan, W., and Gillin, J. C.

(1979). Phase advance of the circadian sleep–wake cycle as an anti-depressant. *Science* 206:710–713.

Weigelin, J. (1868). Versuche uber die harnstoffausscheidung wahrend und mach der muskelthetigkeit. *Arch. Anat. Physiol. Wissensch. Med.*:207–223.

Weitzman, E. D. (1976). Circadian rhythms and episodic hormone secretion in man. *Ann. Rev. Med.* 27:225–243.

Weitzman, E. D., Czeisler, C. A., Coleman, R. M., Spielman, A. J., Zimmerman, J. C., Dement, W. C., Richardson, G. S., and Pollak, C. P. (1981). Delayed sleep phase syndrome: a chronobiologic disorder with sleep onset insomnia. *Archiv. Gen. Psychiatry* 38:737–746.

Weitzman, E. D., Czeisler, C. A., and Moore-Ede, M. C. (1979). Sleep-wake, neuroendocrine and body temperature circadian rhythms under entrained and non-entrained (free-running) conditions in man. In *Biological Rhythms and Their Central Mechanism* (Suda, M., Hayaishi, O., and Nakagawa, H., eds.). New York: Elsevier North-Holland. 199–227.

Weitzman, E. D., Czeisler, C. A., Zimmerman, J. C., Ronda, J. M., and Knauer, R. S. (1981). Chronobiological disorders: analytic and therapeutic techniques. In *Disorders of Sleeping and Waking: Indications and Techniques* (Guilleminault, C., ed.). Menlo Park, Calif.: Addison-Wesley, forthcoming.

Weitzman, E. D., Fukushima, D. K., Nogeire, C., Roffwarg, H., Gallagher, T. F., and Hellman, L. (1971). Twenty-four hour pattern of the episodic secretion of cortisol in normal subjects. *J. Clin. Endocrinol. Metab.* 33:14–22.

Weitzman, E. D., Goldmacher, D., Kripke, D., McGregor, P., Kream, J., and Hellman, L. (1968). Reversal of sleep–waking cycle: effect on sleep stage pattern and certain neuroendocrine rhythms. *Trans. Am. Neurol. Assoc.* 93:153–157.

Weitzman, E. D. and Hellman, L. (1974). Temporal organization of the 24-hour pattern of the hypothalamic-pituitary axis. In *Biorhythms and Human Reproduction* (Ferin, M., Halberg, F., Richart, R. M., and Vandewiele, R. L., eds.). New York: Wiley. 371–395.

Weitzman, E. D., Nogeire, C., Perlow, M., Fukushima, D., Sassin, J., McGregor, P., Gallagher, T. F., and Hellman, L. (1974). Effects of prolonged 3-hour sleep-wake cycle on sleep stages, plasma cortisol, growth hormone and body temperature in man. *J. Clin. Endocrinol. Metab.* 38:1018–1030.

Weitzman, E. D., Perlow, M., Sessin, F. F., Fukushima, E., Buralk, B., and Hellman, L. (1972). Persistency of the 24-hour pattern of episodic cortisol secretion and growth hormone release in blind subjects. *Trans. Am. Neurol. Assoc.* 97:197–199.

Weitzman, E. D., Schaumburg, H., and Fishbein, W. (1966). Plasma 17-hydroxy-corticosteroid levels during sleep in man. *J. Clin. Endocrinol. Metab.* 26:121–127.

Wenger, C. B., Roberts, M. F., Stolwijk, J. A. J., and Nadel, E. R. (1976). Nocturnal lowering of thresholds for sweating and vasodilation. *J. Appl. Physiol.* 41:15–19.

Wenisch, H. J. C. (1976). Retinohypothalamic projection in the mouse: electron microscopic and iontophoretic investigations of hypothalamic and optic centers. *Cell Tissue Res.* 167:547–561.

Went, F. W. (1959). The periodic aspect of photoperiodism and thermoperiodicity. In *Photoperiodism* (Withrow, R. B., ed.). Washington, D.C.: American Association for the Advancement of Science. 551–564.

Wesson, L. G. (1964). Electrolyte excretion in relation to diurnal cycles of renal function. *Medicine* 43:547–592.

Wever, R. (1964). Ein mathematisches modell fur biologische schwingungen. *Z. Tierpsychol.* 21:359–372.

Wever, R. (1969). Autonome circadiane periodik des menschen unter dem einfluss verschiedener beleuchtungsbedingungen. *Pfluegers Arch.* 306:71–91.

Wever, R. (1970a). Zur zeitgeber-starke eines licht-dunkel-wechsels fur die circadiane periodik des menschen. *Pfluegers Arch.* 321:133–142.

Wever, R. (1970b). The effects of electric fields on circadian rhythms in men. *Life Sci. Space Res.* 8:177–187.

Wever, R. (1973). Internal phase angle differences in human circadian rhythms: causes for changes and problems of determinations. *Int. J. Chronobiol.* 1:371–390.

Wever, R. (1974). Influence of light on human circadian rhythms. *Nordic Council Arct. Med. Res. Rep.* 10:33–47.

Wever, R. (1975). The circadian multi-oscillator system of man. *Int. J. Chronobiol.* 3:19–55.

Wever, R. A. (1979). *The Circadian System of Man. Results of Experiments under Temporal Isolation.* New York: Springer-Verlag.

Whishaw, I. Q. (1974). Light avoidance in normal rats and rats with primary visual system lesions. *Physiol. Psychol.* 2:143–147.

Williams, R. L., Agnew, H. W., and Webb, W. B. (1964). Sleep patterns in young adults: an EEG study. *Electroencephalogr. Clin. Neurophysiol.* 17:376–381.

Willoughby, J. O. and Martin, J. B. (1978). The suprachiasmatic nucleus synchronizes growth hormone secretory rhythms with the light-dark cycle. *Brain Res.* 151:413–417.

Wilson, M. M. and Critchlow, V. (1975). Absence of a circadian rhythm in persisting corticosterone fluctuations following surgical isolation of the medial basal hypothalamus. *Neuroendocrinology* 19:185–192.

Winfree, A. (1980). *The Geometry of Biological Time.* New York: Springer-Verlag.

Winfree, A. T. (1971). Corkscrews and singularities in fruitflies: resetting behavior of the circadian eclosion rhythm. In *Biochronometry* (Menaker, M., ed.). Washington, D.C.: National Academy of Sciences. 81–106.

Winget, C. M. (1974). Biorhythms and space experiments with nonhuman primates. In *The Use of Nonhuman Primates in Space* (Winget, C. M., ed.). NASA Conference Publication 005. Moffet Field, Calif.: NASA. 165–170.

Winkler, C. and Potter, A. (1914). *Guide to the Cat's Brain.* Amsterdam: Versluys.

Wolf, A. V. (1950). Osmometric analysis of thirst in man and dog. *Am. J. Physiol.* 161:75–86.

Wolfe, L. K., Gordon, R. D., Island, D. P., and Liddle, G. W. (1966). An analysis of factors determining the circadian pattern of aldosterone excretion. *J. Clin. Endocrinol. Metab.* 26:1261–1266.

Yalow, R. S. (1978). Radioimmunoassay: a probe for the fine structure of biologic systems. *Science* 200:1236–1245.

Yamaoka, S. (1978). Participation of limbic-hypothalamic structures in circadian rhythm of slow wave sleep and paradoxical sleep in the rat. *Brain Res.* 151:255–268.

Yates, F. E. and Maran, J. W. (1974). Stimulation and inhibition of adrenocorticotropin release. In *Handbook of Physiology* (Greep, R. O. and Astwood, E. B., eds.). Baltimore: Williams and Wilkins Company. 367–404.

Yeates, N. T. M. (1949). The breeding season of the sheep with particular reference to its modification by artificial means using light. *J. Agric. Sci.* 39:1–43.

Yochim, J. M. and Spencer, F. (1976). Core temperature in the female rat: effect of ovariectomy and induction of pseudopregnancy. *Am. J. Physiol.* 231:361–365.

Yoshimura, H. (1973). Review of medical researches at the Japanese station (Syowa base) in the Antarctic. In *Polar Human Biology* (Edholm, O. G. and Gunderson, E. K. E., eds.). Great Britain: William Heinemann. 54–65.

Yunis, E. J., Fernandes, G., Nelson, W., and Halberg, F. (1974). Circadian temperature rhythms and aging in rodents. In *Chronobiology* (Scheving, L. E., Halberg, F., and Pauly, J. E., eds.). Tokyo: Iguku Shoin. 358–363.

Zatz, M. (1979). A neuropharmacologic approach to the circadian oscillator regulating rat pineal serotonin N-acetyltransferase activity. In *Biological Rhythms and Their Central Mechanism* (Suda, M., Hayaishi, O., and Nakagawa, H., eds.). New York: Elsevier North-Holland. 149–158.

Zatz, M. and Brownstein, M. J. (1979). Intraventricular carbachol mimics the effects of light on the circadian rhythm in the rat pineal gland. *Science* 203:358–361.

Zeilmaker, G. H. (1966). The biphasic effect of progesterone on ovulation in the rat. *Acta Endocrinol.* (Kbh) 51:461–468.

Zimmerman, E. and Critchlow, V. (1967). Effects of diurnal variation in plasma corticosterone levels on adrenocortical response to stress. *Proc. Soc. Exp. Biol. Med.* 125:658–663.

Zimmerman, N. H. and Menaker, M. (1979). The pineal gland: the pacemaker within the circadian system of the house sparrow. *Proc. Natl. Acad. Sci. USA* 76:999–1003.

Zinn, J. G. (1759). On the sleep of plants. *Hamburgisches Magazin* 22:40–50.

Zucker, I. (1971). Light–dark rhythms in rat eating and drinking behavior. *Physiol. Behav.* 6:115–126.

Zucker, I. and Stephan, S. K. (1973). Light–dark rhythms in hamster eating, drinking, and locomotor behaviors. *Physiol. Behav.* 11:239–250.

Zulley, J. (1979). *Der Einfluss von Zeitbern auf den Schlaf von Menschen.* Frankfurt: Rita G. Fischer.

Zweig, M., Synder, S. H., and Axelrod, J. (1966). Evidence for a nonretinal pathway of light to the pineal gland of newborn rats. *Proc. Natl. Acad. Sci. USA* 56:515–520.

Index